SYNERGY IN SUPRAMOLECULAR CHEMISTRY

SYNERGY IN SUPRAMOLECULAR CHEMISTRY

EDITED BY **TATSUYA NABESHIMA**

CRC Press
Taylor & Francis Group
Boca Raton London New York

CRC Press is an imprint of the
Taylor & Francis Group, an **informa** business

CRC Press
Taylor & Francis Group
6000 Broken Sound Parkway NW, Suite 300
Boca Raton, FL 33487-2742

First issued in paperback 2021

Version Date: 20141008

ISBN 13: 978-1-03-223645-2 (pbk)
ISBN 13: 978-1-4665-9502-6 (hbk)

Publisher's Note
The publisher has gone to great lengths to ensure the quality of this reprint but points out that some imperfections in the original copies may be apparent.

Library of Congress Cataloging-in-Publication Data

Synergy in supramolecular chemistry / editor, Tatsuya Nabeshima.
   p. ; cm.
   Includes bibliographical references and index.
   ISBN 978-1-4665-9800-3 (hardcover : alk. paper)
   I. Nabeshima, Tatsuya, editor.
   [DNLM: 1. Macromolecular Substances--chemistry. 2. Molecular Structure. QD 878]

QD381.8
547'.7--dc23   2014032223

Visit the Taylor & Francis Web site at
http://www.taylorandfrancis.com

and the CRC Press Web site at
http://www.crcpress.com

Contents

Contributors

Dariush Ajami
Scripps Research Institute
La Jolla, California

Markus Albrecht
RWTH Aachen University
Aachen, Germany

Shogo Amemori
Hokkaido University
Sapporo, Japan

Shin Aoki
Tokyo University of Science
Tokyo, Japan

Loïc J. Charbonnière
Laboratoire d'Ingénierie Moléculaire
    Appliquée à l'Analyse
Strasbourg, France

Steven De Feyter
Katholieke Universiteit Leuven
Leuven, Belgium

Amar H. Flood
Indiana University
Bloomington, Indiana

Kei Goto
Tokyo Institute of Technology
Tokyo, Japan

Takeharu Haino
Hiroshima University
Hiroshima, Japan

Akira Harada
Osaka University
Osaka, Japan

Takashi Hashimoto
Sophia University
Tokyo, Japan

Takashi Hayashita
Sophia University
Tokyo, Japan

Keiji Hirose
Osaka University
Osaka, Japan

Yosuke Hisamatsu
Tokyo University of Science
Tokyo, Japan

Kazuya Iseda
Hokkaido University
Sapporo, Japan

Ryo Katoono
Hokkaido University
Sapporo, Japan

Paramjit Kaur
Guru Nanak Dev University
Amritsar, India

Masanori Kitamura
Kanazawa University
Kanazawa, Japan

Hiroyuki Kobayashi
Sophia University
Tokyo, Japan

Kenta Kokado
Hokkaido University
Sapporo, Japan

Yun Liu
Indiana University
Bloomington, Indiana

Hiromitsu Maeda
Ritsumeikan University
Kusatsu, Japan

Hiroyuki Miyake
Osaka City University
Osaka, Japan

Tatsuya Nabeshima
University of Tsukuba
Tsukuba, Japan

Kazuki Sada
Hokkaido University
Sapporo, Japan

Kamaljit Singh
Guru Nanak Dev University
Amritsar, India

Bradley D. Smith
University of Notre Dame
Notre Dame, Indiana

Graeme T. Spence
University of Notre Dame
Notre Dame, Indiana

Kazukuni Tahara
Osaka University
Osaka, Japan

Yoshinori Takashima
Osaka University
Osaka, Japan

Yoshito Tobe
Osaka University
Osaka, Japan

Ali Trabolsi
Centre for Science and Engineering
New York University Abu Dhabi
Abu Dhabi, United Arab Emirates

Hiroshi Tsukube
Osaka City University
Osaka, Japan

Mohd Zulkefeli
University Teknologi MARA
Selangor, Malaysia

Preface

Supramolecular chemistry is a nonconventional field of chemistry concerning the formation of molecular assemblies through non-covalent interactions such as hydrogen bonds, electrostatic interactions, hydrophobic effects, π-π interactions, etc. Thus, supramolecular chemistry provides very different scopes and insights into chemical events, particularly at nano- and meso-scale levels. In addition, the application of supramolecular chemistry to materials science has recently developed significantly to offer a variety of highly functionalized sensors, molecular devices, and polymeric materials.

A very important aspect of recent progress in the supramolecular chemistry concerns the cooperative and synergistic construction and regulation of structures and functions of supramolecules. These supramolecular events are caused by external stimuli and influence circumstances at the molecular level to result in a new structure and function. In particular, biological systems often utilize the cooperative regulation of functions to maintain homeostasis in the body by adjusting the balance of biologically important materials. These sophisticated chemical events do not result from simply combining individual functions of functional molecules that are incorporated in the synergistic molecular systems. These functional units have to structurally and functionally affect each other to cause positive and negative cooperative effects on the functions. Such non-linear-fashioned regulation mechanisms are also very effective and useful in controlling the functions of artificial supramolecules. Hence, cooperative and synergistic chemical events have attracted significant attention from many researchers engaged in organic chemistry, inorganic chemistry, biological chemistry, polymer chemistry, medicinal chemistry, and other related materials sciences. Synergistic supramolecular systems could be developed to amplify the functions and integration of molecular devices in ways that cannot be achieved by conventional single molecules.

The aim of this book is to introduce concepts and various current examples of supramolecular chemistry in terms of cooperativity and synergy, and to survey the recent explosive progress in this field. In this book, the topics are focused on (1) the synergistic and cooperative events in the formation of supramolecular architectures, and (2) synergistic and cooperative control of their properties and functions. The latter category contains at least three categories, supramolecular sensors and devices, supramolecular catalysts, and supramolecular functional assemblies and polymers.

Thus, readers who are engaged in supramolecular chemistry should obtain useful strategies and methods for the design and synthesis of cooperative and synergistic supramolecular systems, which can be applied to highly functional materials of the next generation. We also hope that this book will be useful and informative for graduate students and researchers who are not specialists in supramolecular chemistry but who would like to learn about the frontiers of this field and to realize the ripple effect of cooperative functions on other fields in chemistry.

This book is dedicated to the memory of Professor Hiroshi Tsukube, who made significant contributions to the fields of molecular recognition and supramolecular chemistry. His work and efforts continue to be an inspiration to those working in the field.

Tatsuya Nabeshima
University of Tsukuba

1 Cooperativity and Synergy in Multi-Metal Supramolecular Systems

Tatsuya Nabeshima

CONTENTS

1.1 INTRODUCTION

Cooperative phenomena and synergistic functions in biological systems help to maintain and change pH, ion and molecule concentration, and other factors and also shift equilibria among biologically active chemical species.[1] Such angstrom- and nano-scale chemical events induce subsequent chemical reactions and/or structural changes within biological molecules and molecular assemblies such as proteins, DNA, cells, textures, and even organs.

These processes take place repeatedly and the changes propagate to other biological systems. Consequently, an initial small change at the molecular level is amplified to eventually change structures and functions of a living organism. In other words, information at the molecular level is transferred and amplified to maintain life.

The first step of a cooperative molecular event is to receive an external stimulus, i.e., proton, metal ion, small molecule, light, heat, or pressure to produce conformational and configurational change, formation of self-assembly, or other aggregation. Subsequently, a change of structural, chemical, and/or physical properties (acidity, basicity, reactivity, binding affinity, redox potential, photophysical properties, magnetic properties) of the molecular system occurs and eventually results in cooperative events.

Non-covalent interactions such as hydrogen bonding, $\pi-\pi$ stacking, hydrophobic effect, coordination bonding, cation-π interaction, charge transfer interaction, CH–π interaction, and other events play a key role for controlling these structural changes. These functions based on cooperativity are not simple sums of individual functions. The cooperative functions cannot be generated from simple combinations of individual non-cooperative events. In non-cooperative events, all the constitutional functional units are independent, that is, the units do not communicate with each other.

The cooperative and synergistic events reveal one more very interesting scientific aspect. They are closely related with a unique scientific field known as the *complex system*. It is often said that the complex system exists between chaos and order. This seemingly incomprehensible expression can be understood by employing state function changes familiar to chemists, because chaos and order are created by large positive entropic change and by large negative enthalpic change, respectively.

These metastable states of complex structures such as those observed in biological systems are sustained by correctly responding to external stimuli. Therefore, cooperative control in both artificial and biological systems should be a very useful and effective way to achieve sophisticated functions. There are three types of coopertivity: allosteric, chelate, and interannular. In allostery, conformational change of the allosteric functional molecules is important. In chelate and interannular cooperativities, distance matching of the interacting sites between a chelate and a receptor and also ligating direction are dominant factors.

A coordination bond is one of the most useful tools for the creation of artificial synergistic functional systems because the following properties are available: (1) labile and inert coordination abilities and (2) various geometries and valencies. In addition, metal complexes provide a variety of functions including redox activity, luminescence, catalytic activity, and others for the construction of multi-functional and/or stimuli-responsive molecules with desirable structures. Thus, in this chapter single and multiple metallo-supramolecules that form cooperatively and show cooperative functions are presented.

1.2   SINGLE-METAL SUPRAMOLECULAR SYSTEMS FOR GUEST RECOGNITION

Conformational change is a key factor for allosteric effects on molecular functions. Binding to an effector such as a metal ion or small molecule is the first step in allosteric regulation, which induces a conformational change of the allosteric functional molecule. When the original function of the molecule is enhanced upon effector binding, the effect is called positive allostery. Negative allostery suppresses the function. In reference to molecular recognition as a function of the allosteric system, homotropic and heterotropic are the two terms for describing allosterism. Homotropic means that effector and the guest are the same while heterotropic indicates they are different.

An artificial allosteric ionophore based on complexation with a metal ion as an effector was first synthesized by J. Rebek, Jr.[2] The host **1** consists of a crown ether ring and a 2,2′-bipyridine unit (Figure 1.1). A negative allosteric effect on alkali

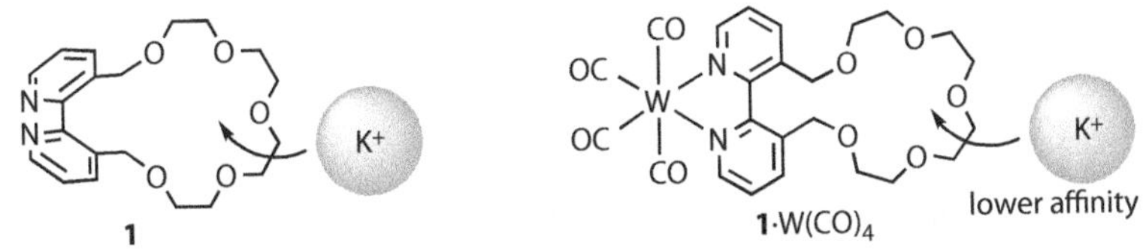

FIGURE 1.1 Negative allosteric effect on alkali metal recognition.

metal binding was observed because crown ring deformation upon complexation with a tungsten ion makes the guest binding unfavorable.

Pseudocrown ethers, whose structures are maintained by coordination bonds instead of covalent bonds like typical crown ethers, are among the most suitable candidates for allosteric regulation of ion binding. A linear podand **2** possessing bipyridine moieties at the ends of the polyether chain was converted easily to the corresponding pseudocrown ether quantitatively by complexation with Cu⁺ (Scheme 1.1).[3] The pseudocrown ether shows a positive allosteric effect on alkali metal ion selectivity in ion transport. The drastic conformational change from a linear to cyclic structure results in a significant macrocyclic effect favorable for ion selectivity.

Pseudocrown ethers **4** with imine functionalities as binding sites for Ni²⁺, Cu²⁺, Zn²⁺ were also synthesized from **3** for recognition of Ba²⁺ in the cavity (Scheme 1.2).[4] Similar crown ether analogues **5** were reported (Figure 1.2).[5] In addition, chiral analogues of **6** were applied to chiral recognition (Figure 1.3).[6]

Switching of helicity of a Cu⁺ complex unit in a pseudomacrocycle **7·Cuᴵ** was achieved by achiral Na⁺ binding (Scheme 1.3).[7] Equilibrium between non-helical and helical forms effectively shifts to the helical form upon Na⁺ binding. This unique helicity control provides a new strategy for controlling helical sense because an achiral guest Na⁺ regulates the chiral structures.

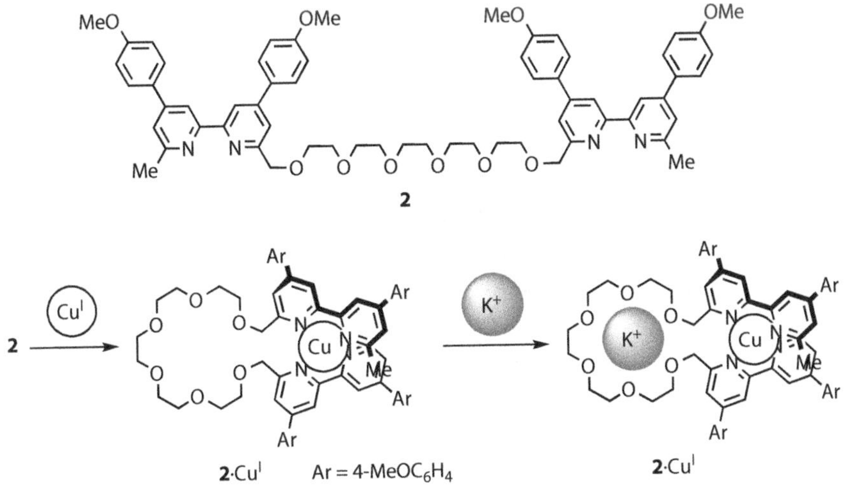

SCHEME 1.1 Allosteric regulation of ion recognition by **2**.

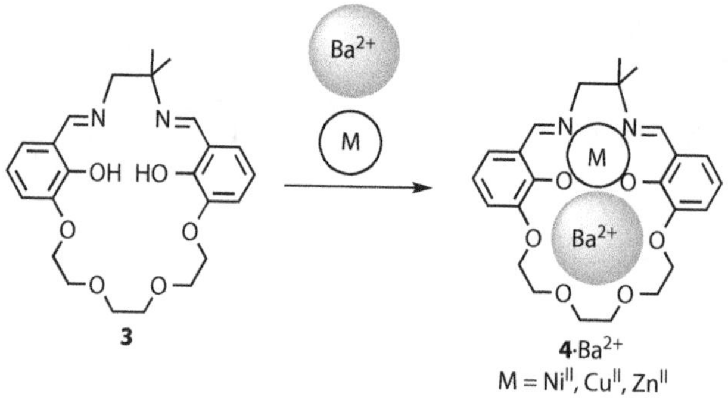

SCHEME 1.2   Ion recognition by **3**.

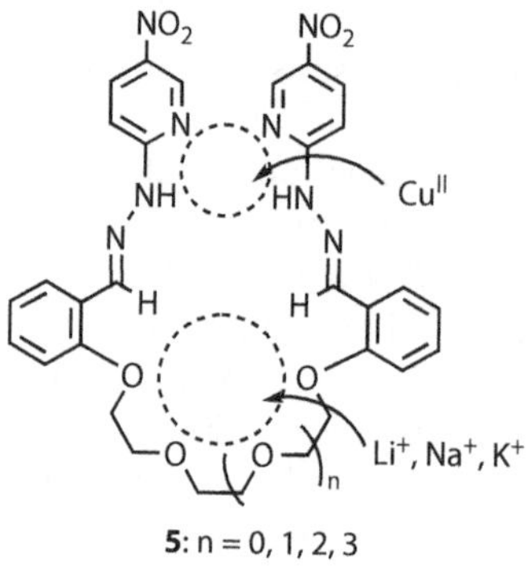

FIGURE 1.2   Allosteric ionophore **5**.

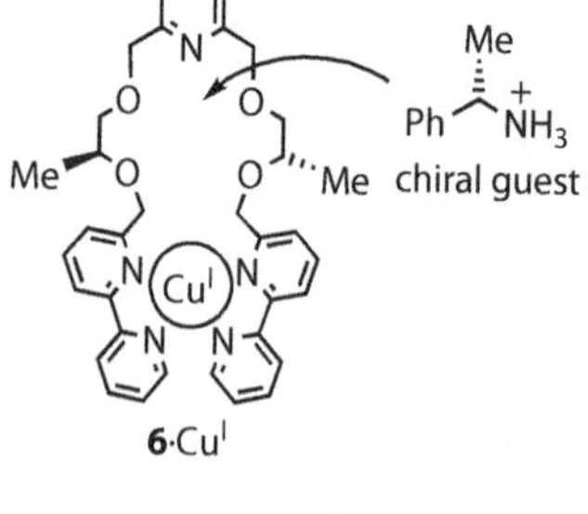

FIGURE 1.3   Chiral pseudocrown ether **6·Cu^I**.

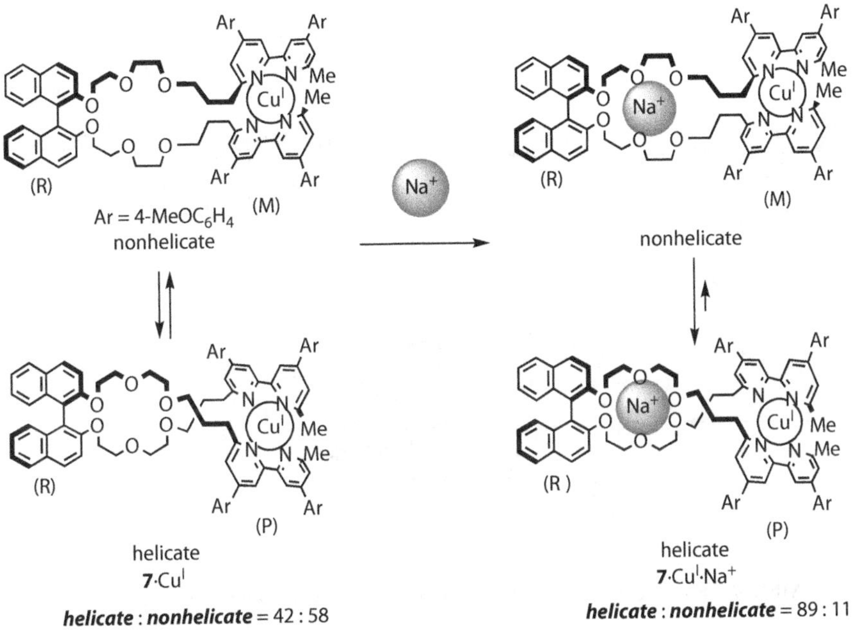

Ar = 4-MeOC$_6$H$_4$

nonhelicate

helicate
7·Cu^I

helicate : nonhelicate = 42 : 58

nonhelicate

helicate
7·Cu^I·Na$^+$

helicate : nonhelicate = 89 : 11

SCHEME 1.3 Control of helical structure of pseudomacrocycle 7·Cu^I.

A larger allosteric efficiency is expected due to larger preorganization effect when a cryptand-like framework is employed. The pseudocryptand 8·Cu^I shows a higher affinity to alkali metal ions and less selectivity than the pseudocrown ether 2·Cu^I.[8] In case of tripodand 9, however, the pseudocryptand formation results in large positive and negative allosteric effects on Cs$^+$ and Na$^+$ binding, respectively (Scheme 1.4).[9]

The macrobicyclic effect and coordination of the polyether oxygen atoms to Cs$^+$ play an important role in the considerable allosteric effects. On the other hand, a CH-π contact was suggested to inhibit Na$^+$ binding due to prohibition of the induced-fit structural change. Tripodand 10 bearing hydroxamic acid units exhibits a homotropic Fe^{3+} binding behavior (Scheme 1.5).[10]

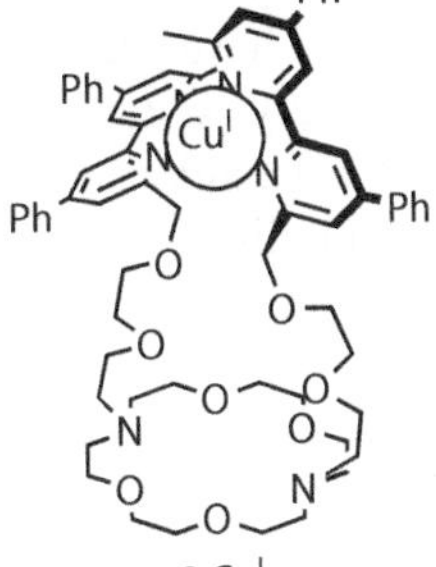

8·Cu^I

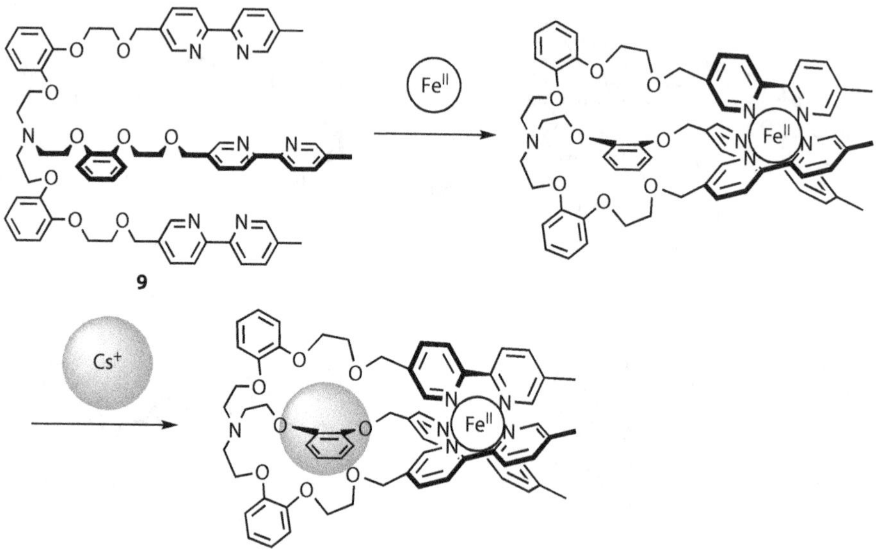

SCHEME 1.4 Positive allosteric effect on ion recognition by **9** and Fe^{II}.

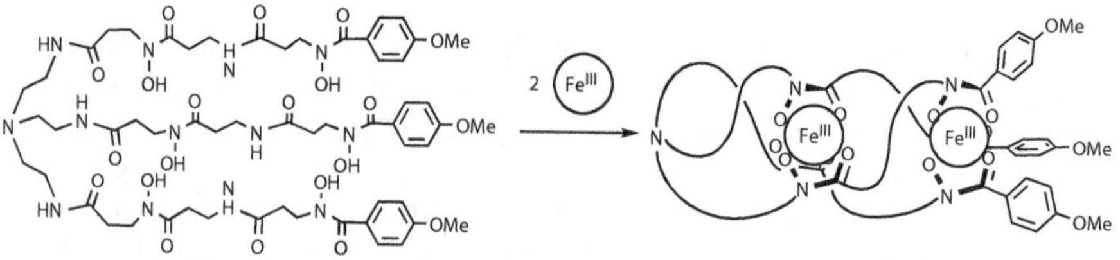

SCHEME 1.5 Cooperative Fe binding by **10**.

When binding sites for anions are introduced into tripodand chains, the pseudocryptand **11**·Fe^{II} acts as an anion receptor.[11] Tripodand **11** consists of an isocyanuric acid core and three chains that have urea moieties as hydrogen bonding sites for anions. X-ray crystallographic analysis indicates that the classical, non-classical, and electrostatic interactions with Fe^{2+}, anion-π interactions, and size effects of the cavities contribute to selective Cl⁻ binding. The Cl⁻ affinity is regulated stepwise by electrochemical redox reactions because the electrostatic interaction depends on the oxidation state of the Fe ion.

A similar Fe pseudocryptand **12**·Fe^{II} shows a strong binding to Cl⁻ due to the strong interaction between the imidazolium moieties and the anion.[12] A conformational

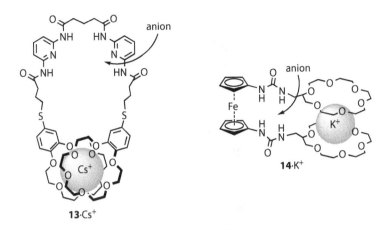

FIGURE 1.4   Biscrown ethers for cooperative anion recognition.

change from linear to cyclic structure works well for regulating anion recognition. Biscrown ether hosts **13**[13] and **14**[14] have urea hydrogen bonding sites (Figure 1.4). Upon complexation with an alkali metal ion, the intramolecular preorganization increases the anion association.

Alkali and heavy metal ions are also utilized as effectors for regulation of anion recognition. A calix[4]arene derivative **15** exhibits drastic stepwise

changes of anion binding by using Na^+ and Ag^+ ions (Figure 1.5).[15] The Ag^+ ion is captured by the bipyridine moieties to give the closed structure considered to be a pseudomacrocycle. Other heavy metal ions such as Ru^{2+} [16], Ba^{2+} [4], and Hg^{2+} [17] are used for cooperative association of receptors bearing macrocyclic-polyether structures with alkali metal or alkaline earth metal ions. Diammonium guest binding of biscrown derivative **16** is also regulated in a negative cooperative fashion (Scheme 1.6).[18]

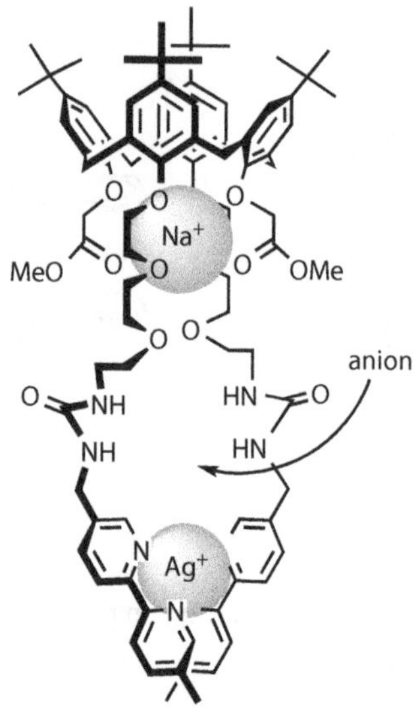

15·Ag⁺·Na⁺

FIGURE 1.5   Multi-step regulation of anion recognition by metal ions.

A biscalix[4]arene host **17** can be applied to homotropic negative cooperativity on alkali metal binding (Figure 1.6).[19] A homotropic positive effect was observed in a multi-binding system **18** which incorporated a porphyrin core and four crown ether moieties (Scheme 1.7).[20] An oligopyrrole host **19** for Ag⁺ ions also shows a homotropic effect on the metal binding (Scheme 1.8).[21]

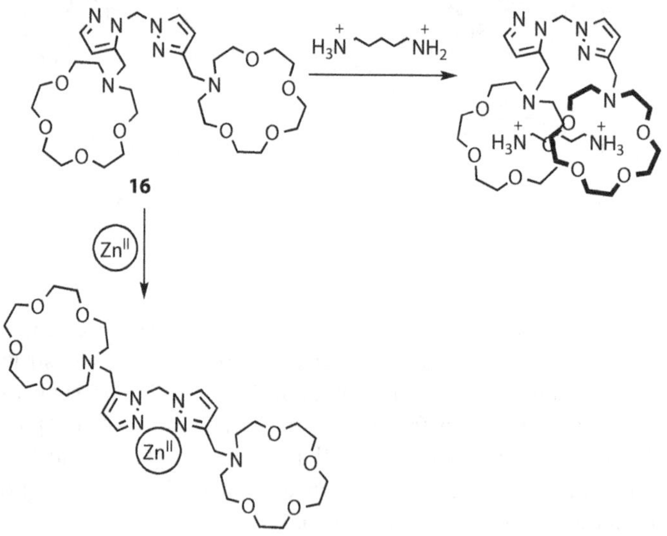

SCHEME 1.6   Inhibition of guest binding by Zn^II.

17

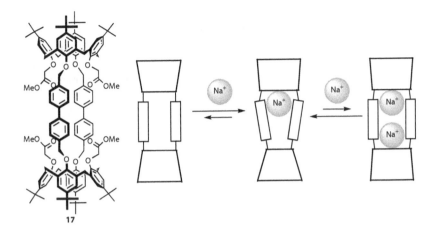

FIGURE 1.6 Biscalix[4]arene **17** for negative ion recognition.

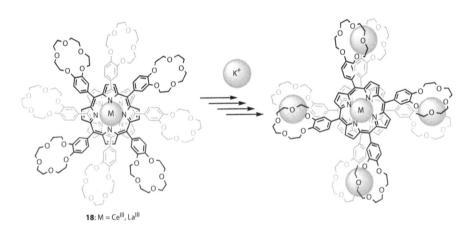

18: M = Ce^{III}, La^{III}

SCHEME 1.7 Ion binding by **18**.

19

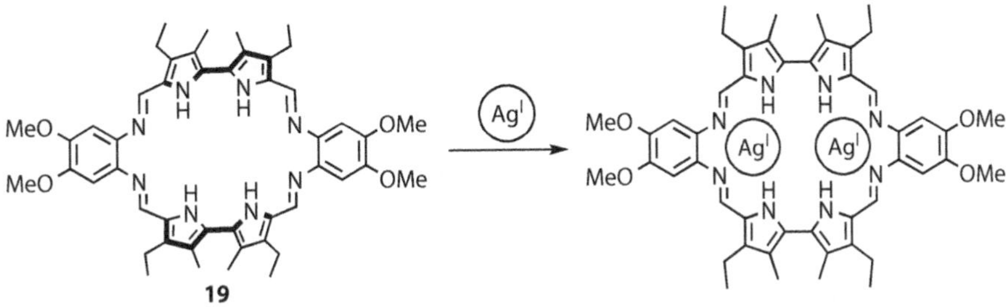

SCHEME 1.8 Ion binding by **19**.

A conformational change induced by the first Ag⁺ binding makes the second Ag⁺ binding more favorable. The calix[4]arene framework is also very useful for the control of salt binding by **20**.[22] As seen in the previous example, the metal ion is considered to act as a cationic effector for anion recognition. In **21**, a Na⁺ ion captured in the multioxygen coordination sites can increase anion affinity through hydrogen bonding.[23]

20·Na⁺

21·Na⁺·Cl⁻

The combination of host conformational change and binding site assembly is an efficient way to enhance alkali-metal-ion recognition. A second-generation ionophore that is formed by complexation with **22** and **23** by multiple hydrogen bonding is recognized more strongly than the host and the first guest molecule.[24]

$R = C_{12}H_{25}$

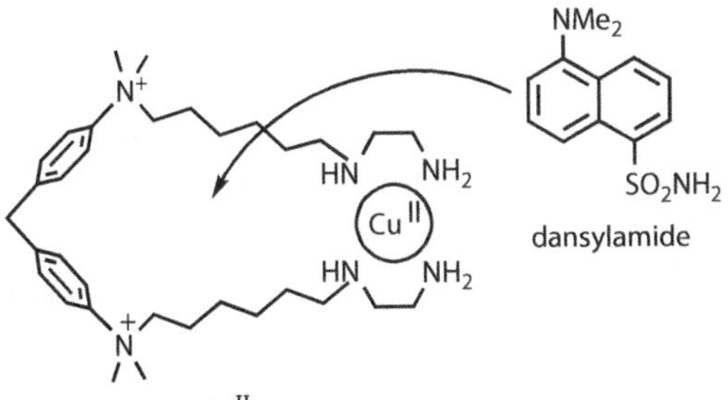

FIGURE 1.7   Guest binding of **24**•CuII.

Molecular recognition can also be well regulated by pseudomacrocycle formation induced by complexation with a metal ion. A Cu²⁺ containing host **24** shows positive cooperativity for dansylamide binding in water (Figure 1.7).[25] Formation of a hydrophobic cavity results in affinity for the much more complicated FMN molecule. Host **25**•Cu¹ transports FMN more efficiently than metal-free host **25** (Figure 1.8).[26a] A similar approach is applied to positive allosteric binding to tryptophane.[26b]

Guest selectivity dependent on a metal ion as an effector was also shown in a pseudomacrocyclic system **26** (Figure 1.9).[27] When the effector is changed from Zn²⁺ and Cu²⁺, the guest selectivity is changed from a naphthalene derivative to a biphenyl one.

Spatial fixation of two calix[5]arene moieties in **27** achieved through Cu⁺ complexation with bipyridines significantly increases their binding affinities for fullerenes C₆₀ and C₇₀ (Figure 1.10).[28] Bishemicarcerand host **28** recognizes a large, less polar bisadamantane derivative due to fixation of the calix[4]arene moieties in the same direction (Figure 1.11).[29]

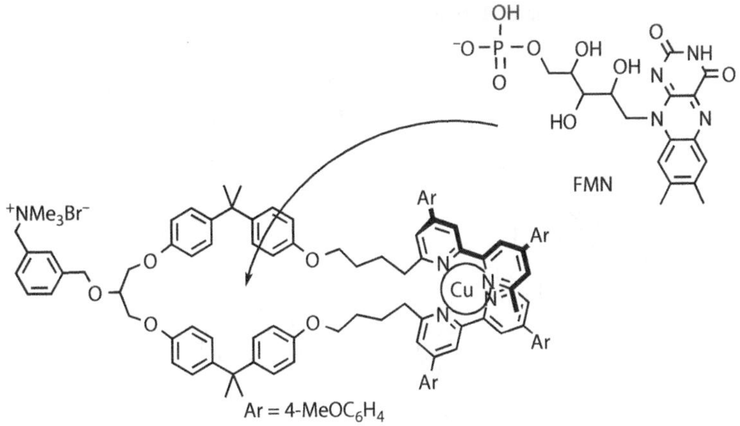

FIGURE 1.8   Guest binding of **25**•Cu¹.

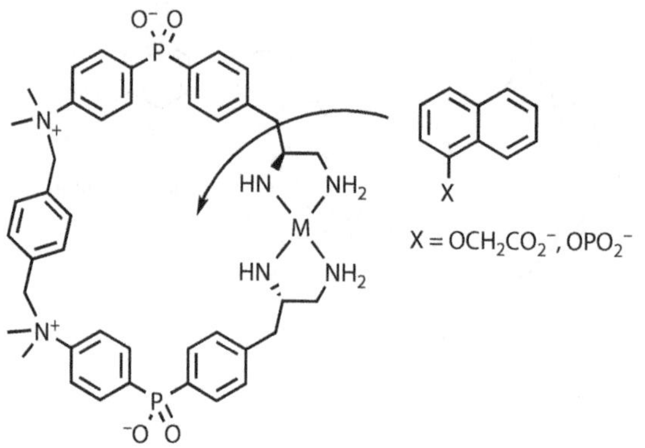

26: M = Zn^{II}, Cu^{II}

FIGURE 1.9   Guest binding of **26**.

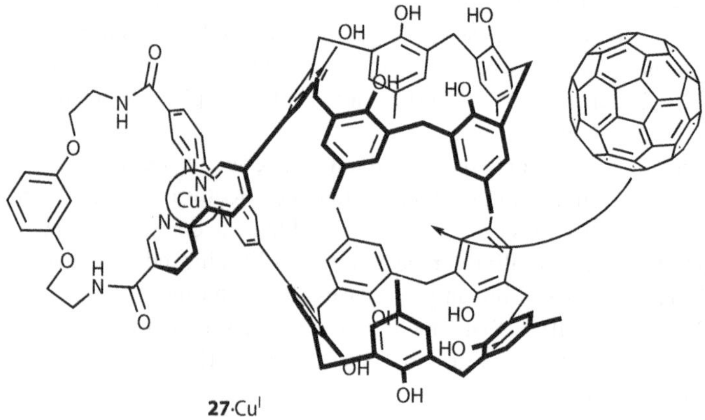

27·Cu^{I}

FIGURE 1.10   Guest binding of **27**•Cu^{I}.

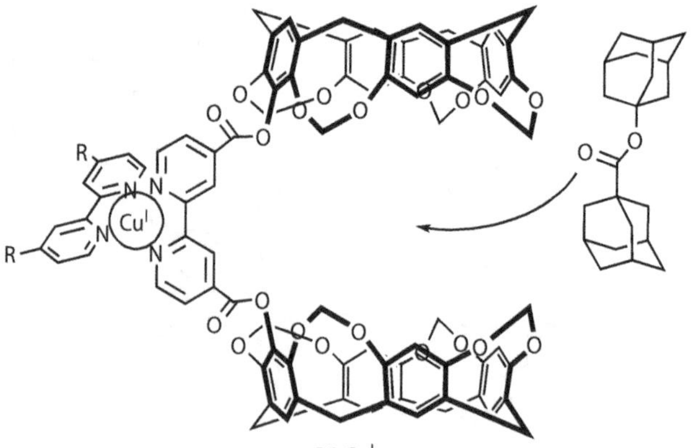

28·Cu^{I}

FIGURE 1.11   Guest binding of **28**•Cu^{I}.

1.3   MULTI-METAL SYSTEMS CONTAINING C = N GROUPS FOR COOPERATIVE COMPLEX FORMATION AND FUNCTIONS

The nitrogen atom of a Schiff base or oxime can coordinate to various metal ions (alkali metal, alkaline earth metal, transition metal, heavy metal ion). In particular, chelates such as salen (bis(salicylidene)ethylenediamine), saloph (bis(salicylidene)-*o*-phenylenediamine), and salamo (bis(salicylideneaminooxy)ethane) afford stable metal complexes, which are utilized for a variety of molecular functions such as catalysis of organic reactions, magnetism, luminescence, etc.[30]

One more interesting feature of such metal complexes is that they can bind to other metal ions, including alkali metal and alkaline earth metal ions by coordination of two negatively charged oxygen atoms in the complex.[31] DFT calculations clearly indicate, which the oxygen atoms are charged more negatively than those of the uncomplexed ligands.

Hetero-multi-nuclear complexes are obtained easily due to this coordination ability. Lanthanide ions and alkali and alkaline earth metal ions can interact with these oxygen atoms. Metallohosts containing salen analogues for cations are designed and constructed on the basis of these binding properties.

Salamo is one of the most useful ligands for the preparation of multi-metal complexes because of its excellent affinity for various metal ions and stability of the ligands toward a C = N cleavage reaction. In addition, salamo metal complexes are usually much more soluble in common organic solvents than the corresponding salen complexes.

The reaction of linear bissalamo **29** with Zn^{2+} quantitatively gives the trinuclear Zn^{2+} complex, which is converted to a variety of heterotrinuclear complexes in a synergistic fashion (Scheme 1.9).[32] The formation of the Zn_3 complex proceeds very cooperatively. The intermediary 1:1 and 1:2 complexes were not observed in the ^1H NMR titration experiment. Only the central Zn was replaced by Ca^{2+} and lanthanide ions.

Although Ca^{2+} selectivity over Mg^{2+} in the transmetalation reaction is extremely high, a lanthanide ion is bound more strongly in the central site to give helical complexes. X-ray crystallographic analysis revealed that the helical structures are more squeezed as the size of the lanthanide ion decreases. In the absence of Zn ions, the lanthanide ions do not interact efficiently with the ligand **29**. This indicates that

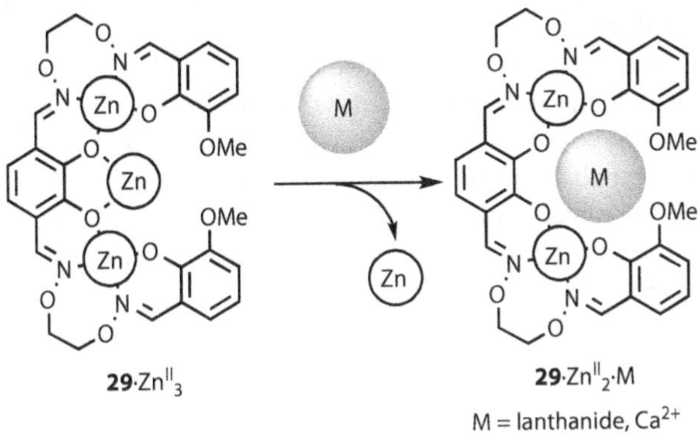

$29 \cdot Zn^{II}_3$ $29 \cdot Zn^{II}_2 \cdot M$

M = lanthanide, Ca^{2+}

SCHEME 1.9   Transmetallization of multi-metal complex $29 \cdot Zn^{II}_3$.

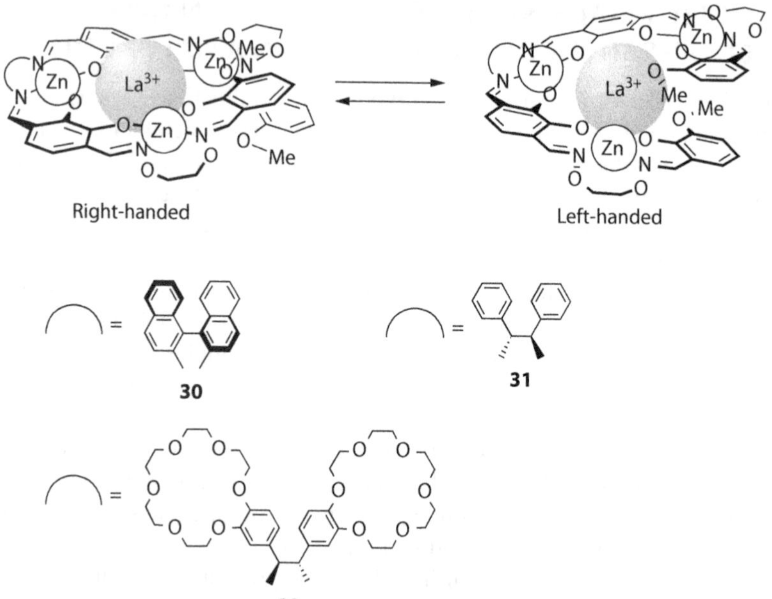

FIGURE 1.12 Multi-metal complexes **30** through **32** with chiral units.

heteronuclear complexes are formed by the synergistic support of complexation with the two Zn ions.

The helical sense of the heteronuclear complexes can be modulated by incorporating a chiral auxiliary into a trisalamo chain. In solution, binaphthyl derivative **30** exclusively gives one helical Zn₃La complex (Figure 1.12).[33] The π–π interaction stabilizes the helical structure. A 1,2-diphenylethylene unit in **31** is not effective for controlling the helical sense in solution.[34] In contrast, the crystals consist of only the left-handed helical complex.

The discrete complexes are arranged helically to yield supercoiled structures. The chiral information of the small chiral part is amplified to result in much larger three-dimensional chiral structures in the crystals.

Helicity of the heterometallic complexes is also regulated by external stimuli. Heterotetranuclear complex **32** has two crown ether rings as a binding site for diammonium guests.[35] According to a molecular leverage motion upon complexation with the guests, the helicity of the complex is changed. The longer guest makes the M-isomer more favorable, while the shorter guest increases the ratio of P-isomer. The multi-step inversion between M and P helicity in heteronuclear complexes of **33** was achieved recently by changing the metal ion (Scheme 1.10).[36]

Most studies on helicity inversion have been based on single-mode regulation, i.e., interconversion between two helically different species. Sequential metal addition results in three-step helicity inversion. This strategy should open new paths for creating different functional compounds with multi-response properties toward different effectors and combinations of effectors.

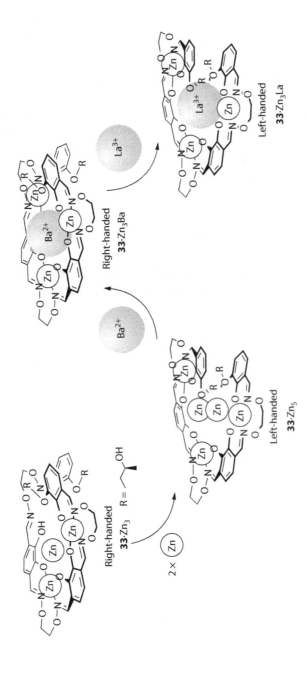

SCHEME 1.10 Multi-step helicity inversion of multi-metal system.

1.4 SYNERGY IN FORMATION OF MULTI-NUCLEAR METAL COMPLEXES POSSESSING TRISALOPH AS LIGAND

Saloph, an analogue of salen, binds to a metal ion to yield the corresponding metal complex, part of which can again interact with another metal ion to afford multi-nuclear complexes. The macrocyclic trimer of saloph known as trisaloph **34**,[37] was expected to yield a trinuclear complex that could bind another metal ion in the central cavity. However, a heptanuclear Zn complex is formed quantitatively by simply mixing the ligand and Zn at room temperature (Scheme 1.11).[38]

Four of the zinc ions are accumulated on the ligand plane. Counter anions of multi-nuclear complexes **35** can control nuclearity finely even though the structural differences among the anions are very small. Acetate anions afford the heptanuclear Zn complex **35** very selectively, while propionate anions result in exclusive formation of the hexanuclear Zn complex **36** (Scheme 1.11). [39]

Interestingly, the Zn_7 complex **35** can be converted to a heteronuclear complex **37** quantitatively by adding one equivalent of La^{3+} ion (Scheme 1.12).[38] Cooperative binding to Zn and La are necessary to form a heteronuclear complex because the metal-free ligand **34** shows no binding affinity for the La^{3+} ion.

Other transition metal ions such as Ni^{2+}, Mn^{2+}, and Cu^{2+} are captured by **34** to produce multi-nuclear metal complexes in high yields.[40] The anions of the heteronuclear Zn_3La complex such as **38** and **39** modulate the structure of the complex.

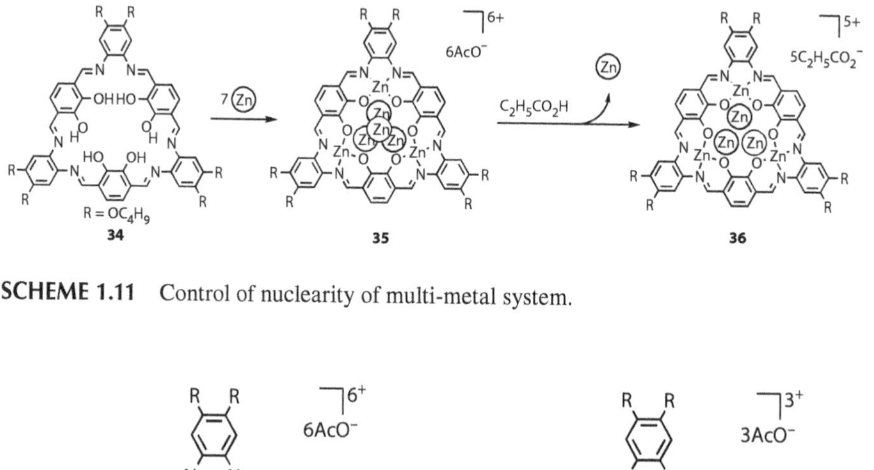

SCHEME 1.11   Control of nuclearity of multi-metal system.

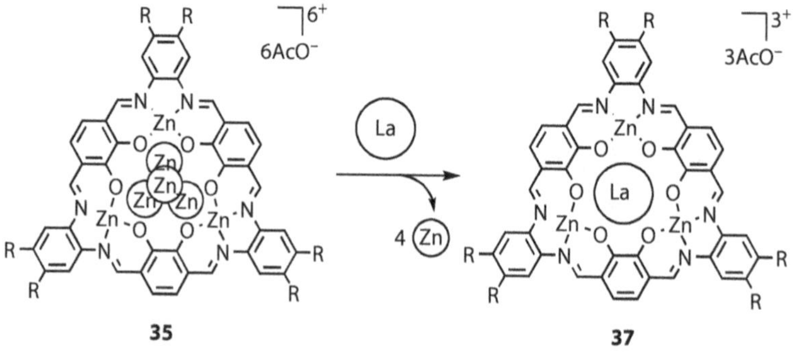

SCHEME 1.12   Transmetallization of multi-metal system.

In particular, 1,3,5-cyclcohexanetricarboxylate yields a C_3 symmetry *uuu*-isomer, probably due to the C_3 symmetric anion structure.[41]

In the homo- and hetero-multi-nuclear metal complexes, the metal ions and counter anions synergistically contribute to the complex formation and the unique structures of the discrete complexes and their assemblies at the meso scale.[42]

R = OC$_4$H$_9$

38   **39**

1.5 CONCLUSION AND PERSPECTIVE

Metal coordination is one of the most useful ways to construct molecular structures cooperatively and control functions because specific bond angles and various functions are available by appropriate selection of metal ions and ligands. Even in mono-metal molecular systems, the metal coordination strategy provides a variety of responsive, cooperative, and synergistic functional molecules. In multi-metal systems, each metal ion often works cooperatively to form unique complexes.

Although some multi-metal complexes can be utilized as hosts, efficient catalysts and functional materials, hetero multi-metal complexes are expected to show more sophisticated functions based on their more complex structures. In biological systems, multi-metal centers provide very high catalytic activities and selectivities that are not achieved by singular metal systems. Combinations of homo- and hetero-metal systems produce more elegant functional systems as seen in photosynthetic cascades. In artificial molecules and molecular assemblies, such combinatorial strategies should create a novel and useful variety of cooperative and synergistic supramolecular functional systems.

REFERENCES AND NOTES

1. Reviews on cooperativity: (a) C. R. Cantor and P. R. Schimmel, in *Biophysical Chemistry*, W. H. Freeman, New York, 1980, pp. 849–1371. (b) M. F. Perutz, *Mechanisms of Cooperativity and Allosteric Regulation in Proteins*, Cambridge University Press, Cambridge, 1989. (c) J. Rebek, Jr., *Acc. Chem. Res.* 17 (1984): 258–264. (d) I. Tabushi, *Pure. Appl. Chem.* 60 (1988): 581–586. (e) C. A. Hunter and H. L. Anderson, *Angew. Chem. Int. Ed.* 48 (2009): 7488–7499. (f) T. Nabeshima, *Bull. Chem. Soc. Jpn.* 83 (2010): 969–991. (g) G. Ercolani and L. Schiaffino, *Angew. Chem. Int. Ed.* 50 (2011): 1762–1768. (h) C. Kremer and A. Lützen, *Chem. Eur. J.* 19, (2013): 6162–6196.
2. J. Rebek, Jr., J. E. Trend, R. V. Wattley and S. Chakravorti, *J. Am. Chem. Soc.* 101 (1979): 4333–4337.
3. T. Nabeshima, T. Inaba, and N. Furukawa, *Tetrahedron Lett.* 28 (1987): 6211–6214.

4. C. J. van Staveren, J. van Eerden, F. C. J. M. van Veggel et al., *J. Am. Chem. Soc.* 110 (1988): 4994–5008.
5. H. Sakamoto, J. Ishikawa, H. Nakagami et al., *Chem. Lett.* (1992): 481–484.
6. Y. Habata, J. S. Bradshaw, X. X. Zhang et al., *J. Am. Chem. Soc.* 119 (1997): 7145–7146.
7. T. Nabeshima, A. Hashiguchi, T. Saiki et al., *Angew. Chem. Int. Ed.* 41 (2002): 481–484.
8. T. Nabeshima, T. Inaba, T. Sagae et al., *Tetrahedron Lett.* 31 (1990): 3919–3922.
9. T. Nabeshima, Y. Yoshihira, T. Saiki et al., *J. Am. Chem. Soc.* 125 (2003): 28–29.
10. S. Blanc, P. Yakirevitch, E. Leize et al., *J. Am. Chem. Soc.* 119 (1997): 4934–4944.
11. T. Nabeshima, S. Masubuchi, N. Taguchi et al., *Tetrahedron Lett.* 48 (2007): 1595–1598.
12. (a) V. Amendola, M. Boiocchi, B. Colasson et al., *Angew. Chem. Int. Ed.* 45 (2006): 6920–6924. (b) K. Sato, Y. Sadamitsu, S. Arai et al., *Tetrahedron Lett.* 48 (2007): 1493–1496.
13. T. Nabeshima, T. Hanami, S. Akine et al., *Chem. Lett.* (2001): 560–561.
14. F. Otón, A. Tárraga, M. D. Velasco et al., *Dalton Trans.* (2005): 1159–1161.
15. T. Nabeshima, T. Saiki, J. Iwabuchi et al., *J. Am. Chem. Soc.* 127 (2005): 5507–5511.
16. P. D. Beer and A. S. Rothin, *J. Chem. Soc., Chem. Commun.* (1988): 52–54.
17. C. J. Baylies, T. Riis-Johannessen, L. P. Harding et al., *Angew. Chem. Int. Ed.* 44 (2005): 6909–6912.
18. J. C. Rodríguez-Ubis, O. Juanes, and E. Brunet, *Tetrahedron Lett.* 35 (1994): 1295–1298.
19. T. Nabeshima, T. Saiki, K. Sumitomo et al., *Tetrahedron Lett.* 45 (2004): 6761–6763.
20. A. Robertson, M. Ikeda, M. Takeuchi et al., *Bull. Chem. Soc. Jpn.* 74 (2001): 883–888.
21. J. L. Sessler, E. Tomat, and V. M. Lynch, *J. Am. Chem. Soc.* 128 (2006): 4184–4185.
22. (a) H. Murakami and S. Shinkai, *J. Chem. Soc., Chem. Commun.* (1993): 1533–1535. (b) J. Scheerder, J. P. M. van Duynhoven, J. F. J. Engbersen et al., *Angew. Chem. Int. Ed. Engl.* 35 (1996): 1090–1093.
23. M. J. Deetz, M. Shang, and B. D. Smith, *J. Am. Chem. Soc.* 122 (2000): 6201–6207.
24. T. Nabeshima, T. Takahashi, T. Hanami et al., *J. Org. Chem.* 63 (1998): 3802–3803.
25. H.J. Schneider and D. Ruf, *Angew. Chem. Int. Ed. Engl.* 29 (1990): 1159–1160.
26. (a) T. Nabeshima, A. Hashiguchi, S. Yazawa et al., *J. Org. Chem.* 63 (1998): 2788–2789. (b) T. Nabeshima and Hashiguchi, *Tetrahedron Lett.* 43 (2002): 1457–1459.
27. F. Wang and A. W. Schwabacher, *J. Org. Chem.* 64 (1999): 8922–8928.
28. T. Haino, Y. Yamanaka, H. Araki et al., *Chem. Commun.* (2002): 402–403.
29. A. Lützen, O. Haß, and T. Bruhn, *Tetrahedron Lett.* 43 (2002): 1807–1811.
30. Selected reviews on salen complexes: (a) A. W. Kleij, *Chem. Eur. J.* 14 (2008): 10520–10529. (b) H. L. C. Feltham and S. Brooker, *Coord. Chem. Rev.* 253, (2009): 1458–1475. (c) S. Akine and T. Nabeshima, *Dalton Trans.* (2009): 10395–10408.
31. (a) I. Ramade, O. Kahn, Y. Jeannin et al., *Inorg. Chem.* 36 (1997): 930–936. (b) L. Carbonaro, M. Isola, P. L. Pegna et al., *Inorg. Chem.* 38 (1999): 5519–5525. (c) D. Cunningham, P. McArdle, M. Mitchell et al., *Inorg. Chem.* 39 (2000): 1639–1649.
32. (a) S. Akine, T. Taniguchi, T. Saiki et al., *J. Am. Chem. Soc.* 127 (2005): 540–541. (b) S. Akine, T. Taniguchi, and T. Nabeshima, *J. Am. Chem. Soc.* 128 (2006): 15765–15774.
33. (a) S. Akine, T. Matsumoto, and T. Nabeshima, unpublished results. (b) T. Nabeshima and M. Yamamura, *Pure Appl. Chem.* 85 (2013): 763–776.
34. S. Akine, T. Matsumoto, and T. Nabeshima, *Chem. Commun.* (2008): 4604–4606.
35. S. Akine, S. Hotate, and T. Nabeshima, *J. Am. Chem. Soc.* 133 (2011): 13868–13871.
36. S. Akine, S. Sairenji, T. Taniguchi et al., *J. Am. Chem. Soc.* 135 (2013): 12948–12951.
37. S. Akine, T. Taniguchi, and T. Nabeshima, *Tetrahedron Lett.* 42 (2001): 8861–8864.
38. T. Nabeshima, H. Miyazaki, A. Iwasaki et al., *Chem. Lett.* 35 (2006): 1070–1071.
39. M. Yamamura, H. Miyazaki, M. Iida et al., *Inorg. Chem.* 50 (2011): 5315–5317.
40. T. Nabeshima, H. Miyazaki, A. Iwasaki et al., *Tetrahedron* 63 (2007): 3328–3333.
41. M. Yamamura, M. Iida, K. Kanazawa et al., *Bull. Chem. Soc. Jpn.* (2014): 334–340.
42. M. Yamamura, M. Sasaki, M. Kyotani et al., *Chem. Eur. J.* (2010): 10638–10643.

2 Hierarchically Assembled Titanium(IV) Helicates

Markus Albrecht

CONTENTS

2.1 INTRODUCTION

Helicates are defined as metallo-supramolecular coordination compounds in which linear ligands wrap around several metal ions in a helical fashion.[1] The helicate term was introduced by J. M. Lehn in 1987,[2] and from that point the chemistry of helicates became a prominent subdiscipline of supramolecular chemistry. Initially, helicates were considered simple model compounds explaining mechanistic aspects of self-assembly. Recently, special properties of the oligonuclear complexes became more important for developing novel materials based on the helicate motif.[3]

Helicates are formed in cooperative metal-directed self-assembly processes, usually from linear ligand strands possessing several metal coordination sites and appropriate numbers of metal ions (Scheme 2.1, top). As an alternative, helicates are obtained in a hierarchical process by initial formation of mononuclear complexes connected by non-covalent linkers (e.g., metal ions) in a subsequent step. Thus, in the

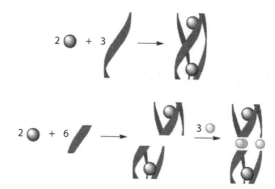

SCHEME 2.1 Self-assembly of triple stranded dinuclear helicate from three linear ligand strands and two metal ions (top). Hierarchical self-assembly of helicates by initial formation of mononuclear complexes followed by metal coordination of two units to several bridging metal ions (bottom).

hierarchical self-assembly of helicates, synergistic effects between the formation of the non-covalently linked ligand and the terminal metal complexes leads to metallo-supramolecular structures.

Hierarchical processes provide facile routes to supramolecular complexes[4] and allow easy functionalization of aggregates by modification of the starting ligand systems. A representative example of such a hierarchically assembled helicate with catechol phosphane ligands and bridging trans-palladium(II)dibromide is shown in Figure 2.1.[5]

In this chapter, the hierarchical self-assembly of titanium catecholate helicates will be described and functionalized compounds and stereochemical aspects will be discussed. In addition, related systems based on other metals and other ligands (8-hydroxyquinoline) will be introduced briefly. In the described examples, self-assembly relies on the synergism between binding of lithium cations to salicylate units and a series of other metal ions to catecholates or hydroxyquinolates.

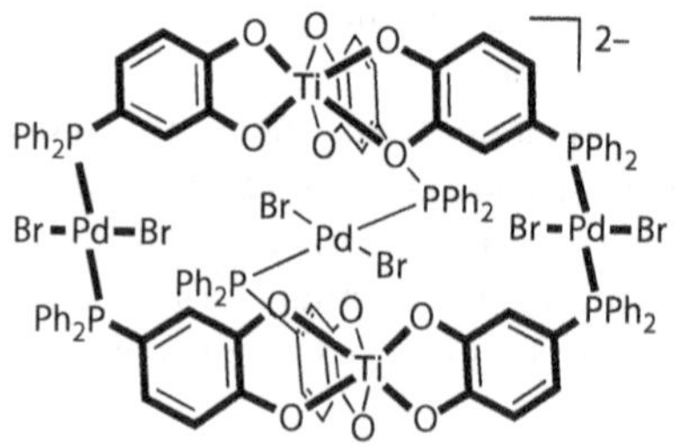

FIGURE 2.1 Hierarchically assembled triple stranded helicate described by Raymond and Wong.

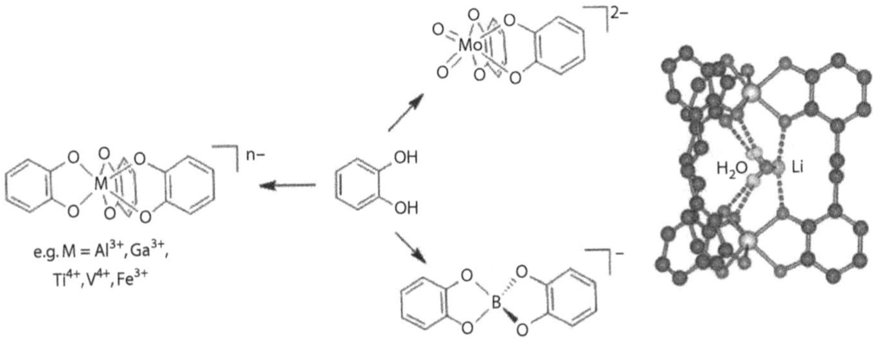

SCHEME 2.2   Catechol complexes with different ligand:metal ratios and solid state structure of a dinuclear titanium(IV) helicate with ethylene-bridged dicatechol ligands.

The chemistry of catecholate metal coordination compounds is well explored. The dianionic ligand forms trisligand complexes with a series of metal cations.[6] Bis- and monoligand compounds are found as well but are less common.[7]

2.2   LITHIUM-CONTROLLED ASSEMBLY OF CATECHOL TITANIUM HELICATES: SOLID STATE, SOLUTION, AND GAS PHASE STUDIES

In the early 1990s, covalently linked biscatechol ligands were introduced by us[8] and others[9] as appropriate ligands for the self-assembly of dinuclear helicates. The resulting complexes were investigated intensely with a focus on different aspects of their chemistry.

As a representative example, the solid state structure of the corresponding ethylene-bridged complex is shown in Scheme 2.2 and exhibits the special role of one lithium counter ion that stabilizes the helicate in concert with two molecules of water as a template.[10] Later, bisimine-bridged ligands were also used to examine this class of coordination compounds. In this case, the helicates can be formed by simply mixing a diamine, 2,3-dihydroxybenzaldehyde, a source of titanium(IV) ions, and an alkali metal carbonate as a base.[11] This procedure in principle resembles a hierarchical process with an initial imine condensation followed by helicate self-assembly, although the order of events cannot be predicted definitively.

A more defined hierarchical reaction sequence in which metal coordination occurs first followed by bridging of the monomeric complex units by three (or two) ligand strands is the focus of this chapter and is shown in Scheme 2.3. Reaction of the ligand with the metal affords monomeric complexes that, in the presence of lithium cations, are in equilibrium with the homochiral dimer.

A series of dinuclear triple-stranded lithium-bridged titanium helicates has been characterized by x-ray diffraction studies. Figure 2.2 shows a representative example of the structure of the anionic dinucelar complex [Li3{(1)6Ti2}]−. Two

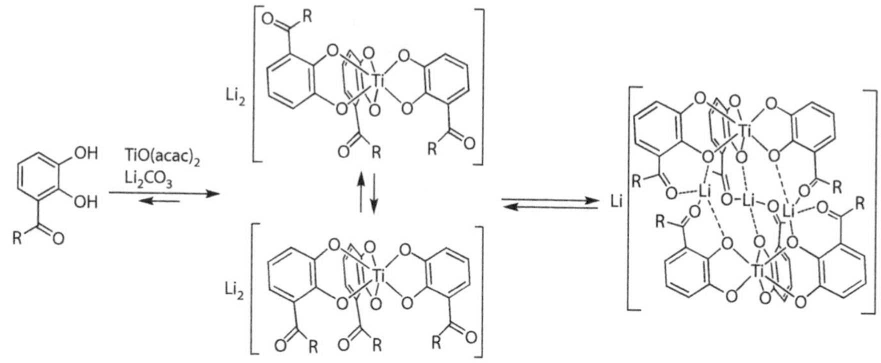

SCHEME 2.3 Hierarchical assembly of triple lithium-bridged dinuclear titanium(IV) helicates by initial formation of mononuclear triscatecholates that are in equilibrium with dinuclear lithium bridged helicates.

tris(carbaldehydecatecholate) titanium(IV) complex units are present. Both show syn orientations of the aldehyde units and adopt distorted octahedral geometries at the metal centers. Each ligand of one of the complex units coordinates through a salicylate moiety (aldehyde oxygen and internal catecholate oxygen) to a lithium ion bridging to the second titanium(IV) complex.

The lithium cation adopts a distorted tetrahedral geometry. This enforces the relative stereochemistry at the two triscatecholate complexes. Dimerization is possible only if both units are homochiral, resulting in the $\Lambda\Lambda$ or $\Delta\Delta$ complex, respectively. Thus, the structure can be described as a helicate in which three lithium bis-salicylates act as ligand spacers (Figure 2.2).[12]

Crystallization leads only to the dinuclear titanium(IV) helicates. However, the monomer–dimer equilibrium in solution can be investigated easily by using different NMR techniques. In the dimer, the protons of R near the carbonyl unit of one complex moiety are neighbored by an aromatic catecholate of the second and thus experience an anisotropic shift to high field. In the monomers, the enantiomeric Λ and Δ complexes equilibrate very quickly, generating only one signal of CH_2 groups at the substituent R. In the dimer, the stereochemistry is locked and no fast racemization

FIGURE 2.2 Structure of monoanionic dinuclear titanium(IV) complex $[Li_3\{(1)_6Ti_2\}]^-$ in crystal. Left: structural representation. Center: side view. Right: top view along Ti–Ti axis.

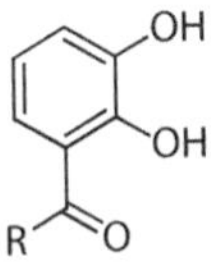

Aldehyde: R = H: **1**-H$_2$
Ketone: R = Alkyl, benzyl: **2**-H$_2$
Ester: R = OR': **3**-H$_2$ (R' = Me: **a**, R' = Et: **b**)

FIGURE 2.3 Selected aldehyde-, keto-, or ester-substituted catechol ligands.

takes place and separated diastereotopic protons of the methylene units may be observed. DOESY NMR allows an unambiguous assignment of the monomeric and dimeric species.

Dinuclear triple lithium-bridged helicates are obtained with 3-aldehyde, 3-keto, or 3-ester substituted ligands. Most substituents allow the observation of the mono-mer–dimer equilibrium in solution (Figure 2.3). The equilibrium between the mono-meric and lithium-bridged dimeric species depends on various factors.

First, the equilibrium is influenced strongly by the nature of the carbonyl group at the ligand. The tendency to form the dimer increases in the order of aldehyde < ketone < ester. This is due to the increasing donor ability of the carbonyl units and the corresponding tendency to bind the lithium cations within the dimer. The dimerization constant K_{dim} of the aldehyde in methanol at room temperature was determined to be $K_{dim} = 10$ M^{-1} for the methylketone (R = CH$_3$), $K_{dim} = 3715$ M^{-1}, and for the ester (R = OCH$_3$) $K_{dim} = 25600$ M^{-1}.

Second, the size of the substituent R influences the formation of the dimer versus the monomer. The substituent of one complex is located near a catecholate of the second. It thus destabilizes the dimer if the group becomes too bulky. This is exem-plified at room temperature in methanol for the ketone series with R = Me ($K_{dim} = 3715$ M^{-1}), Et ($K_{dim} = 785$ M^{-1}), or iPr ($K_{dim} = 3.6$ M^{-1}).

Third, lithium ions in the dimer are embedded well within the complex. To remove the bridging ions, the solvent must be able to solvate this kind of cation. Highly polar solvents dissolve lithium cations well and thus destabilize (disrupt) the dimer. Non-polar solvents are not able to compete with the salicylate binding site for lithium and the dimer will be favored. At room temperature, the following series of dimerization constants was obtained for the aldehyde derivative Li[Li$_3${(**1**)$_6$Ti$_2$}]:

$$\text{DMSO-d}_6, \text{D}_2\text{O (only monomer)} < \text{CD}_3\text{OD } (K_{dim} = 10 \text{ M}^{-1})$$
$$< \text{THF-d}_8 \ (K_{dim} = 950 \text{ M}^{-1})$$
$$< \text{acetone-d}_6 \ (K_{dim} = 1330 \text{ M}^{-1})$$

Specific K_{dim} values may be obtained by selecting the appropriate solvent mixtures.

Finally, the central metal of the triscatecholate unit plays a crucial role as well for the equilibrium of the monomeric and dimeric complexes. Figure 2.4 depicts helicates with two titanium(IV) or gallium(III) complex units for comparison. Due to the different charges of the central metal ions, the lithium-bridged titanium(IV) complexes possess a total charge of 1 and the corresponding gallium(III) complex a charge of 3. Due to the higher charge of the latter, the embedded lithium cat-ions are attracted far more strongly and the gallium dimer is more stable than that of titanium. In the case of the aldehyde system, the switch from titanium(IV) to

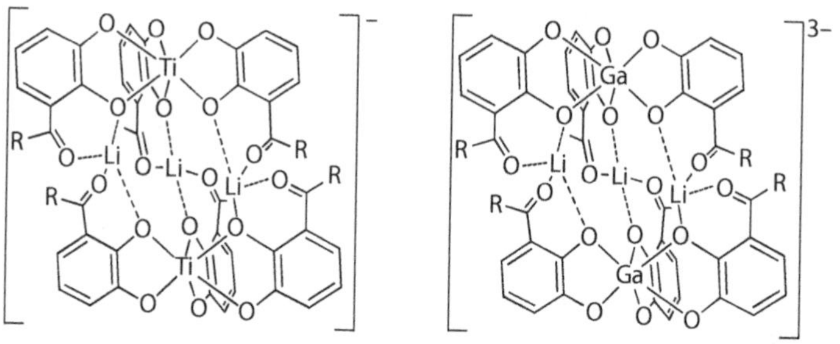

FIGURE 2.4 Comparison of monoanionic dinuclear titanium(IV) and trisanionic gallium(III) helicates.

gallium(III) enhances the dimerization constants in methanol-d_4 from $K_{dim} = 10$ M^{-1} to $K_{dim} = 200000$ M^{-1}.

In addition to the control mechanisms for the dimerization process, its thermodynamics were investigated as well. Van't Hoff plots for the dimerization behavior of the aldehyde-derived dinuclear titanium(IV) or gallium(III) helicates revealed that the dimerization is both enthalpically and entropically driven. It is expected that enthalpy change is favorable due to the formation of the salicylate–lithium cation complex formation. Scheme 2.4 presents an explanation for the initial "unusual" entropic driving force of the formation of one dimer from two monomers. Considering all species that participate in the reaction (including solvents) reveals that lithium binding within the dinuclear helicate liberates many solvent molecules that are bound initially to the lithium cations, leading to a higher level of entropy of the system.

The lithium-controlled hierarchical assembly of the helicates, the monomer–dimer equilibrium, and the dynamic processes can be investigated as crystals or in solution. Gas phase studies can be performed by electrospray ionization mass spectrometry (ESI MS). The measurements help explain the solution behaviors of the compounds. However, gas phase reactivity can be studied as well by collision-induced decay (CID) experiments. ESI MS revealed the following findings.

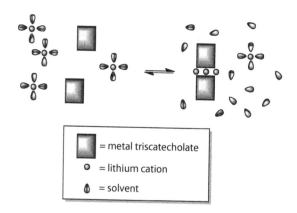

= metal triscatecholate

= lithium cation

= solvent

SCHEME 2.4 Monomer–dimer equilibrium showing involvement of solvent molecules.

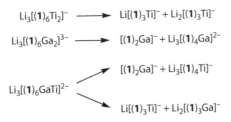

SCHEME 2.5 Gas phase decompositions of $[Li_3\{(\mathbf{1})_6Ti_2\}]^-$, $[Li_3\{(\mathbf{1})_6Ga_2\}]^{3-}$, and $[Li_3\{(\mathbf{1})_6TiGa\}]^{2-}$ shown by CID experiments.

First, directly after dissolution of the complexes, the dimer is the observed species that then generates more and more of the monomer until equilibrium is reached. Second, mixing two similar lithium-bridged complexes bearing different R substituents leads to a slow exchange of ligands. The exchange of monomeric triscatecholate complex units is fast while ligand scrambling based on dissociation of the complexes is much slower.

Third, in CID experiments, the triple lithium-bridged helicates $[Li_3\{(\mathbf{1})_6Ti_2\}]^-$ and $[Li_3\{(\mathbf{1})_6Ga_2\}]^{3-}$ and the heterodinuclear $[Li_3\{(\mathbf{1})_6TiGa\}]^{2-}$ (obtained within a mixture by mixing the corresponding homodinuclear complexes in a 1:1 ratio) were mass selected in the gas phase and studied in MS/MS experiments using argon as the collision gas. The observed initial decompositions are summarized in Scheme 2.5.

The titanium complex fragmentizes symmetrically, leading to triscatecholate complexes (the neutral species cannot be observed by MS). In case of the gallium helicate alone, an asymmetric decomposition to a biscatecholate and a species involving four catecholates results. This is probably due to the high thermodynamic gas phase stability of gallium biscatecholate bearing only a single negative charge.

For the heterodinuclear $[Li_3\{(\mathbf{1})_6TiGa\}]^{2-}$, both decomposition pathways are observed. On the one hand, symmetric decomposition into the triscatechol gallium(III) and triscatechol titanium(IV) species occurs. Conversely, the asymmetric formation of a tetracatecholate titanium(IV) and a biscatecholate gallium(III) species takes place. Formation of a dicatecholate titanium derivative and a tetracatecholate gallium species was not observed.[12]

2.2.1 HIERARCHICALLY ASSEMBLED DINUCLEAR TITANIUM(IV) HELICATES AS DENDRIMERS

An understanding of the fundamental principles and control mechanisms of the hierarchical formation of dinuclear helicates is a prerequisite to modifying such systems further and obtaining more highly functionalized derivatives. In a first attempt, dendritic side groups were attached to catecholate ligands. The **3c** through **e**-H$_2$ esters were prepared to represent dendritic arms of generation 0 (G0), generation 1 (G1), and generation 2 (G2); see Figure 2.5.

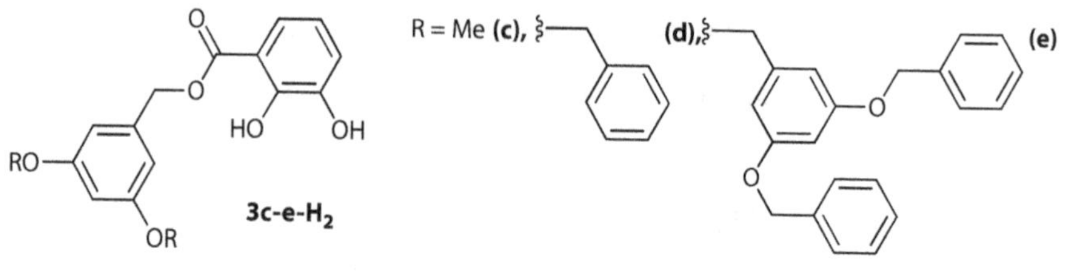

FIGURE 2.5 Ligands **3c-e-H$_2$** bearing dendritic side chains for hierarchical assembly of dendrimers with dinuclear titanium helicate cores.

All ligands **3c** through **e-H$_2$** form coordination compounds with titanium(IV) ions that possess dendrimer structures. Due to the non-polar Frechet-type dendritic substituents, the compounds are soluble only in solvents with low polarities (mixtures of methanol and chloroform). Consequently, the dimeric complexes are observed as the major species in solution. Figure 2.6 shows the structural formula of Li[Li$_3${(3e)$_6$Ti$_2$}] and the structure derived from x-ray diffraction of the anion of the G0 dendrimer [Li$_3${(3c)$_6$Ti$_2$}]$^-$. The central triple lithium-bridged part and the dendritic substituents decorating its periphery may be seen.

As a curiosity, DOSY NMR did not allow the separation of the data sets of the monomeric and the dimeric complexes. This is ascribed to an interlocking of the dendritic side chains upon formation of the dinuclear helicates. Based on this, the size of the dendrimer does not change but its density does.[13]

2.2.2 STEREOCHEMISTRY OF HIERARCHICALLY ASSEMBLED TITANIUM(IV) HELICATES: A STEREOCHEMICAL SWITCH

Introduction of catechol ligands that bear stereocenters at the substituent Rs of the catechol ligands should help induce specific stereochemistry at the complexes (Λ versus Δ). For this purpose, the ligands **3f-H$_2$** and **3g-H$_2$** were prepared and submitted to a coordination study (Figure 2.7). As noted above, the monomeric units racemized quickly while the stereochemistry in the dimers is locked.

If the stereochemical information at the substituent is located far from the complex units, stereocontrol is low. Consequently, ligand **3f-H$_2$** results in the helicate Li[Li$_3${(3f)$_6$Ti$_2$}] with a diastereomeric excess (de) of 20%. In Li[Li$_3${(3g)$_6$Ti$_2$}], the stereocenters are much closer to the helicate and in methanol only one enantiomerically pure diastereoisomer is observed.[14]

Recording the circular dichroism (CD) spectra of Li[Li$_3${(3g)$_6$Ti$_2$}] in either methanol or DMSO results in two nearly opposite spectra as shown in Figure 2.8. The strong transitions around 350 and 410 nm can be assigned to ligand-to-metal charge transfer bands. Based on earlier CD spectroscopic investigations of chiral triscatechol titanium(IV) complexes, it appears that in DMSO the metal complexes possess Λ configuration and in methanol a Δ configuration.[15] The stereochemistry at the substituent is fixed as S. However, it induces different configurations at the metal complex units based on the solvent.[14]

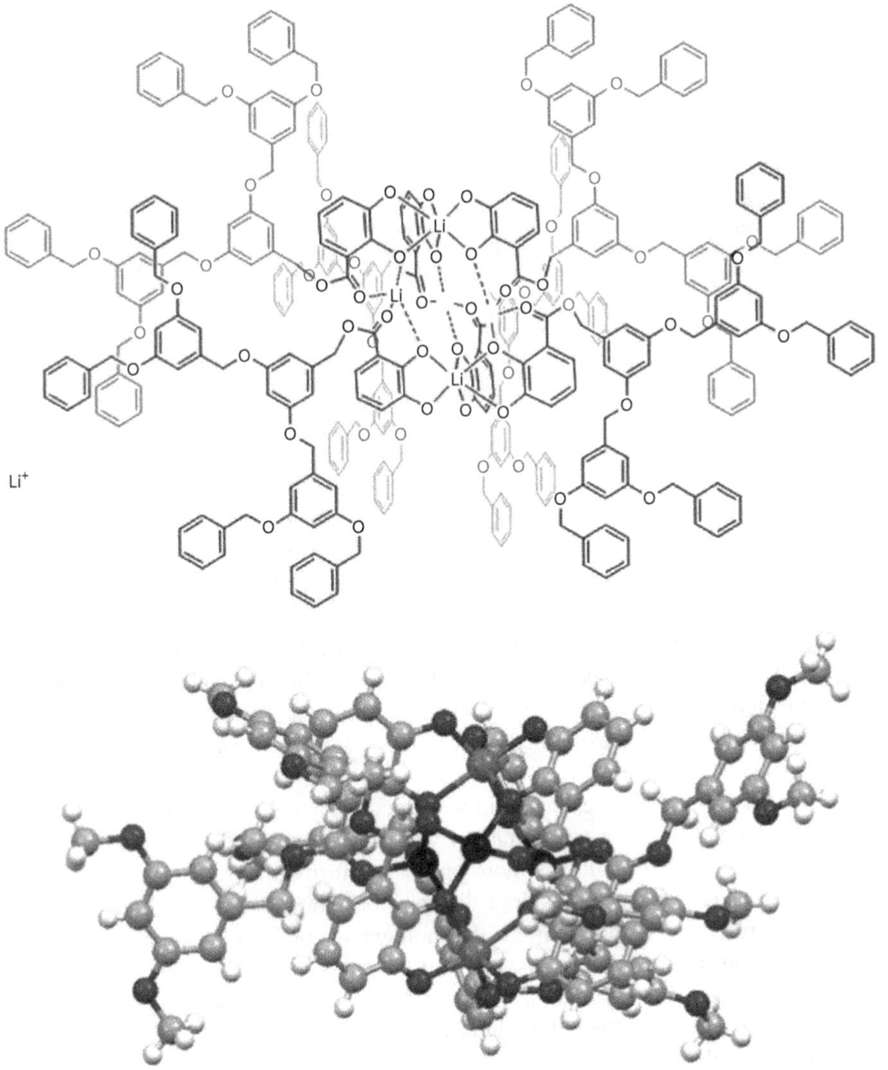

FIGURE 2.6 G2 dendrimer Li[Li$_3${(3e)$_6$Ti$_2$}] (top) and x-ray structure of G0 dendrimer [Li$_3${(3c)$_6$Ti$_2$}]$^-$ (bottom).

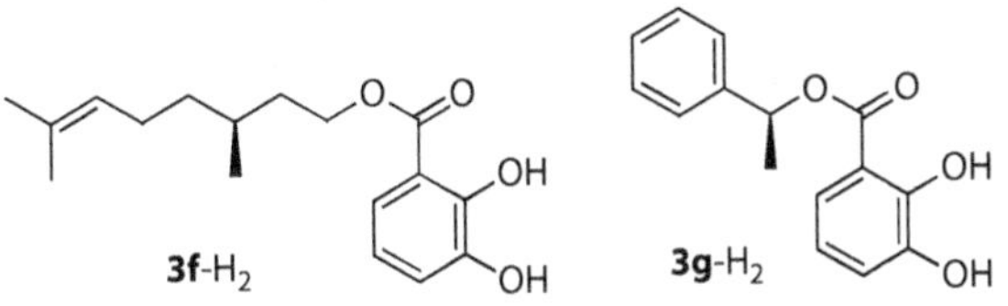

3f-H$_2$ **3g-H$_2$**

FIGURE 2.7 Enantiomerically pure ligands for stereoselective hierarchical assembly of helicates.

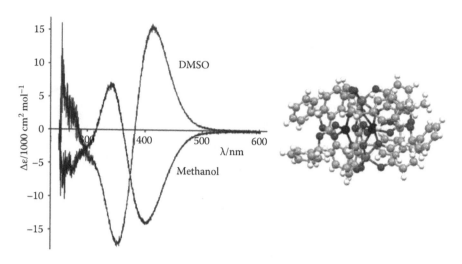

FIGURE 2.8   CD spectra of Li[Li$_3${(**3g**)$_6$Ti$_2$}] in methanol and in DMSO and crystal structure of dimer showing S-configured side chains and Λ configuration at metal (*Source:* M. Albrecht, E. Isaak, M. Baumert et al., *Angew. Chem. Int. Ed.* **2011**, *50*, 2850. With permission.)

The observed unusual solution behavior of Li[Li$_3${(**3g**)$_6$Ti$_2$}] is due to the preferences of the monomeric species in DMSO and the dimeric species in methanol. Thus, the stereochemical induction of the ester substituent must be different in the two species. This can be explained by different conformational orientations of the side chains.

It is well known that an ester group with secondary α-carbon centers at the alkoxide component adopts a conformation with the CH unit syn to the C = O carbonyl moiety.[16] Thus, this unit is more or less conformationally rigid. As depicted in Figure 2.9, the ester groups adopt different orientations in the monomeric and dimeric coordination compounds. In the monomer, repulsion between the electron pairs of the internal catecholate oxygen and the carbonyl oxygen atom takes place, leading to an outward orientation of the carbonyl (and thus of the α-CH). In the dimer, the carbonyl (and the α-CH) have to point inward to chelate lithium together with the catecholate oxygen.

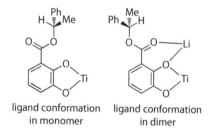

ligand conformation   ligand conformation
in monomer   in dimer

FIGURE 2.9   Various orientations of ester substituent in mononuclear Li$_2$[(**3g**)$_3$Ti] or dinuclear Li[Li$_3${(**3g**)$_6$Ti$_2$}].

The ester has to turn by 180 degrees upon transformation of the monomer into the dimer and vice versa. This switches the orientations of the large phenyl and small methyl groups. The steric pressure of the phenyl moiety onto the twisted titanium chelate complex reverses its direction, leading to a preference of the opposite configuration. This is an impressive example in which the control of rotation around a single bond influences the stereochemistry of a complex molecular aggregate.[14]

2.3 HIERARCHICALLY ASSEMBLED DINUCLEAR DOUBLE-STRANDED HELICATES UTILIZING BORON(III) AND CIS-DIOXOMOLYBDENUM(VI) AS CENTRAL IONS

Titanium(IV) and gallium(III) ions form triscatecholate coordination compounds. Substitution of those central metal ions by others should allow the formation of related hierarchically assembled lithium-bridged helicates possessing different structural features. The use of boron(III) instead of gallium(III) leads to the formation of double-stranded lithium helicates $Li[Li_2\{(1/3)_2B_2]$ with the ligands $1-H_2$, $3a-H_2$, and $3b-H_2$. In those complexes boron(III) occupies the catecholate binding sites and has a near-tetrahedral geometry. The lithium ions are coordinated by the salicylate units as observed in the corresponding triple-stranded titanium(IV) complexes (see Figure 2.10).[17]

The complexes are soluble only in solvents like chloroform and only the dimer can be observed by NMR spectroscopy. In solution, the boron(III) complexes show some interesting fluorescence properties.[17]

Cis-molybdenum(VI) dioxide forms biscatecholate coordination compounds. The third position is already blocked by the two cis-oriented oxo ligands. The reaction of bis(acetylacetonate) cis-dioxomolybdenum with carbonyl-substituted catecholates in the presence of lithium carbonate results in the formation of bis catecholate cis-dioxomolybdenum(VI) complexes that are in equilibrium with their dimers.

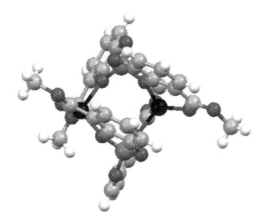

FIGURE 2.10 Structure of $[(3a)_2B_2]^-$ in crystal.

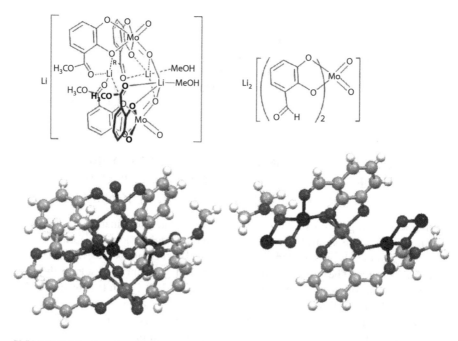

FIGURE 2.11 Results of x-ray crystal structure determinations of $[Li_3\{(3a)_4(MoO_2)_2\}\cdot$ 2 MeOH]$^-$ (left) and unit of $[Li_2\{(1)_2(MoO_2)\}\cdot$ DMF]$_n$ (right).

However, the dimerization constant is low and even in non-polar acetone-d_6, K_{dim} values were 12 M^{-1} (with **1**), 30 M^{-1} (with **2**, R = Me), and 100 M^{-1} (**3a**), respectively. Crystallization from methanol and diffraction studies revealed the structure of Li[Li$_3$\{(**3a**)$_4$(MoO$_2$)$_2$\}] (Figure 2.11, left). Two biscatecholate cis-dioxomolybdenum complexes are bridged by three lithium cations. One is found in a bis-salicylate environment, as observed earlier. The two other lithium cations bind only to one of the salicylates bridging to an oxo ligand of the second complex unit. Those two lithiums are coordinated also by one molecule of methanol each.[17]

The dimerization constants of the cis-dioxomolybdenum(VI) complexes are very low. Consequently, crystallizing the complex of the aldehyde ligand 1 from a highly polar solvent led to the formation of a coordination polymer in the solid that consisted of lithium-bridged monomeric units Li$_2$\{(**1**)$_2$(MoO$_2$)\}.

These observations indicate that the principle of hierarchical self-assembly of helicates is applicable not only to octahedral trischelate complexes. The reaction is more versatile and can generate new compounds with interesting physical (B: fluorescence) or structural (MoO$_2$) features.[17]

2.4 LITHIUM-CONTROLLED ASSEMBLY OF HELICATES WITH 8-HYDROXYQUINOLINE-BASED LIGANDS

To expand the system, catecholate was substituted by 8-hydroxyquinolinate as a ligand unit. To produce an 8-hydroxyquinoline ligand to form coordination compounds like the catechol **3a**-H$_2$ forms, ligand **4**-H was prepared. This ligand forms

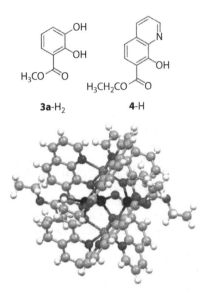

3a-H$_2$ 4-H

FIGURE 2.12 Comparison of catechol ligand **3a**-H$_2$, hydroxquinoline **4**-H, and molecular structure of [Li$_3${(**4**)$_6$Ni$_2$}]$^+$ in crystal.

trischelate complexes with divalent metal ions like cobalt(II), nickel(II), and zinc(II). In the presence of lithium cations, they result in hierarchically formed triple lithium-bridged helicates. The compounds are characterized in solution and in the gas phase. As an example, the nickel(II) complex [Li$_3${(**4**)$_6$Ni$_2$}]$^+$ is shown in Figure 2.12. It reveals structural features similar to those of the catechol compounds Li[Li$_3${(**1**–**3**)$_6$Ti$_2$}].[17]

Interestingly, ligand **4**-H forms neutral trischelates with gallium(III) or aluminum(III) ions. However, the addition of lithium salts in this case does not lead to the formation of hierarchically assembled helicates. This shows that the electrostatic interaction of the monomeric complex units with the lithium cations is crucial for the formation of triple lithium-bridged dinuclear complexes.[17]

2.5 CONCLUSION

This chapter describes in detail the hierarchical self-assembly of lithium-bridged helicates that are in equilibrium with the corresponding monomeric building blocks in solution. Structural and thermodynamic principles of the reaction were investigated in solid, solution, and gas phases and are well understood. The scope and the limits of the process related to metal ion or ligand variations were also discussed. In addition, it was shown that the peripheries of the complexes can be decorated to obtain novel dendrimers or observe some unprecedented stereochemical effects. The work is now on a level to be utilized for the development of functional materials or in systems chemistry.

REFERENCES AND NOTES

1. Reviews: (a) C. Piguet, G. Bernardinelli, and G. Hopfgartner, *Chem. Rev.* 1997, *97*, 2005. (b) M. Albrecht, *Chem. Rev.* 2001, *101*, 3457. (c) M. J. Hannon and L. J. Childs, *Supramol. Chem.* 2004, *16*, 7. (d) M. Albrecht, *Angew. Chem. Int. Ed.* 2005, *44*, 6448.

2. J. M. Lehn, A. Rigault, J. Siegel et al., *Proc. Natl. Acad. Sci. USA* 1987, *84*, 2565.

3. G. Maayan and M. Albrecht, Eds., *Metallofoldamers: Supramolecular Architectures from Helicates to Biomimetics*, Wiley, Chichester, 2013.

4. Covalent hierarchical approaches to supramolecular aggregates: (a) J. R. Nitschke and J. M. Lehn, *Proc. Natl. Acad. Sci. USA* 2003, *100*, 11970. (b) J. R. Nitschke, M Hutin, and G. Bernardinelli, *Angew. Chem. Int. Ed.* 2004, *43*, 6724. (c) P. Mal, B. Breiner, K. Rissanen et al., *Science* 2009, *324*, 1697.

5. (a) X. Sun, D. W. Johnson, D. Caulder et al., *Angew. Chem. Int. Ed.* 1999, *38*, 1303. (b) X. Sun, D. W. Johnson, D. Caulder et al., *J. Am. Chem. Soc.* 2001, *123*, 2752. (c) X. Sun, D. W. Johnson, K. N. Raymond et al., *Inorg. Chem.* 2001, *40*, 4504. (d) A. K. Das, A. Rueda, L. R. Falvello et al., *Inorg. Chem.* 1999, *38*, 4365. (e) J. Heinicke, N. Peulecke, K. Karaghiosoff et al., *Inorg. Chem.* 2005, *44*, 2137.

6. (a) B. A. Borgias, S. R. Cooper, Y. B. Koh et al., *Inorg. Chem.* 1984, *23*, 1009. (b) B. A. Borgias, S. J. Barclay, and K. N. Raymond, *J. Coord. Chem.* 1986, *15*, 109.

7. M. Albrecht, S. F. Franklin, K. and N. Raymond, *Inorg. Chem.* 1994, *33*, 5785.

8. M. Albrecht, *Chem. Soc. Rev.* 1998, *27*, 281.

9. (a) D. L. Caulder and K. N. Raymond, *Angew. Chem. Int. Ed.* 1997, *36*, 1440. (b) M. Meyer, B. Kersting et al., *Inorg. Chem.* 1997, *36*, 5179. (c) E. J. Enemark and T. D. P. Stack, *Angew. Chem. Int. Ed.* 1995, *34*, 996. (d) E. J. Enemark and T. D. P. Stack, *Inorg. Chem.* 1996, *35*, 2719.

10. M. Albrecht and S. Kotila, *Angew. Chem. Int. Ed.* 1996, *35*, 1208.

11. M. Albrecht, I. Janser, S. Kamptmann et al., *Dalton Trans.* 2004, 37.

12. M. Albrecht, S. Mirtschin, M. de Groot et al., *J. Am. Chem. Soc.* 2005, *127*, 10371.

13. (a) M. Albrecht, M. Baumert, H. D. F. Winkler et al., *Dalton Trans.* 2010, 7220. (b) M. Baumert, M. Albrecht, H. Winkler et al., *Synthesis* 2010, 953.

14 M. Albrecht, E. Isaak, M. Baumert et al., *Angew. Chem. Int. Ed.* 2011, *50*, 2850.

15. M. Albrecht, I. Janser, J. Fleischhauer et al., *Mendeleev Commun.* 2004, *14*, 250.

16. K. Omata, K. Kotani, K. Kabuto et al., *Chem. Commun.* 2010, 3610.

17. M. Albrecht, M. Fiege, M. Baumert et al., *Eur. J. Inorg. Chem.* 2007, 609.

3 Supramolecular Hosts and Catalysts Formed by Synergistic Molecular Assemblies of Multinuclear Zinc(II) Complexes in Aqueous Solutions

Shin Aoki, Mohd Zulkefeli, Masanori Kitamura, and Yosuke Hisamatsu

CONTENTS

3.1 INTRODUCTION

Supramolecular strategies have proven powerful methodologies for designing well-defined structures for molecular recognition, sensing, and storage, electronic devices, reaction catalysts, drug delivery systems, and other applications.[1] Multicomponent self-assembly via metal–ligand coordination is of considerable interest in the supramolecular chemistry field.

Supramolecular chemistry in aprotic solvents is currently a well-developed field, as described in many reviews and books.[1] In contrast, self-assembly, molecular recognition, and catalytic reactions in nature take place largely in aqueous media. Therefore, the design and synthesis of supramolecular complexes for the recognition and sensing of biological macromolecules and for catalytic reactions of biorelevant substrates in aqueous solution are of great interest.[2]

After iron, the zinc ion (Zn^{2+}) is the second most ubiquitous metal in natural biological systems. More than 300 molecular structures that involve the functionality of Zn^{2+} proteins and Zn^{2+} enzymes have been reported.[3] The role of Zn^{2+} in Zn^{2+}-containing proteins can be classified into three categories: (1) structural factors in zinc finger proteins and zinc(II) enzymes such as alcohol dehydrogenase, (2) catalytic factors as in zinc(II) enzymes, including carbonic anhydrase and alcohol dehydrogenase, and (3) co-catalytic factors as found in some aminopeptidases such as the enzyme from *Aeromonas proteolytica* (Scheme 3.1).

Kimura and coworkers established that Zn^{2+} complexes of macrocyclic polyamines such as Zn^{2+}-cyclen **1** $(ZnL^1)^{2+}$ are stable in aqueous solutions at neutral pH (Scheme 3.2) and are good models for Zn^{2+} enzymes such as carbonic anhydrase, alkaline phosphatase, class II aldolases, and related enzymes (cyclen = 1,4,7,10-tetraazacyclododecane).[4–7] These Zn^{2+} complexes form 1:1 complexes (1-X$^-$ complexes) with various anions, including carboxylates,[5] imides such as thymine,[6] phosphates,[7] and thiolates.[7,8]

Recognition of phosphate monoester dianions by **1** $(ZnL^1)^{2+}$ through O$^-$Zn^{2+} coordination results in the formation of a 1:1 complex **2** (Scheme 3.3), the dissociation constants (K_d) of which are in the millimolar order in aqueous solution at pH 7.4 (K_d = 0.8 mM when R = 4-nitrophenyl).[7] It was also found that 1 recognizes imide-containing compounds such as thymidine (T) and forms a 1:1 complex **3**, whose K_d values are also in the millimolar order (K_d = 0.3 mM for 1-(T$^-$) complex at pH 8).[6]

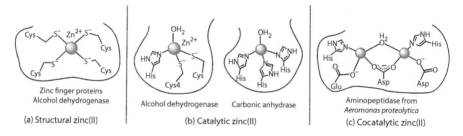

Zinc finger proteins
Alcohol dehydrogenase

(a) Structural zinc(II)

Alcohol dehydrogenase   Carbonic anhydrase

(b) Catalytic zinc(II)

Aminopeptidase from
Aeromonas proteolytica

(c) Cocatalytic zinc(II)

SCHEME 3.1   The role of Zn^{2+} in Zn^{2+}-containing proteins.

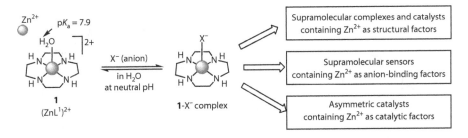

SCHEME 3.2 Recognition of anions by Zn^{2+}-cyclen complex (**1**) in aqueous solution and its applications.

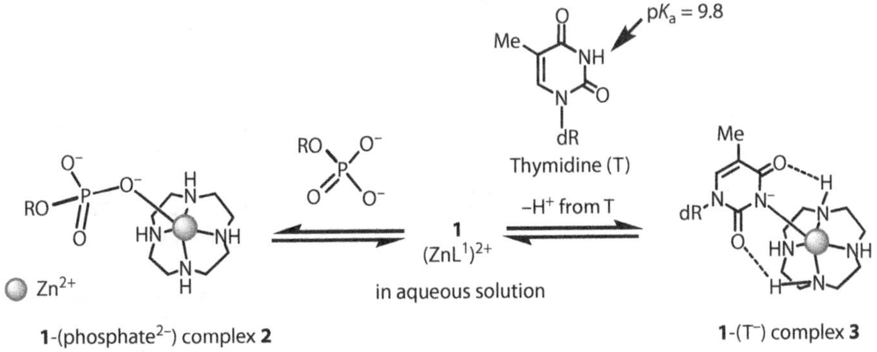

SCHEME 3.3 Complexation of Zn^{2+}-cyclen complex (**1**) with phosphate dianion and imide anion of deprotonated thymidine in aqueous solution.

In this review, we consider various applications involving the anion complexations of polymeric derivatives of **1** in the construction of supramolecular complexes and catalysts that utilize Zn^{2+} ions as structural factors and supramolecular sensors (and receptors) containing Zn^{2+} ions as anion-binding factors (Scheme 3.2).[9] Various supramolecular complexes of different structures and functionalities can be prepared from the same Zn^{2+} complex or ligand by synergistic assembly with various building blocks.

3.2 MOLECULAR RECOGNITION BY MULTINUCLEAR ZINC(II) COMPLEXES IN AQUEOUS SOLUTION AND SUPRAMOLECULAR COMPLEXES FORMED BY SELF-ASSEMBLY OF TRIMERIC ZINC COMPLEX WITH CYANURIC ACID

Multinuclear Zn^{2+} complexes **4** (p-Zn_2L^2)$^{4+}$, **5** (m-Zn_2L^3)$^{4+}$, and **8** (p,p-Zn_3L^4)$^{6+}$ were designed and synthesized for use in the recognition of dinucleotide thymidylthymidine (TpT) to form a 1:1 complex **6**, trinucleotide thymidylthymidylthymidine (TpTpT) to form **9**, and polythymidine (polyT) (Scheme 3.4).[4,10,11] These Zn^{2+} complexes effectively destabilize DNA double helices upon binding to T units in DNA and break the hydrogen bonding between T and adenine (A).[10]

This function was utilized to inhibit the photo[2 + 2]cycloaddition reaction of TpT that produces a *cis-syn* cyclobutane-type thymine dimer (T[*c,s*]T) that is a typical

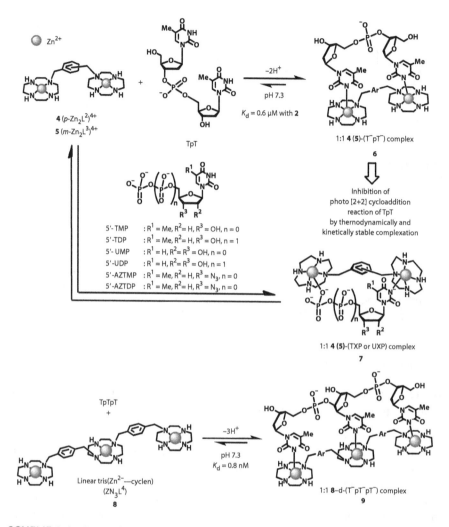

SCHEME 3.4 Recognition of dinucleotide TpT, trinucleotide TpTpT and thymine (uridine) nucledotides by multinuclear Zn^{2+}-cyclen complexes.

photolesion of DNA (Scheme 3.4 top).[11] In addition, these ditopic Zn^{2+} complexes $(4)^{4+}$ and $(5)^{4+}$ form 1:1 complexes **7** with thymine (uridine) nucleotides (TXP, UXP, and AZTXP, where AZT = azidothymidine) whose K_d values are in the micromolar to nanomolar order.[12,13]

It was reported that a dimeric Zn^{2+}–cyclen 4 $(Zn_2L^2)^{4+}$ binds to barbital, a dimide compound, to form the 1:1 complex **10** $[(Zn_2L^2)-(Bar^{2-})]^{2+}$ (Scheme 3.5).[14] A trimeric $(Zn^{2+}$–cyclen) complex having a 1,3,5-trimethylbenzene core **11** (tris(Zn^{2+}–cyclen), $(Zn_3L^5)^{6+}$) was synthesized for use in recognizing cyanuric acid (CA), a tri-imide compound for the formation of a 1:1 complex **12** (Scheme 3.5).

Instead, x-ray examination of fine crystal structures obtained from an aqueous solution of $(11)^{6+}$ and CA revealed that two types of self-assembled complexes were

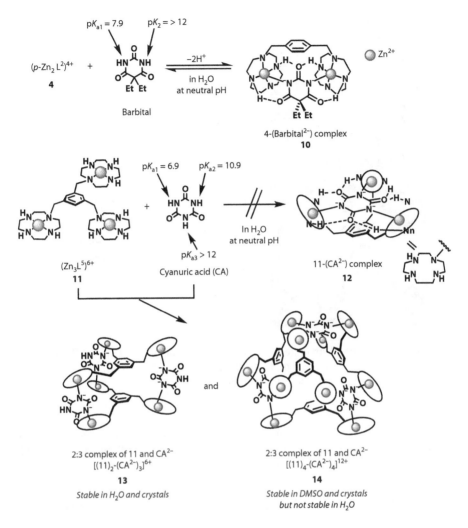

SCHEME 3.5 Formation of supramolecular complexes (**13** and **14**) by self-assembly of tri-meric Zn^{2+}-cyclen complex (**11**) and cyanuric acid.

produced. One is a 2:3 complex of (**11**)$^{6+}$ and the dianionic form of CA (CA^{2-}), **13** [(Zn_3L^5)$_2$–(CA^{2-})$_3$]$^{6+}$, which has a sandwich-like structure formed through N-Zn^{2+} coordination bonds, as evidenced by x-ray crystal structure analysis (Figure 3.1a).[15] The second is a 4:4 complex of (**11**)$^{6+}$ and trianionic CA (CA^{3-}), **14** [(Zn_3L^5)$_4$–(CA^{3-})$_4$]$^{12+}$ isolated as crystals from a strongly alkaline (pH 11.5) aqueous solution (Figure 3.1b). The 2:3 complex (**13**)$^{6+}$ was stable in both neutral aqueous solution and in the crystalline state. The 4:4 complex (**14**)$^{12+}$ was stable only in DMSO and in crystals and was unstable in water.

These two supermolecules are equivalent in terms of acid–base equilibrium. The direction of the disproportion of (**14**)$^{12+}$ is favored in the crystal, possibly due to its stability resulting from the intramolecular hydrogen bond network. Figure 3.1

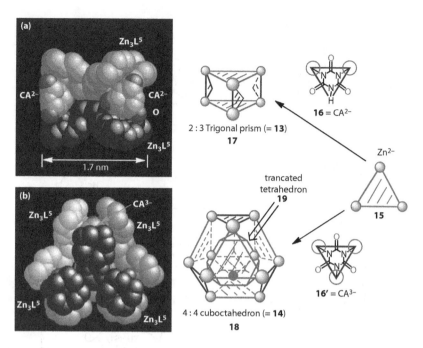

FIGURE 3.1 (a) Crystal structures of 2:3 complex of **11** and CA^{2-}, **13** [(Zn$_3$L^5)$_2$–(CA^{2-})$_3$]$^{6+}$ (side view). (b) The 4:4 complex of **11** and CA^{3-}, **14** [(Zn$_3$L^5)$_4$–(CA^{2-})$_4$]$^{12+}$ viewed along a C_2-symmetric axis with schematic presentations.

illustrates (**13**)$^{6+}$ and (**14**)$^{12+}$, both of which have highly symmetrical structures formed by the assembly of a triangle **15** that presents (**11**)$^{6+}$ with a smaller triangle representing CA (**16** for CA^{2-} or **16′** for CA^{3-}) in 2:3 (**17** for **13**) and 4:4 (**18** for **14**) ratios. The inner hollow structure is visualized as a truncated tetrahedron **19**.

3.3 SUPRAMOLECULAR CAGE FORMED BY SELF-ASSEMBLY OF TRIMERIC ZINC COMPLEX WITH TRITHIOCYANURIC ACID

For the construction of more robust supramolecular cages in aqueous solutions at neutral pH, we replaced the CA of (**14**)$^{12+}$ with trithiocyanuric acid (TCA; Scheme 3.6).[16] An expected advantage of TCA over CA was that the three imide deprotonation constants of TCA (pK$_a$ values of 5.1, 8.2, and 11.7) were about two orders of magnitude lower than those for CA (pK$_a$s of 6.9, 10.9, and >12). For this reason, the trianionic TCA^{3-} might stabilize at near-neutral pH.

Another advantage may be that TCA^{3-} tends to aromatize to a 1, 3, 5-triazine structure, and the anions would be more localized on the exocyclic sulfur atoms that would favor stronger Zn–S$^-$ interactions. In fact, the slow evaporation of a mixture of **11** (Zn$_3$L^5)$^{6+}$ and TCA in aqueous solution at pH 8 yielded a 4:4 complex (**20**)$^{12+}$ [(Zn$_3$L^5)$_4$–(TCA^{3-})$_4$]$^{12+}$ as crystals, whose structure was verified by x-ray structure analysis (Figure 3.2). The exterior frame is visualized as a twisted cuboctahedron (Figure 3.2b).

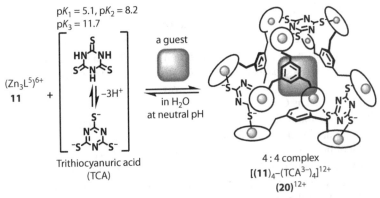

SCHEME 3.6 A 4:4 supramolecular assembly $(20)^{12+}$ from **11** and trithiocyanuric acid (TCA^{3-}).

An interesting property of $(20)^{12+}$ is that lipophilic organic compounds may be encapsulated within its inner cavity. An x-ray structure analysis of the adamantane (ADM)-including $(20)^{12+}$ disclosed that ADM is trapped in the inner cavity (Figure 3.2a). The encapsulation of guest molecules in aqueous solution was observed by the upfield shifts of their ^1H NMR signals due to shielding of the magnetic field formed by the surrounding four benzene rings of Zn_3L and the four triazine rings of TCA^{3-}.

The encapsulated guest molecules are generally size- and shape-matching hydrophobic molecules that are neutral, anions, or cations such as ADM, nitrophenols, ibuprofen, diadamantane, or tetra(n-propyl)ammonium (TPA). Analysis of typical potentiometric pH titrations strongly indicated that the 4:4 complex $(20)^{12+}$

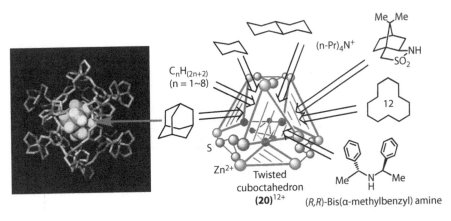

FIGURE 3.2 (a) X-ray crystal structure of adamantane-inclusion complex of $(20)^{12+}$. (b) Schematic presentation of $(20)^{12+}$, a twisted cuboctahedron, and representative hydrophobic guest compounds.

yield exceeds 95% in the presence of a guest molecule at neutral pH. Although the guest-including $(20)^{12+}$ is kinetically stable on the NMR timescale, the guest molecules can be displaced slowly by other guests at room temperature.

The molecular recognition of hydrocarbon guests by $(20)^{12+}$ in water was examined in detail.[17] On the basis of ^1H NMR data, gas chromatography (GC) measurements, and crystal structure analysis, it was confirmed that single molecules of hydrocarbons such as $C_nH_{(2n+2)}$ (n = 1 to 8), cis- and trans-decalines, cyclododecane, and (R,R)-bis(α-methylbenzenzyl)amine ((R,R)-BMBA) may be incorporated as guests. Computational simulations of $(20)^{12+}$ guest complexes using Amber* of the MacroModel indicate that $(20)^{12+}$ has a flexible structure originating from somewhat flexible rotation about the C–S⁻ bonds of the $(Zn^{2+})_3$–(TCA^{3-}) unit.

The $\Delta SASA$ values $(SASA_G + SASA_H) - (SASA_{HG})$, in which $(SASA_G)$, $(SASA_H)$, and $(SASA_{HG})$ are the surface-accessible areas[18] of the guest, host, and host–guest complex, respectively, were calculated as parameters to evaluate surface complementarity between guest molecules and the inner cavity of the $(20)^{12+}$. The result was a possible explanation for the order of stability determined by the guest replacement experiments (water solvent in this case). It is assumed that many water molecules would be present around the outer surface and the inner cavity wall of the cage $(20)^{12+}$ (see 21 in Scheme 3.7).

After a hydrophobic guest molecule becomes encapsulated, water molecules in the inner cavity are excluded to the outside (22). It was concluded that the order of stability of the host–guest complexes of encapsulated guest molecules such as cis-decalin, adamantane, trans-decalin, and cyclododecane is parallel to the order of the $\Delta SASA$ values for these guest molecules, i.e., the greater $\Delta SASA$ value, the more stable the host–guest complex.

In addition, the storage and release activities of volatile molecules by $(20)^{12+}$ in aqueous solution and the solid state were examined.[17,19] Time-dependent changes in the concentrations of propane and n-butane under ambient conditions were monitored by ^1H NMR spectra and GC after sparging these gases in a solution of $(20)^{12+}$

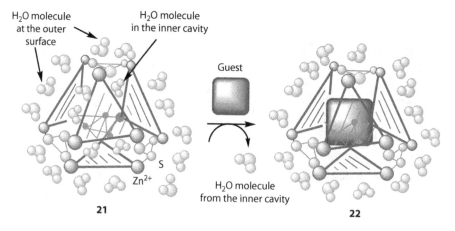

SCHEME 3.7 Expulsion of H_2O molecules in inner cavity of $(20)^{12+}$ to outer surface upon encapsulation of guest molecule, resulting in decreases of *SASA* values.

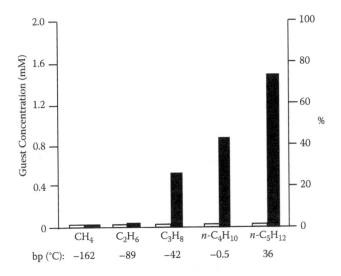

bp (°C):   −162   −89   −42   −0.5   36

FIGURE 3.3 Storage of methane, ethane, propane, *n*-butane, and *n*-pentane in aqueous solutions containing $(20)^{12+}$. Concentrations of guest molecules were determined by gas chromatography after incubation of sample solutions for 24 hr in D_2O (3 mL) in the absence (open bars) and presence (closed bars) of $(20)^{12+}$ (2 mM) at pD 7.4 and 25°C for 10 hr in glass bottles under ambient pressure (1 atm).

(2 mM) in D_2O at 20°C. The concentrations of methane, ethane, propane, *n*-butane, and *n*-pentane in aqueous solution of $(20)^{12+}$ (2 mM) after incubation for 24 hr are shown in Figure 3.3 and indicate that propane, *n*-butane, and *n*-pentane may be stored in aqueous solutions of $(20)^{12+}$.

3.4   SUPRAMOLECULAR TRIGONAL PRISMS FROM LINEAR MULTINUCLEAR ZN–CYCLEN COMPLEXES AND TRITHIOCYANURIC ACID FOR COMPLEXATION WITH DNA DOUBLE HELIX

The above principle of molecular recognition and self-assembly was applied to the preparation of prismatic molecular assemblies.[20,21] The self-assembly of bis(Zn^{2+}–cyclen) **4** (p–Zn_2L^2)$^{4+}$ and linear tris(Zn^{2+}–cyclen) **8** (p,p–Zn_3L^4)$^{6+}$ with TCA^{3-} afforded the 3:2 complex $(23)^{6+}$ [(p–Zn_2L^2)$_3$–(TCA^{3-})$_2$]$^{6+}$ and the 3:3 complex $(24)^{9+}$ [(p,p–Zn_2L^3)$_3$–(TCA^{3-})$_3$]$^{9+}$, respectively, in neutral aqueous solutions (Scheme 3.8).[20]

The crystal structure of $(24)^{9+}$ formed by the 3:3 assembly of $(8)^{6+}$ with TCA is shown in Figure 3.4 and its exterior can be represented as a trigonal prism as shown in Scheme 3.8. The distance between two adjacent TCA^{3-} $(24)^{9+}$ molecules is about 1.2 nm. Of the three TCA^{3-} units, the two terminal units bind to Zn^{2+} ions through S–Zn^{2+} coordination bonds (2.31 Å on average). The TCA^{3-} takes an aromatic triazine form. It is interesting to note that the central TCA^{3-} unit has a less aromatic structure with the anion localized more on the imide Ns that are major donors to Zn^{2+} (the Zn^{2+}–N^- distance is 2.07 Å on average) and supplementary Zn^{2+}–S coordination (2.88 Å on average) bonds.

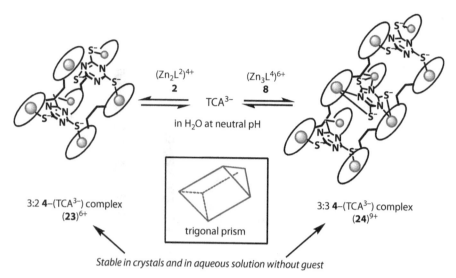

3:2 **4**–(TCA^{3-}) complex
(23)$^{6+}$

trigonal prism

3:3 **4**–(TCA^{3-}) complex
(24)$^{9+}$

Stable in crystals and in aqueous solution without guest

SCHEME 3.8 Trigonal prism structures of supermolecules (**23**)$^{6+}$ and (**24**)$^{9+}$ formed by 3:2 and 3:3 assembly of **2** and **8** with TCA in aqueous solution.

Figure 3.4 also indicates that the three phenyl groups from three molecules of (**8**)$^{6+}$ are assembled in very close proximity.

A tetrakis(Zn^{2+}–cyclen) complex **25** (Zn$_4$L^6)$^{8+}$ was synthesized (Scheme 3.9).[21] Compound (**25**)$^{8+}$ formed a 3:4 complex (**26**)$^{12+}$ [(Zn$_4$L^6)$^{8+}$–(TCA^{3-})$_4$]$^{12+}$ with TCA^{3-} that was very stable at submicromolar concentrations at neutral pH, as evidenced by ^{1}H NMR titration, potentiometric pH and UV titrations, and mass spectrometry (MS) measurements.

Potentiometric titrations, UV spectrophotometric titrations, ^{1}H NMR titrations, and ESI MS supported the conclusion that (**26**)$^{12+}$ is thermodynamically and kinetically (on the NMR timescale) stable and is formed quantitatively in micromolar concentrations at neutral pH. In Figure 3.5, the distribution of (**26**)$^{12+}$ is compared with that for (**24**)$^{9+}$ at [Zn$_4$L^6]$_{total}$ = [Zn$_3$L^2]$_{total}$ = 0.6 μM based on the results of the potentiometric pH titrations, showing that (**26**)$^{12+}$ is much more stable than (**24**)$^{9+}$ at micromolar concentrations. Also, (**26**)$^{12+}$ does not dissociate into starting building blocks even in the presence of strong Zn^{2+} binding anions such as phosphates and double-stranded DNA.

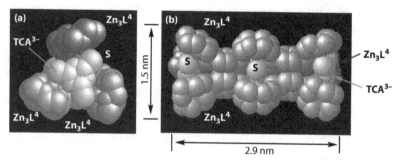

FIGURE 3.4 Trigonal prism complex formed by 2:3 self-assembly (**24**)$^{9+}$ of linear **8** (*p,p*-Zn$_3$L^4)$^{6+}$ with TCA. (a) Top view. (b) Side view.

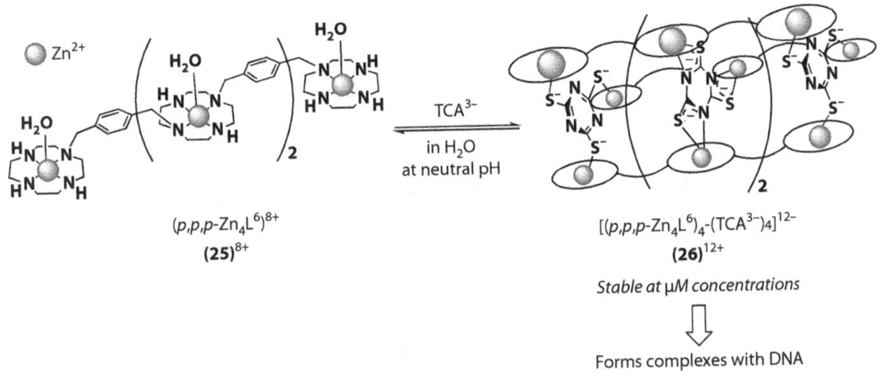

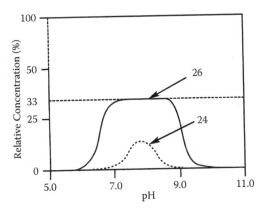

SCHEME 3.9 Extremely stable trigonal prism $(26)^{12+}$ produced from $(25)^{8+}$ and TCA^{3-}.

Interestingly, $(26)^{12+}$ forms a complex with double-stranded DNA and inhibits the interaction of DNA with ethidium bromide (EB, a DNA intercalator) and DAPI (4'6-diamidino-2-phenylindole) and pentamidine (4,4'-[pentane-1,5-diylbis(oxy)] dibenzenecarboximidamide)—both are DNA minor groove binders. The results of an EB displacement experiment of calf thymus DNA (ctDNA) with the trigonal prism $(26)^{12+}$ and polyacrylamide gel electrophoresis (PAGE) of DNA with $(26)^{12+}$ suggested that $(26)^{12+}$ interacts strongly with ctDNA. The concentration for the 50% complexation (IC_{50}) of $(26)^{12+}$ with DNA was estimated to be 2 to 4 μM. The melting temperature (T_m) of ctDNA was not affected by $(26)^{12+}$.[21]

The hypothesis for the interaction mode of $(26)^{12+}$ with ctDNA is presented in Scheme 3.10. It is assumed that the ionic interaction of $(26)^{12+}$ and double-stranded DNA is important for achieving complexation, as shown in **27**, disturbs the interaction of DNA with EB, DAPI, and pentamidine. It should be noted that $(26)^{12+}$ induces the dissociation of EB–DNA (**28a**), DAPI–DNA (**28b**), and spermine–DNA (**28c**) complexes, while $(26)^{12+}$ has only a negligible effect on the T_m of ctDNA.

FIGURE 3.5 Comparison of pH-dependent formation of supramolecular trigonal prisms, $(24)^{9+}$ (at $[p,p\text{-}Zn_3L^4]_{total} = [TCA]_{total} = 0.6$ μM) and $(26)^{12+}$ (at $[p,p,p\text{-}Zn_4L^6]_{total} = 0.6$ μM and $[TCA]_{total} = 0.8$ μM), based on results of potentiometric pH titrations at 25°C with $I = 0.1$ (NaNO$_3$).

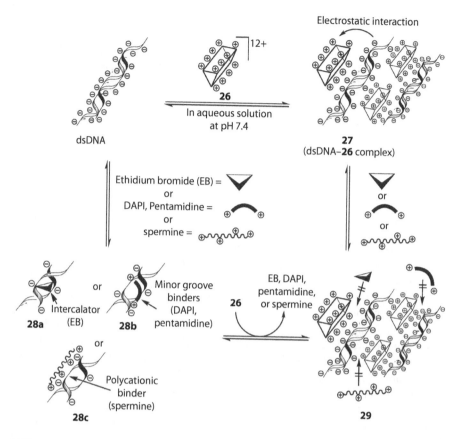

SCHEME 3.10 Interaction mode of trigonal prism $(26)^{12+}$ with ctDNA.

Therefore, it was hypothesized that $(26)^{12+}$ and double-stranded DNA aggregate mainly due to electrostatic interactions to inhibit the complexation of DNA with the DNA intercalator (EB), minor groove binders (DAPI and pentamidine), and a polycationic DNA binder (spermine) as illustrated in **29** in Scheme 3.10, although $(26)^{12+}$ does not affect the T_m values of DNA. The results of dynamic light scattering (DLS) measurement of $(26)^{12+}$ with pUC18 supported this hypothesis. This result should be useful in future attempts to design stable supramolecular complexes and study their applications in the areas of molecular recognition of biological molecules and drug design.

3.5 RECOGNITION AND SENSING OF D-*MYO*-INOSITOL 1,4,5-TRISPHOSPHATE BY SUPRAMOLECULAR LUMINESCENCE SENSOR FORMED BY 3:1 COMPLEXATION OF DIMERIC ZN²⁺ COMPLEX WITH RU²⁺

It is well known that D-*myo*-inositol 1,4,5-trisphosphate (Ins(1,4,5)P$_3$) is an important second messenger in intracellular signal transduction, which induces the release of Ca^{2+} from intracellular Ca^{2+} stores such as the endoplasmic reticulum

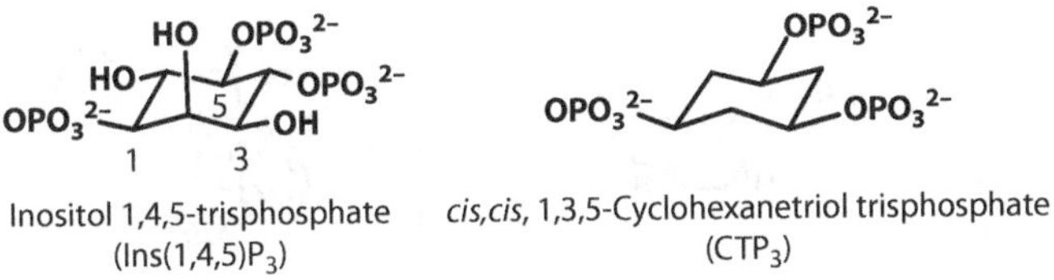

HO   OPO₃²⁻
HO┐ ╱ ╱OPO₃²⁻
OPO₃²⁻┘ └ 5 └OH
     1   3

OPO₃²⁻
OPO₃²⁻⌐ ╱ ╱OPO₃²⁻

Inositol 1,4,5-trisphosphate   cis,cis, 1,3,5-Cyclohexanetriol trisphosphate
(Ins(1,4,5)P₃)   (CTP₃)

SCHEME 3.11   Structures of inositol 1,4,5-trisphosphate (Ins(1,4,5)P₃) and *cis,cis*-1,3,5-cyclohexanetriol trisphosphate (CTP₃).

(ER; Scheme 3.11).[22] Reports indicate that the Ins(1,4,5)P₃ receptors (InsP₃R) are intracellular channel proteins that mediate the release of Ca²⁺ from ER and regulate a number of processes such as cell proliferation and death.

The intracellular concentration of Ins(1,4,5)P₃ is generally in the nanomolar range and InsP₃R cooperatively responds to subtle (nanomolar) changes in concentration. In addition, intracellular Ins(1,4,5)P₃ converts rapidly into derivatives that cannot activate the Ca²⁺ channel.[22] Therefore, the development of novel receptors that bind tightly and rapidly to Ins(1,4,5)P₃ would be highly desirable.

The results of a potentiometric pH titration for a mixture of tris(Zn²⁺–cyclen) **8**[9a] (Zn₃L⁵)⁶⁺ and *cis,cis*-cyclohexanetriol trisphosphate (CTP₃), a model compound for Ins(1,4,5)P₃, suggested the formation of a 1:1 complex of (**8**)⁶⁺ and CTP₃ **30** whose dissociation constant was determined to be 10 nM at neutral pH (Scheme 3.12).[23]

Since **8** has a low molecular extinction coefficient and is non-luminescent, a new supramolecular complex (**32**)¹⁴⁺ was designed and synthesized as a luminescence sensor for Ins(1,4,5)P₃ and CTP₃ by the ruthenium(II)-templated assembly of three molecules of bis(Zn²⁺–cyclen) containing a 2,2′-bipyridyl (bpy) linker (**31**)⁴⁺ based on the assumption that luminescence emissions from the Ru(bpy)₃ core **34** would change upon interactions with trisphosphates (Scheme 3.13).[23]

Single-crystal x-ray diffraction analysis of a racemic mixture of (**32**)¹⁴⁺ (a mixture of Δ and Λ forms) showed that three of the six Zn²⁺–cyclen units are oriented to face the opposite side of the molecule of (**32**)¹⁴⁺ and the three apical ligands (Zn²⁺-bound HO⁻) of each of the three (Zn²⁺–cyclen) units are located on the same face (Figure 3.6).

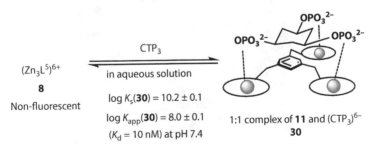

$$ \text{(Zn}_3\text{L}^5)^{6+} \xrightarrow[\text{in aqueous solution}]{\text{CTP}_3} $$

8

Non-fluorescent

$\log K_s(\mathbf{30}) = 10.2 \pm 0.1$

$\log K_{app}(\mathbf{30}) = 8.0 \pm 0.1$ 1:1 complex of **11** and (CTP₃)⁶⁻

$(K_d = 10 \text{ nM})$ at pH 7.4   **30**

SCHEME 3.12   Complexation of trimeric Zn²⁺-cyclen complex (**11**) with CTP₃.

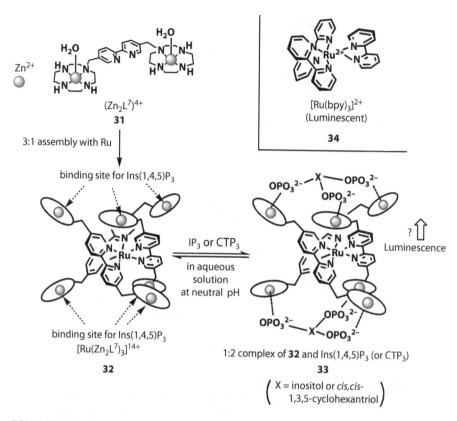

$(Zn_2L^7)^{4+}$

31

$[Ru(bpy)_3]^{2+}$
(Luminescent)

34

3:1 assembly with Ru

binding site for Ins(1,4,5)P$_3$

IP$_3$ or CTP$_3$

in aqueous
solution
at neutral pH

? Luminescence

binding site for Ins(1,4,5)P$_3$
$[Ru(Zn_2L^7)_3]^{14+}$

32

1:2 complex of **32** and Ins(1,4,5)P$_3$ (or CTP$_3$)

33

$\left(\begin{array}{c} X = \text{inositol or } cis,cis\text{-} \\ 1,3,5\text{-cyclohexantriol} \end{array}\right)$

SCHEME 3.13 Synthesis of supramolecular luminescence sensor of Ins(1,4,5)P$_3$, (**32**)$^{14+}$, by 3:1 assembly of **31** $(Zn_2L^7)^{2+}$ with Ru^{2+}.

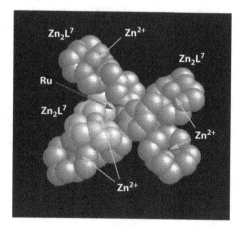

FIGURE 3.6 Crystal structure of (**32**)$^{14+}$ [Ru(Zn$_2$L^7)$_3$]$^{14+}$.

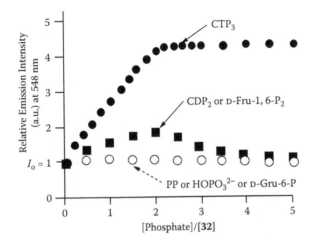

FIGURE 3.7 Luminescence response of $(32)^{14+}$ (10 μM) at 584 nm to CTP_3 (●), monophosphates such as phenyl phosphate and inorganic phosphate, and D-glucose-6-phosphate (○), and diphosphates, including CDP_3 and D-Fru-1,6-P_2 (■) at pH 7.4 (10 mM HEPES with I = 0.1 (NaNO₃)) and 25°C (excitation at 300 nm). I_0 is the emission intensity of $(32)^{14+}$ at 584 nm in absence of guest.

Luminescent titrations of $(32)^{14+}$ suggest that it forms a 1:2 complex **33** with CTP_3 in aqueous solution at neutral pH. In the absence of guest molecules, $(32)^{14+}$ (10 μM) has an emission maximum at 610 nm at pH 7.4 and 25°C (excitation at 300 nm). The addition of 2 equivalents of CTP_3 induced a 4.2-fold enhancement in the emission of $(32)^{14+}$ at 584 nm, as shown in Figure 3.7 and Figure 3.8, suggesting the formation of 1:2 complex **33** from $(32)^{14+}$ with CTP_3. It should be noted that $(32)^{14+}$ is the first chemical sensor to be developed that directly responds to CTP_3 and IP_3 and discriminates these triphosphates from monophosphates (phenylphosphate (PP), D-glucose-6-phosphate (D-Glu-6-P) and HPO_4^{2-}), and diphosphates (*trans*-1,2-cyclohexanediol bisphosphate (CDP_2) and D-fructose-1,6-diphosphate (D-Fru-6-P)).

The crystal structure study of $Ins(1,4,5)P_3$ complexed with the $InsP_3R$ binding core and the binding site of $Ins(1,4,5)P_3$ 3-kinase (IPK) suggests that two phosphates at the 4 and 5 positions (P4 and P5) of $Ins(1,4,5)P_3$ are important for complexation at the binding sites of these proteins, and the 1 phosphate (P1) and other hydroxyl groups are less important for complexation.[24]

Thus, it was hypothesized that chiral ditopic Zn^{2+} complexes would be effective for the cooperative and selective recognition of P4 and P5 of $Ins(1,4,5)P_3$. Based on this information about molecular recognition by dimeric Zn^{2+}–cyclen complexes noted above (Scheme 3.4), chiral ditopic Zn^{2+} complexes $[(S,S)$-**35**$]^{14+}$ and $[(R,R)$-**35**$]^{14+}$ (Scheme 3.14) were synthesized from L- and D-tartaric acid or D-mannitol, and their abilities to recognize P4 and P5 of $Ins(1,4,5)P_3$ were examined.[25]

As shown in Scheme 3.14, the absolute configuration of $Ins(1,4,5)P_3$ at $C4$ and $C5$ is R, allowing us to assume that (R,R)-*trans*-1,2-cyclohexanediol bisphosphate **36** $((R,R)$-1,2,-$CDP_2)$ could serve as a simple model for $Ins(1,4,5)P_3$. We also envisioned that $[(S,S)$-**35**$]^{14+}$ and $[(R,R)$-**35**$]^{14+}$ might be used to discriminate *trans*-1,2-CDP_2

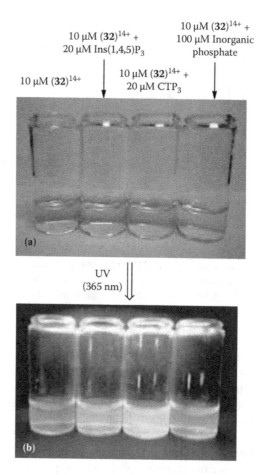

FIGURE 3.8 (a) Solutions of 10 μM (**32**)$^{14+}$, 10 μM (**32**)$^{14+}$ + 20 μM Ins(1,4,5)P$_3$, 10 μM (**32**)$^{14+}$+ 20 μM CTP$_3$, and 10 μM (**32**)$^{14+}$ + 100 μM inorganic phosphate (left to right) at pH 7.4 (10 mM HEPES with I = 0.1 (NaNO$_3$)) and 25°C (excitation at 300 nm). (b) Luminescence emission from 10 μM (**32**)$^{14+}$, 10 μM (**32**)$^{14+}$ + 20 μM Ins(1,4,5)P$_3$, 10 μM (**32**)$^{14+}$+ 20 μM CTP$_3$, and 10 μM (**32**)$^{14+}$ + 100 μM inorganic phosphate (left to right) at pH 7.4 excited by UV light at 365 nm.

from its regioisomers such as *cis*-1,3-CDP$_2$ **37** and *trans*-1,4-CDP$_2$ **38** that denote P1 + P5 and P1 + P4 of Ins(1,4,5)P$_3$, respectively.

The x-ray crystal structure of [(*S,S*)-**35**]$^{14+}$ was determined, as shown in Figure 3.9, which disclosed that the distance between the two Zn^{2+} ions is about 6.8 Å. The distances between two phosphorus atoms of *trans*-1,2-CDP$_2$, *cis*-1,3-CDP$_3$, and *trans*-1,4-CDP$_3$ calculated by density functional theory at MPW1PW91/6-31++G level are 6.7, 8.4, and 8.9 Å, respectively. Therefore, it was expected that the two Zn^{2+} ions of [(*S,S*)-**35**]$^{14+}$ could accommodate the two phosphate groups of *trans*-1,2-CDP$_2$, as well as two phosphate groups at P4 and P5 of Ins(1,4,5)P$_3$.

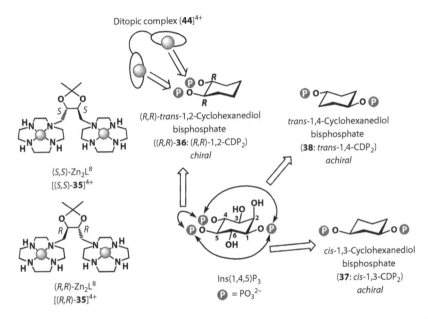

SCHEME 3.14 Complexation of chiral bis(Zn^{2+}-cyclen) complexes ((S,S)-**35**) and (R,R)-**35** with cyclohexanediols as models for Ins(1,4,5)P$_3$.

^{1}H NMR and ^{31}P NMR titrations (and Job's plots) of [(S,S)-**35**]$^{14+}$ with optically pure **36** (*trans*-1,2-CDP$_2$) at neutral pH strongly indicated a 1:1 stoichiometry for complexation with both (S,S)-1,2-CDP$_2$ and (R,R)-1,2-CDP$_2$. The complexation of [(S,S)-**35**]$^{14+}$ ((S,S)-Zn$_2$L^8) with (S,S)-**36**, (R,R)-**36**, **37**, and **38** was studied by isothermal titration calorimetry (ITC) at pH 7.4 (50 mM HEPES with $I = 0.1$ (NaNO$_3$)) and 25°C. The resulting log K_{app} values of 6.9 ± 0.1 and 7.0 ± 0.1 for a 1:1 complex of [(S,S)-**35**]$^{14+}$–(S,S)-**36** for [(S,S)-**34**]$^{14+}$–(R,R)-**36**), respectively, indicated negligible chiral discrimination of (S,S)- and (R,R)-**36** by [(S,S)-**35**]$^{14+}$.

The results of ESI-MS and ITC experiments for Ins(1,4,5)P$_3$ with (S,S)- and [(R,R)-**35**]$^{14+}$ at 25°C and pH 7.4 (50 mM HEPES with $I = 0.1$ (NaNO$_3$)) allowed us

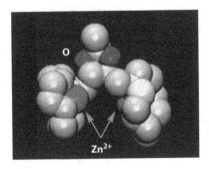

FIGURE 3.9 Crystal structure of [(S,S)-**35**]$^{14+}$. (Space-filling drawings of Zn^{2+}-bound (NO$_3$)$^-$ and external anions are omitted.)

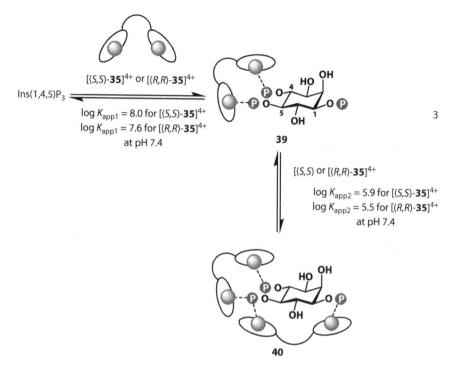

SCHEME 3.15   Proposed binding scheme of $(35)^{4+}$ with Ins(1,4,5)P$_3$.

to conclude that (S,S)- and $[(R,R)$-**35**$]^{14+}$ form 2:1 complexes with Ins(1,4,5)P$_3$ (**40** in Scheme 3.15), in which the first molecule of $(35)^{4+}$ cooperatively recognizes P4 and P5 of Ins(1,4,5)P$_3$ (in **39**), whose affinity (K_d = 10–25 nM) is comparable to that for InsP$_3$R; the second molecule of $(35)^{4+}$ binds to P1 and P5 (**40**).

Although the log K_{app} values for first complexations of Ins(1,4,5)P$_3$ with (S,S)- and $[(R,R)$-**35**$]^{14+}$ were nearly identical, the differences in thermodynamic parameters (enthalpy and entropy changes) suggest different binding modes for the two combinations.[25] These results provide important information about the design and synthesis of chemical receptors, sensors, and inhibitors for Ins(1,4,5)P$_3$ and related compounds such as PtdIns(4,5)P$_2$, and for the design of supramolecular complexes using phosphate–metal coordination bonds.

3.6   SUPRAMOLECULAR PHOSPHATASE FORMED BY SELF-ASSEMBLY OF BIS(ZN^{2+}–CYCLEN) COMPLEX, CYANURIC ACID, AND COPPER FOR SELECTIVE HYDROLYSIS OF PHOSPHATE MONOESTER IN AQUEOUS SOLUTION

The phosphorylation and dephosphorylation of proteins and enzymes are important processes in intracellular regulation. The dephosphorylation of phosphoserine and phosphothreonine residues in proteins is catalyzed by protein phosphatases that

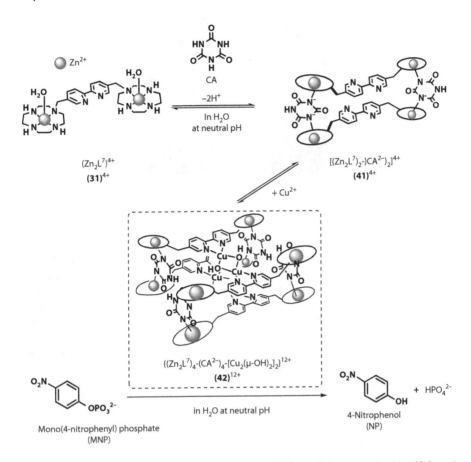

SCHEME 3.16 Formation of 4:4:4 supermolecule $(42)^{12+}$ via 2:2 supermolecule $(41)^{4+}$ and hydrolysis of MNP. The μ–$Cu_2(OH)_2$ center at the bottom of $(42)^{12+}$ is omitted for clarity.

catalyze the reverse reactions of protein kinases.[26] It has been reported that alkaline phosphatase (AP) from *Escherichia coli* contains Zn^{2+}–Zn^{2+} centers, and kidney bean purple acid phosphatase (KBPAP) and Ser/Thr protein phosphatase I contain Zn^{2+}–Fe^{3+} centers.[4]

Although chemical models for phosphatases have been developed,[27,28] artificial compounds that catalyze the hydrolysis of phosphonic acid monoesters such as mono(4-nitrophenyl)phosphate (MNP) are limited, and most contain two metal-macrocyclic polyamine complexes connected by covalent bonds.[29]

We found that the aforementioned dizinc complex **31** $(Zn_2L^7)^{4+}$, cyanuric acid (CA), and Cu^{2+} assemble automatically to form a 4:4:4 complex in aqueous solution as the result of metal–ligand coordination, π–π stacking, and hydrogen bond formation (Scheme 3.16).[29] A dianion of CA (CA^{2-}) and $(31)^{4+}$ assemble in a 2:2 ratio to give $(41)^{4+}$ and the addition of Cu^{2+} (or Cu^{+}) to $(41)^{4+}$ induces the 2:4 complexation of $(41)^{4+}$ and Cu^{2+} to afford $(42)^{12+}$, which is a 4:4:4 complex of $(31)^{4+}$, CA^{2-}, and Cu^{2+}.

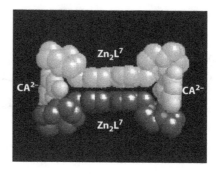

FIGURE 3.10 X-ray crystal structure of (**41**)$^{4+}$. (Space-filling drawings of all external anions are omitted.)

The x-ray crystal structure analysis of the fine colorless crystals obtained by the slow evaporation of a 1:1 mixture of (**31**)$^{4+}$ and CA in aqueous solution at pH 7.4 provided evidence for the formation of a 2:2 complex, **41** [(Zn$_2$L^7)$_2$-(CA^{2-})$_2$]$^{4+}$. The complex has a sandwich-like structure stabilized by four N(CA)–Zn^{2+} coordination bonds and π–π stacking interactions between two bpy units (Figure 3.10) in which two pyridine rings have a trans-configuration in the solid state.[30]

The addition of 2 equivalents of copper ions (Cu(NO$_3$)$_2$, CuI, or CuBr) to an aqueous solution of (**41**)$^{4+}$ produced a pale blue solution, from which blue crystals were obtained. An x-ray crystal structure analysis indicated a 2:4 assembly of (**41**)$^{4+}$ and Cu, namely, a 4:4:4 complex of (**31**)$^{4+}$, CA^{2-} and Cu, as shown in Figure 3.11 and designated **42** {(Zn$_2$L^7)$_4$-(CA^{2-})$_4$-[μ–Cu$_2$(OH)$_2$]$_2$}$^{12+}$. This structure is stabilized by N$^-$–Zn^{2+} coordination bonds between Zn^{2+} and CA^{2-}, π–π stacking between bpy units, and hydrogen bonds between two CA^{2-} function to stabilize (**42**)$^{12+}$.

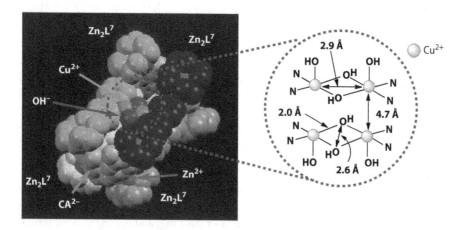

FIGURE 3.11 Crystal structure of (**42**)$^{12+}$ (diagonal view of space-filling model). In dashed circle, crystal diagram shows representative coordination bond lengths in μ–Cu$_2$(OH)$_2$ center. (All NO$_3^-$ anions and H$_2$O molecules are omitted for clarity.)

Importantly, $(42)^{12+}$ contains two $\mu\text{-Cu}_2(\text{OH})_2$ units in the center of the structure (Figure 3.11, right) and each copper has a square pyramidal structure including one water molecule (or I⁻) at the apical position. The Cu–Cu and O–O distances in one $\mu\text{-Cu}_2(\text{OH})_2$ unit are about 2.9 Å and 2.6 Å, respectively, and the distances between two Cu ions of $\mu\text{-Cu}_2(\text{OH})_2$ upside and downside are 4.7 Å. Analysis of the potentiometric pH titration curve and UV/Vis titrations of $(31)^{4+}$ with CA, and then with Cu^{2+}, suggested the formation of $(41)^{4+}$ and $(42)^{12+}$ at micromolar order concentrations.[30]

The central $\mu\text{-Cu}_2(\text{OH})_2$ cores of $(42)^{12+}$, as disclosed by x-ray structure, resemble the catalytic centers of dinuclear metalloenzymes such as AP and KBPAP, although the metal ions they contain are different. Indeed, $(42)^{12+}$ (10 μM) accelerates the selective hydrolysis of MNP (100 μM) in aqueous solution at pH 7.4 (10 mM HEPES with $I = 0.1$ (NaNO₃)) and 37°C to give 4-nitrophenol (NP) and inorganic phosphate (Figure 3.12). Negligible or very low hydrolysis of MNP occurred in the presence of $(31)^{4+}$, CA^{2-}, Cu^{2+}, and $\text{Cu}(2,2'\text{-bpy})_2$, indicating that all of the three components of $(42)^{12+}$ are required for the hydrolysis. More interestingly, it has been suggested that $(42)^{12+}$ works as a catalyst for the hydrolysis of MNP and the highest hydrolysis yield is obtained at pH 7.4.

In this supramolecular system, Zn^{2+} ions act as structural factors for the assembly of these three components and Cu^{2+} ions function as catalytic (possibly co-catalytic) factors for the hydrolysis of MNP. Cu^{2+} is required for the hydrolytic activity of $(42)^{12+}$. Other metal cations such as Zn^{2+}, Co^{2+}, Ni^{2+}, Mn^{2+}, Fe^{2+}, and Fe^{3+} exhibit negligible activity. It should also be noted that $(42)^{12+}$ negligibly catalyzes the hydrolysis of tris(4-nitrophenyl) phosphate (TNP), bis(4-nitrophenyl) phosphate (BNP), and carboxylic acid derivatives such as 4-nitrophenyl acetate (NA) and L-leucine-4-nitroanilide (Leu-NA).

A kinetic study indicates that MNP hydrolysis by $(42)^{12+}$ obeys Michaelis-Menten kinetics. The V_{max} (maximum velocity in NP production from MNP promoted by $(42)^{12+}$), K_{m} (Michaelis constant) value, and k_{cat} values for $(42)^{12+}$ were $(9.5 \pm 0.2) \times$

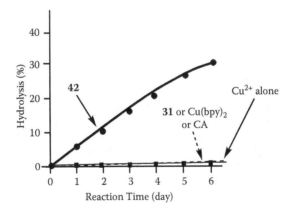

FIGURE 3.12 Time course for hydrolysis yields for MNP by $(42)^{12+}$ (bold curve with closed circle), $(31)^{4+}$, $\text{Cu}(2,2'\text{-bpy})_2$ complex, CA (plain curve for $(31)^{4+}$, $\text{Cu}(\text{bpy})_2$ and CA, Cu^{2+} (closed circle).

10^{-3} μM·min^{-1}, 470 ± 20 μM, and (9.5 ± 0.2) × 10^{-4} min^{-1}, respectively. We know that HPO$_4{}^{2-}$, a hydrolysis product of MNP, inhibits alkaline phosphatase (AP) production.[33] It was postulated that HPO$_4{}^{2-}$ inhibits MNP hydrolysis by a mixed-type inhibition mechanism and its K_i was estimated to be about 70 μM.

One-pot and two-step hydrolysis of bis(4-nitrophenyl)phosphate (BNP) to inorganic phosphate by a combined use of natural phosphatase (phosphodiesterase I (PDE) from *Crotalus atrox* (EC 3.1.4.1)) and (**42**)$^{12+}$ was successful, suggesting the biocompatibility of supramolecular catalysts for the chemoenzymatic synthesis and bioorthogonal reactions[31] in flasks, living cells, and/or bodies.

3.7 SUMMARY

As described in the introduction, natural biological systems utilize Zn^{2+} as structural, catalytic, and co-catalytic factors at the appropriate sites under appropriate conditions. This review summarizes some of the methodologies for constructing supramolecular complexes, sensors, and catalysts by the synergistic and cooperative molecular assembly of multinuclear Zn^{2+} complexes with organic molecules and/ or other metal compounds in which Zn^{2+} ions are used as structural, catalytic, and anion recognition factors.

We should note that various supramolecular complexes that exhibit these factors have dramatically different structures and functionalities can be produced from the same Zn^{2+} complex by utilizing various building blocks. For example, TCA reacts with tris(Zn^{2+}–cyclen) **11** (Zn$_3$L^5)$^{6+}$ to afford a cuboctahedral cage (**20**)$^{12+}$, while a mixture of TCA and a linear tris(Zn^{2+}–cyclen) **8** (Zn$_3$L^4)$^{6+}$ or tetrakis(Zn^{2+}–cyclen) **25** (Zn$_4$L^6)$^{8+}$ yields trigonal prismatic complexes (**24**)$^{12+}$ and (**26**)$^{12+}$. In other cases, a combination of (Zn$_2$L^7)$^{4+}$ with Ru^{2+} affords the luminescence Ins(1,4,5)P$_3$ sensor, (**32**)$^{12+}$, while a combination with CA and Cu^{2+} gives a supramolecular phosphatase (**42**)$^{12+}$. This body of information should be useful in the future design of stable, biologically active, and biocompatible supramolecular hosts and catalysts and their efficient syntheses from readily available building blocks or intermediates.

REFERENCES AND NOTES

1. Reviews: (a) Fujita. M., Ed. *Molecular Self-Assembly: Organic versus Inorganic Approaches*, Springer, Heidelberg, 2000. (b) Prins, L., Reinhoudt, D., and Timmerman, D. N. P. *Angew. Chem. Int. Ed.* 2001, *40*, 2382–2426. (c) Tanaka, K., Okada, T., and Shionoya, M. In *Redox Systems under Nanospace Control*, Hirano, T., Ed. Springer, Heidelberg, 2006, pp. 155–165. (d) Nabeshima, T. and Akine, S. In *Redox Systems under Nanospace Control*, Hirano, T., Ed. Springer, Heidelberg, 2006, pp. 167–178. (e) Oshovsky, G. V., Reinhoudt, D. N., and Verboom, W. *Angew. Chem. Int. Ed.* 2007, *46*, 2366–2393. (f) van Leeuwen, P. W. *Supramolecular Catalysis*, Wiley-VCH, New York, 2008. (g) Nabeshima, T. and Akine, S. *Chem. Rec.* 2008, *8*, 240–251. (h) Kumar, A., Sun, S. S., and Lees, A. J. *Coord. Chem. Rev.* 2008, *252*, 922–939. (i) Northrop, B. H., Zheng, Y. R., Chi, K. W. et al. *Acc. Chem. Res.* 2009, *42*, 1554–1563. (j). Yoshizawa, M., Klosterman, J. K., and Fujita, M. *Angew. Chem. Int. Ed.* 2009, *48*, 3418–3438. (k) Beeletskaya, I., Tyurin, V. S., Tsivadze, A. Y. et al. *Chem. Rev.* 2009, *109*, 1659–1713. (l) Steed, J. W. and Atwood, J. L. *Supramolecular Chemistry*, 2nd Ed., John Wiley & Sons,

Wiltshire, 2009. (m) Cragg, P. J. *Supramolecular Chemistry: From Biological Inspiration to Biomedical Applications*, Springer, New York, 2010. (n) Traboisi, A., Khashab, N., Fahrenbach, A. C. et al. *Nat. Chem.* 2010, *2*, 42–49. (o) Ma, Z. and Moulton, B. *Coord. Chem. Rev.* 2011, 1623–1641. (p) Safont-Sempere, M. M., Fernandez, G., and Wurthner, F. *Chem. Rev.* 2011, *255*, 5784–5814. (q) Hargrove, A. E., Nieto, S., Zhang, T. et al. *Chem. Rev.* 2011, *211*, 6603–6782. (r) Shinoda, S. *Chem. Soc. Rev.* 2013, *42*, 1825–1835. (s) Tashiro, S. and Shionoya, M. *Chem. Lett.* 2013, *42*, 456–462.

2. (a) Prins, L. J., Reinhoudt, D. N., Timmerman, P. *Angew. Chem., Int. Ed.* 2001, *40*, 2382–2426. (b) Atwood, J. L., Leonard, J. B., and Agoston, J. *Science* 2002, *296*, 2367–2369. (c) Kubik, S., Reheller, C., and Stüwe, S. *J. Inclusion Phenom. Macrocyclic Chem.* 2005, *52*, 137–187. (d) Schmuck, C. *Angew. Chem., Int. Ed.* 2007, *46*, 5830–5833. (e) Oshovsky, G. V., Reinhoudt, D. N., and Verboom, W. *Angew. Chem., Int. Ed.* 2007, *46*, 2366–2393.

3. (a) Barrett, A. J., Rawlings, N. D., and Woessner, J. F., Eds. *Handbook of Proteolytic Enzymes*, Academic Press, London, 2000. (b) Lipscomb, W. N. and Styräter, N., *Chem. Rev.* 1996, *96*, 2375–2433. (c) Coleman, J. E. *Curr. Opin. Chem. Biol.* 1998, *2*, 222–234. (d) Aoki, S. and Kimura, E. In *Comprehensive Coordination Chemistry II*, Vol. 8, Que, L., Jr. and Tolman, W. B., Eds., Elsevier, Amsterdam, 2003, pp. 601–640. (e) Averill, B. A. In *Comprehensive Coordination Chemistry II*, Vol. 8, Que, L., Jr. and Tolman, W. B. Eds., Elsevier, Amsterdam, 2003, pp. 641–676. (f) Wenston, J. *Chem. Rev.* 2005, *105*, 2151–2174. (g) Cleland, W. W. and Hengge, A. C. *Chem. Rev.* 2006, *106*, 3252–3278. (h) Auld, D. S., *BioMetals* 2001, *14*, 271–313.

4. Reviews: (a) Kimura, E. and Koike, T., In *Comprehensive Supramolecular Chemistry*, Vol. 10, Reinhoudt, D. N., Ed., Pergamon, Tokyo, pp 429–444. (b) Kimura, E., Koike, T., and Shionoya, M. In *Structure and Bonding: Metal Sites in Proteins and Models*, Vol. 89, Sadler, J. P., Ed., Springer, Berlin, 1997, pp 1–28. (c) Kimura, E. *Acc. Chem. Res.* 2001, *34*, 171–179. (d) Kimura, E. and Aoki, S. *BioMetals* 2001, *14*, 191–204. (e) Aoki, S. and Kimura, E. *Rev. Mol. Biotech.* 2002, *90*, 129–155. (f) S. Aoki, E. Kimura, *Chem. Rev.* 2004, *104*,769–787. (g) Kimura, E. *Bull. Jpn. Soc. Coord. Chem.* 2012, *59*, 26–47. (h) Timmons, J. C. and Hubin, T. J. *Coord. Chem. Rev.* 2010, *254*, 1661–1685.

5. (a) Kimura, E., Ikeda, T., Shionoya, M. et al. *Chem., Int. Ed. Engl.* 1995, *34*, 663–664. (b) Vargova, Z., Kotek, J., Rudovsky, J. et al. *J. Inorg. Chem.* 2007, 3974–3987.

6. (a) Shionoya, M., Kimura, E., Shiro, M. *J. Am. Chem. Soc.* 1993, *115*, 6730–6737. (b) Aoki, S., Honda, Y., and Kimura, E. *J. Am. Chem. Soc.* 1998, *120*, 10018–10026. (c) Gasser, G., Belousoff, M. J., Bond, A. M. et al. *Inorg. Chem.* 2007, *46*, 1665–1674. (d) Kaletas, B. K., Joshi, H. C., van der Zwan, G. et al. *J. Phys. Chem. A.* 2005, *109*, 9443–9455.

7. (a) Kimura, E., Aoki, S., Koike, T., et al. *J. Am. Chem. Soc.* 1997, *119*, 3068–3076. (b) Aoki, S., Iwaida, K., Hanamoto, N. et al. *J. Am. Chem. Soc.* 2002, *124*, 5256–5258. (c) Aoki, S., Kagata, D., Shiro, M. et al. *J. Am. Chem. Soc.* 2004, *126*, 9129–9139. (d) Aoki, S., Jikiba, A., Takeda, K. et al. *J. Phys. Org. Chem.* 2004, *17*, 489–497. (e) Bhuyan, M., Katayev, E., Stadlbauer, S. et al. *Eur. J. Org. Chem.* 2011, 2807–2817.

8. Reviews of anion recognition and sensing by other types of Zn^{2+} complexes: (a) Tamaru, S. and Hamachi, I. In *Structure and Bonding: Recognition of Anions*, Vol. 129, Springer, Berlin, 2008, pp. 95–125. (b) Kinoshita E., Kinoshita-Kikuta, E., and Koike, T. *Neuromethods: Protein Kinase Technologies*, Vol. 68, Springer, New York, 2012, pp. 13–34.

9. Reviews of chiral Zn^{2+}–cyclen complex catalysts mimicking natural aldolase reactions for asymmetric carbon–carbon formation reactions: (a) Itoh, S., Kitamura, M., Yamada, Y., and Aoki, S. *Chem. Eur. J.* 2009, *15*, 10570–1058. (b) Itoh, S., Tokunaga, T., Sonoike, S. et al. *Chem. Asian J.* 2013, *8*, 2125–2135. (c) Itoh, S., Tokunaga, T., Kurihara, M.

et al. *Tetrahedron Asym.* 2013, *24*, 1583–1590. (d) Sonoike, S., Itakura, T., Kitamura, M. et al. *Chem. Asian J.* 2012, *7*, 64–74. (e) Mlynarski, J. and Baś, S. *Chem. Soc. Rev.* 2014, 43, *2*, 577–587.

10. (a) Kimura, E., Kikuchi, M., Kitamura, H. et al. *Chem. Eur, J.* 1999, *5*, 3113–3123. (b) Kikuta, E., Aoki, S., and Kimura, E. *J. Am. Chem. Soc.* 2001, *123*, 7911–7912.

11. (a) Aoki, S., Sugimura, C., and Kimura, E. *J. Am. Chem. Soc.* 1998, *120*, 10094–10102. (b) Yamada, Y. and Aoki, S. *J. Biol. Inorg. Chem.* 2006, *11*, 1007–1023.

12. Aoki, S. and Kimura, E. *J. Am. Chem. Soc.* 2000, *122*, 4542–4548.

13. Reviews: (a) Svoboda, J. and Konig, B. *Chem. Rev.* 2006, *106*, 5413–5430. (b) Zing, Z., Torreiro, A. A. J., Belousoff, M. J. et al. *Chem. Eur, J.* 2009, *15*, 10988–10996. (c) Timmons, J. C. and Hubin, T. J. *Coord. Chem. Rev.* 2010, *254*, 1661–1685.

14. (a) Koike, T., Takashige, M., Kimura, E. et al. *Chem. Eur. J.* 1996, *2*, 617–623. (b) Fujioka, H., Koike, T., Yamada, N. et al. *Heterocycles* 1996, *42*, 775–787.

15. Aoki, S., Shiro, M., Koike, T. et al. *J. Am. Chem. Soc.* 2000, *122*, 577–584.

16. Aoki, S., Shiro, M., and Kimura, E. *Chem. Eur. J.* 2002, *8*, 929–939.

17. Aoki, S., Suzuki, S., Kitamura, M. et al. *Chem. Asian J.* 2012, *7*, 944–956.

18. Houk, K. N., Leach, A. G., Kim, S. P. et al. *Angew. Chem. Int. Ed.* 2003, *42*, 4872–4897.

19. Reviews of gas encapsulation by supramolecular complexes: (a) Horike, S., Shimomura, S., and Kitagawa, S. *Nature Chem.* 2007, *1*, 695–704. (b) Seo, J., Sakamoto, H., Matsuda, R. et al. *J. Nanosci. Nanotechnol.* 2010, *10*, 3–20.

20. Aoki, S., Zulkefeli, M., Shiro, M. et al. *Proc. Natl. Acad. Sci. USA* 2002, *99*, 4894–4899.

21. Zulkefeli, M., Sogon, T., Takeda, K. et al. *Inorg. Chem.* 2009, *48*, 9567–9578.

22. (a) Reitz, A. B. *Inositol Phosphates and Derivatives,* American Chemical Society, Washington, 1991. (b) Miyawaki, A., Furuichi, T., Ryou, Y. et al. *Proc. Natl. Acad. Sci. USA* 1991, *88*, 4911–4915. (c) Luzzi, V., Sims, C. E., Soughayer, J. S. et al. *J. Biol. Chem.* 1998, *273*, 28657–28662. (d) Uchiyama, T., Yoshikawa, F., Hishida, A. et al. *J. Biol. Chem.* 2002, *277*, 8106–8113. (e) Bockhoff, I., Tareilus, E., Strolmann, J. et al. *EMBO J.* 1990, *9*, 2453–2458.

23. Aoki, S., Zulkefeli, M., Shiro, M. et al. *J. Am. Chem. Soc.* 2005, *127*, 9129–9139.

24. (a) Bosanac, I., Alattia, J. R., Mal, T. K. et al. *Nature* 2002, *420*, 696–700. (b) González, B., Schell, M. J., Letcher, A. J. et al. *Mol. Cell.* 2004, *15*, 689–701.

25. Kitamura, M., Nishimoto, H., Aoki, K. et al. *Inorg. Chem.* 2010, *49*, 5316–5327.

26. (a) Woodgett, J. *Protein Kinase Functions,* Frontiers in Molecular Biology Series, Oxford University Press, Oxford, 2000. (b) Denu, J. M., Stuckey, J. A., Saper, M. A. et al. *Cell* 1996, *87*, 361–364.

27. Reviews: (a) Garcia-Espana, E., Diaz, P., Llinares, J. M., Bianchi, A. *Coord. Chem. Rev.* 2006, *250*, 2952–2986. (b) Natale, D. and Mareque-Rivas, J. C. *Chem. Commun.* 2008, 425–437. (c) Mewis, R. E. and Archibald, S. J. *Coord. Chem. Rev.* 2010, *254*, 1686–1712.

28. Recent examples: (a) Bonomi, R., Selvestrel, F., Lombardo, V. et al. *J. Am. Chem. Soc.* 2008, *130*, 15744–15756. (b) Taran, O., Medrano, F., and Yatsimirsky, A. K. *Dalton Trans.* 2008, 6609–6618. (c) Bazzicalupi, C., Bencini, A., Bonaccini, C. et al. *Inorg. Chem.* 2008, *47*, 5473–5484. (d) Piovezan, C., Jovito, R., Bortoluzzi, A. J. et al. *Inorg. Chem.* 2010, *49*, 2580–2582. (e) Bonomi, R., Scrimin, P., and Mancin, F. *Org. Biomol. Chem.* 2010, *8*, 2622–2626.

29. (a) Williams, N. H., Takasaki, B., Wall, M. et al. *Acc. Chem. Res.* 1999, *32*, 485–493. (b) Vance, D. H. and Czarnik, A. W. *J. Am. Chem. Soc.* 1993, *115*, 12165–12166. (c) Koike, T., Inoue, M., Kimura, E. et al. *J. Am. Chem. Soc.* 1996, *118*, 3091–3099. (d) Hettich, R. and Schneider, H. J. *J. Am. Chem. Soc.* 1997, *119*, 5638–5647.

30. Zulkefeli, M., Suzuki, A., Shiro, M. et al. *Inorg. Chem.* 2011, *50*,10113–10123.

31. Sletten, E. M. and Bertozzi, C. R. *Acc. Chem. Res.* 2011, *44*, 666–676.

4 Supramolecular Assemblies Based on Interionic Interactions

Hiromitsu Maeda

CONTENTS

4.1 INTRODUCTION

Organic molecules with well-designed geometries and appropriate substituents can act as building blocks for providing dimension-controlled organized structures.[1] Ordered arrangements of such molecules can facilitate the formation of soft materials[2] such as liquid crystals[3] and supramolecular gels[4] that are extremely useful due to their abilities to change their bulk structures according to external conditions. One promising class of building blocks consists of π-electronic molecules, with their highly planar structures enabling efficient stacking. In addition, they often exhibit optical absorption in the visible region, resulting in the ability to fabricate functional electronic materials.

It is known that molecular assemblies can be constructed through a variety of non-covalent interactions. In contrast to electronically neutral molecules, charged species form ion pairs both in solution and in bulk states via interionic electrostatic interactions. Therefore, the specific geometries and electronic states of ionic species are highly important for determining the properties of ion pairs and their resulting assemblies that may be fabricated by synergetic effects of non-covalent interactions.

Among the many available ionic species, charged organic structures can afford numerous ion pairs because of their extensive diversity. For example, bulky geometries of both cation and anion constituents enable effective production of ionic liquids

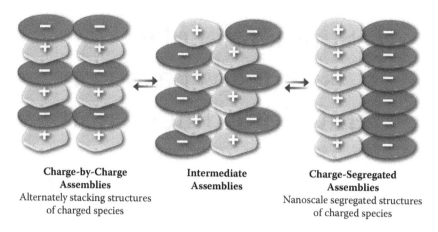

Charge-by-Charge   **Intermediate**   **Charge-Segregated**
   Assemblies   **Assemblies**   **Assemblies**
Alternately stacking structures   Nanoscale segregated structures
    of charged species   of charged species

FIGURE 4.1   Conceptual diagram of charge-based assemblies.

by discouraging crystallization by weakening ionic interactions.[5] Conversely, carefully designed ionic species can provide dimension-controlled assemblies as liquid
crystals and related materials.[3b,6]

Most of the ion-based materials reported thus far have been fabricated using
either cationic or anionic components as the main building blocks of the assemblies,
accompanied by corresponding counter ions for charge compensation, or by using
components possessing charged moieties in their side chains. To effectively utilize
the great potential of these ordered assemblies, a new strategy is required for fabricating ion-based materials utilizing both cations and anions as crucial building blocks.

To achieve effective stacking, a geometry consisting of planar positively and
negatively charged species would enable the formation of dimension-controlled
assemblies consisting of both cationic and anionic components in ordered arrangements. The appropriate design of planar charged species could provide a variety of functional materials by controlling their interactions (Figure 4.1). Here,
we define *charge-by-charge assembly* as the formation of an organized structure comprising alternately stacking positively and negatively charged species.
Charge-segregated assembly, on the other hand, results from combining the
appropriate cations and anions to yield electrostatic repulsion between species
of the same charge.

From a practical view, partial contributions of charge-by-charge and charge-
segregated assemblies would be observed in the intermediate assembling modes.
These ordered structures comprising charged components (electron-deficient and
electron-rich species) exhibit potential for use as organic semiconductors. In particular, charge-segregated assemblies may enable high charge carrier density by decreas-
ing the electrostatic repulsion between species of the same charge.

Compared to working with planar cations, it is not easy to prepare planar anionic
structures for use as building blocks in stacking assemblies because excess electrons
encourage electrophilic attacks on the anions. One promising strategy for preparing

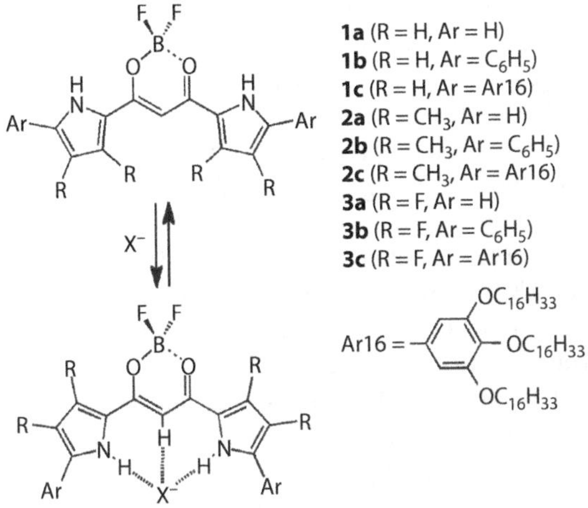

1a (R = H, Ar = H)
1b (R = H, Ar = C₆H₅)
1c (R = H, Ar = Ar16)
2a (R = CH₃, Ar = H)
2b (R = CH₃, Ar = C₆H₅)
2c (R = CH₃, Ar = Ar16)
3a (R = F, Ar = H)
3b (R = F, Ar = C₆H₅)
3c (R = F, Ar = Ar16)

FIGURE 4.2 Dipyrrolyldiketone BF₂ complexes and their anion-binding mode.

planar anions is combining planar electronically neutral anion receptor molecules with inorganic anions such as halides to produce a variety of planar anionic species as receptor–anion complexes. Therefore, the design and synthesis of π-electronic systems that efficiently bind anions would be extremely useful.

As anion-responsive π-conjugated molecules,[7] dipyrrolyldiketone boron complexes (**1** through **3** ⇒ **1–3** Figure 4.2)[8,9] demonstrated anion-binding behavior with inversion of the pyrrole rings to form planar receptor–anion complexes. Such anion receptors have been shown to be suitable motifs for fabricating supramolecular assemblies such as thermotropic liquid crystals,[9d,j,m–o] supramolecular gels,[9b,h] and amphiphilic organized structures[9c] in anion-free form. The results are due to their fairly planar conformations and abilities to introduce various substituents to induce additional non-covalent interactions. As these building blocks are anion-responsive molecules, the resulting structures exhibit anion-responsive behaviors[10] and fascinatingly, the formation of ion-pairing assemblies of receptor–anion complexes and counter cations.

4.2 SOLID-STATE ION-BASED ASSEMBLED STRUCTURES

Before discussing ion-based soft materials, the solutions and solid-state anion-binding behaviors of dipyrrolyldiketone boron complexes are introduced briefly. The center boron-bridged 1,3-propanedione moiety between the two pyrrole rings is effective in affording suitable electronic states that exhibit UV/vis absorption and emission maxima in the visible region, as observed at 432 and 451 nm, respectively, for **1a** in CH₂Cl₂, for example.[9a]

Upon the addition of anions as tetrabutylammonium (TBA⁺) salts, N–H···X⁻ and bridging C–H···X⁻ interactions can be identified from the ¹H NMR chemical shifts, as demonstrated for a series of anion receptors. UV/vis absorption and emission spectra were also altered in the presence of anions, suggesting their potential as colorimetric and fluorescent anion sensors. Anion-binding constants (K_a) of, for example, α-phenyl **1b** in CH_2Cl_2 were estimated to be 30,000, 2,800, 210,000, and 72,000 M^{-1} for Cl^-, Br^-, $CH_3CO_2^-$, and $H_2PO_4^-$, respectively.[9b]

In the solid state, **1b** afforded planar [1 + 1]-type pentacoordinated Cl^-/Br^- complexes by using pyrrole NH, bridging CH, and aryl *ortho*-CH units (Figure 4.3a). Planar **1b** · Cl^-/Br^- anions stacked along with tetrapropylammonium cations (TPA$^+$) to form charge-by-charge columnar structures with distances between the halides of 8.54 and 8.67 Å for Cl^- and Br^-, respectively, and those between the receptor–halide complexes of 7.29 and 7.44 Å, respectively.[9b,d] Similarly, β-substituted **2b** and **3b** exhibited the formation of [1 + 1]-type complexes.

In the crystal state, both **2b** · Cl^- and **3b** · Cl^- formed columnar structures with counter cations, with the TPA$^+$ salt of **2b** · Cl^- showing alternately stacking cationic and anionic components. The TBA$^+$ salt of **3b** · Cl^- formed a columnar assembly comprising a pair of **3b** · Cl^- and a pair of TBA$^+$ in a row (Figure 4.3b and c).[9h]

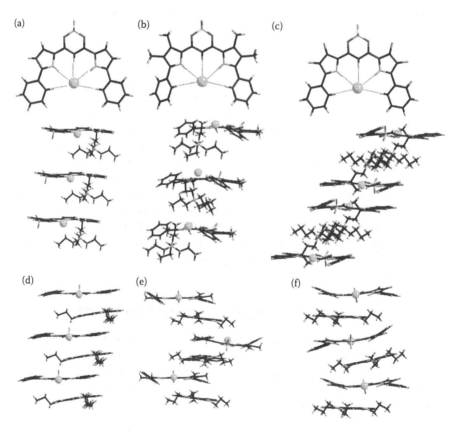

FIGURE 4.3 Top and side-packing view from single-crystal X-ray analysis of (a) **1b** · Cl^- TPA$^+$, (b) **2b** · Cl^--TPA$^+$, (c) **3b** ↔ Cl^-TBA$^+$, (d) **1b** · Cl^--TATA$^+$, (e) **2b** · Cl^--TATA$^+$, and (f) **3b** · Cl^--TATA$^+$ (reproduced from cif files CCDC-646480, 781980, 78198, 745780, 853090, and 853091, respectively). Stacking structures only are shown in (d)–(f).

For evaluation of the effect of the cation geometry, a planar 4,8,12-tripropyl-4,8,12-triazatriangulenium cation (TATA$^+$)[11] was employed in place of the tetraalkylammonium cations. Ion pair **1b** · Cl$^-$-TATA$^+$ formed a charge-by-charge columnar structure like **1b** · Cl$^-$-TPA$^+$, but with a smaller distance between the **1b** · Cl$^-$ units (6.85 and 7.29 Å for TATA$^+$ and TPA$^+$ salts, respectively; see Figure 4.3d).[9f]

A similar tendency in the crystal packing was observed for the corresponding Br$^-$ complexes. Furthermore, β-methyl **2b** · Cl$^-$-TATA$^+$ exhibited a packing mode similar to **2b** · Cl$^-$-TPA$^+$ (Figure 4.3e), whereas β-fluorinated **3b** · Cl$^-$-TATA$^+$ formed a columnar assembly by alternate stacking of **3b** · Cl$^-$ and TATA$^+$ (Figure 4.3f), in contrast to **3b** · Cl$^-$-TBA$^+$.

The Cl···Cl distances of 12.43 and 10.57 Å for **2b** · Cl$^-$-TATA$^+$ and **3b** · Cl$^-$-TATA$^+$, respectively, along the column are much longer than the 6.85 Å observed for **1b** · Cl$^-$-TATA$^+$. Smaller overlaps between receptor–anion complexes and TATA$^+$ in **2b** · Cl$^-$-TATA$^+$ and **3b** · Cl$^-$-TATA$^+$ implied that the β-substituents of the receptor molecules interfered with the formation of stable charge-by-charge columnar structures.[9k]

4.3 ANION-RESPONSIVE SUPRAMOLECULAR GELS TO ION-BASED SOFT MATERIALS

Modifications at the peripheries of anion receptor molecules make it possible to stabilize stacking structures in soft materials. In fact, introduction of aliphatic chains at the α-aryl rings (**1c**, **2c**, and **3c**) enabled the formation of dimension-controlled organized structures exhibiting anion-responsive behaviors. In general, among the assembled states of π-conjugated molecules, gel materials, especially those that respond to external stimuli, are of great interest as soft materials that consist of solvent and a small amount of gelating species.[4] Supramolecular gels are constructed from nanoscale structures such as fibers, tubes, and sheets formed from smaller molecular assemblies. The ability to modulate supramolecular gels using chemical control is extremely attractive because a large variety of potential additives are available and may be included in the assemblies as building subunits.[10]

Aliphatic anion receptor **1c** was found to form a supramolecular gel from octane (10 mg mL^{-1}; Figure 4.4a(i)) with a gel-to-solution transition temperature of 27.5°C. The stacking structure of **1c** in the octane gel exhibited split absorption bands with λ_{max} values of 525 and 555 nm, along with a shoulder at 470 nm. This was in contrast to the single peak at 493 nm obtained for a dilute solution (0.01 mM) of the dispersed monomer. The octane gel exhibited emission at 654 nm (excited at 470 nm), which was red-shifted compared to that of the diluted condition (0.01 mM; λ_{em} = 533 nm excited at 493 nm).

An ordered structure stabilized by non-covalent interactions between the π-electronic moieties and their substituents in a gel was observed using scanning electron microscopy (SEM) and atomic force microscopy (AFM). The addition of a solid-state TBA salt of Cl$^-$ (10 equivalents) to the octane gel resulted in a gradual transition to the solution state, beginning from the areas close to where the solid salts were added (Figure 4.4a(ii)). After the receptor molecules in the gel formed

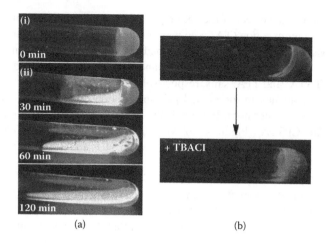

(a)   (b)

FIGURE 4.4 (a)(i) Supramolecular gel of **1c** from octane (10 mg mL^{-1}); (a)(ii) time-dependent transition to solution at 20°C upon the addition of TBACl (10 equivalents); (b) supramolecular gel of **3c** from octane (10 mg mL^{-1}) in the absence (top) and presence (bottom) of TBACl (1 equivalent). Photographs were obtained under UV$_{365\,nm}$ light.

complexes with the anions, TBA$^+$ concertedly approached the receptor–anion complexes to form ion pairs that were soluble in octane. The transition of the gel of **1c** was in contrast to the crystal state of **1b**·Cl$^-$-TPA$^+$ that was constructed because of its lower solubility in hydrocarbon solvents.[9b]

For the β-substituted derivatives, although β-methyl **2c** could not form a gel, β-fluorinated **3c** afforded an octane gel (Figure 4.4b, top) whose transition temperature of 38°C is higher than that of **1c** (27.5°C). In contrast to **1c**, the addition of TBACl did not disrupt the octane gel of **3c** (Figure 4.4b, bottom). The UV/vis absorption and fluorescence bands differed from those of the anion-free state.

The introduction of electron-withdrawing moieties has been shown to induce dipoles in receptors and anionic complexes, resulting in the formation of stable stacking organized structures.[9h] It is noteworthy that in the gel systems discussed in this chapter, intermolecular hydrogen bonding was not the main driving force for assembly and gelation of receptors in the anion-free form. Therefore, anion binding of the receptors was able to modulate the gel state rather than just decompose it.

The assembled structures of **1c** and **3c** are typical examples of supramolecular gels that use π–π interactions as the main force for aggregation. Thus, as observed for **3c**, anions as additives do not always act as inhibitors, but sometimes act as building units for the construction of soft materials. From this view, both the structural modifications of anion receptors and the choices of appropriate anions and cations as guest species for harnessing the fascinating properties of ion-based assemblies have been the foci of investigation. In particular, as observed in the solid-state assemblies (Figure 4.3), counter cations are essential for determining the states of assemblies.

To achieve dimension-controlled assemblies comprising positively and negatively charged species, introduction of planar rather than sterical cations

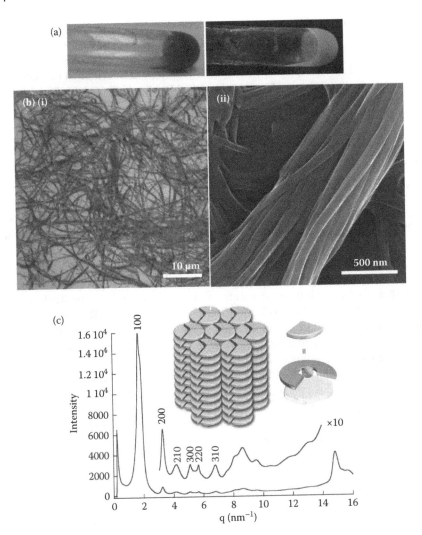

FIGURE 4.5 (a) Photographs of gel of **1c**·Cl⁻TATA⁺ under visible (left) and UV$_{365 nm}$ (right) light; (b)(i) OM and (ii) SEM images of octane xerogel of **1c**·Cl⁻-TATA⁺ at 20°C; (c) synchrotron XRD pattern of octane xerogel of **1c**·Cl⁻TATA⁺ at 25°C along with proposed Col$_h$ model based on charge-by-charge assembly.

has been observed to be effective. In contrast to the addition of TBACl, the combination of **1c** (10 mg mL⁻¹) with Cl⁻ as a TATA⁺ salt (1 equivalent) afforded an opaque gel of **1c**·Cl⁻TATA⁺ (Figure 4.5a). The gel–solution transition temperature of **1c**·Cl⁻TATA⁺ at 35°C was ~8°C higher than that for **1c**, also suggesting that the TATA salt stabilized the gel.

Optical microscopy (OM), SEM, and AFM analyses of the xerogel of **1c**·Cl⁻-TATA⁺ showed a sub-micrometer scale entangled fibril morphology (Figure 4.5b). Furthermore, synchrotron X-ray diffraction (XRD) analysis of **1c**·Cl⁻-TATA⁺ fibers as a xerogel revealed the structure as a hexagonal columnar (Col$_h$) phase with $a =$ 4.25 nm and $c = 0.73$ nm based on a trimeric assembly ($Z = 3.0$ for $\rho = 1$; Figure 4.5c). The c value of 0.73 nm corresponded to the distance ascribed to alternately stacking ion pairs of 1c·Cl$^-$ and TATA$^+$, strongly suggesting the contribution of a charge-by-charge assembly. The circular trimeric assembly in a single disk unit was also consistent with the fan-like geometry of 1c·Cl$^-$.[9f]

4.4 ION-BASED THERMOTROPIC LIQUID CRYSTALS

Peripheral modification of the receptor molecules may allow the existence of stacking structures in soft materials such as thermotropic liquid crystals. Differential scanning calorimetry (DSC) analysis of the octane xerogel of 1c·Cl$^-$-TATA$^+$ suggested the formation of a mesophase as observed in the phase transitions at 88 and 42°C upon first cooling from the isotropic liquid state (Iso) and at 44 and 96°C upon second heating.

Cooling from Iso afforded larger focal conic domains in polarized optical microscopy (POM) images, along with dark domains (Figure 4.6a(i)), suggesting that discotic columnar structures were well aligned perpendicularly to the substrates. Upon cooling to 70°C from Iso, synchrotron XRD analysis showed relatively sharp peaks due to a Col$_h$ phase with $a = 4.64$ nm and $c = 0.73$ nm based on a tetrameric assembly ($Z = 3.58$ for $\rho = 1$; Figure 4.6b(i)), with shear-driven alignment of the columnar structure.

An optical response of the mesophase was observed in POM images upon application of an electric field.[9f] Conversely, β-methyl 2c and β-fluorinated 3c with TATACl exhibited POM textures (Figure 4.6a(ii)) as mesophases, with transition temperatures of 149/34°C (first cooling) and 40/153°C (second heating) and 145/38°C (first cooling) and 45/146°C (second heating), respectively.

Synchrotron XRD analysis of the mesophases of 2c · Cl$^-$-TATA$^+$ and 3c · Cl$^-$-TATA$^+$ revealed Col$_h$ structures with $a = 4.99$ nm and $c = 0.37$ nm ($Z = 2.05$ for $\rho = 1$) and $a = 4.92$ nm and $c = 0.37$ nm ($Z = 1.98$ for $\rho = 1$), respectively (Figure 4.6b(ii)). The periodicity of 0.37 nm that differed from that of 1c · Cl$^-$-TATA$^+$ (0.73 nm) was comparable to the ordinary π–π stacking distance. This observation strongly suggests that local stacking structures of identically charged planes were formed, resulting in Col$_h$ mesophases. A contribution of charge-segregated assembly may result from the distorted π-conjugated units due to the β-substituents observed in the single-crystal X-ray structures (Figures 4.3e and f).[9k]

Furthermore, time-of-flight (TOF) electrical conductivity measurements of 1c ↔ Cl$^-$-TATA$^+$, 2c · Cl$^-$-TATA$^+$, and 3c · Cl$^-$-TATA$^+$ indicated ambipolar charge carrier transport behaviors with well-balanced values at high mobilities (10^{-2} to 10^{-3} cm^2 V^{-1} s^{-1}) for both holes and electrons without special purification procedures. The highest value of zero-field limit mobility (μ ($E = 0$)) was observed for the positive charge in 3c · Cl$^-$-TATA$^+$ (0.11 cm^2 V^{-1} s^{-1}). The value could be considered a result of the partial contributions of charge-segregated assemblies with distinct arrays of species with the same charge. In the case of the negative charge carriers, 3c · Cl$^-$-TATA$^+$ exhibited almost equivalent values of mobility (5×10^{-3} cm^2 V^{-1} s^{-1} at $E = 4 \times 10^3$ V cm^{-1}) at 100 to 140°C, with negligible electric field dependence.

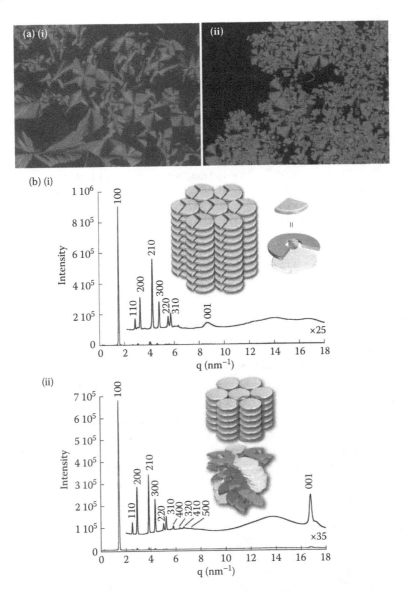

FIGURE 4.6 (a) POM images of (i) **1c**·Cl⁻-TATA⁺ and (ii) **2c**·Cl⁻-TATA⁺ at 70 and 150°C, respectively, upon cooling from Iso; (b) synchrotron XRD patterns of (i) **1c**·Cl⁻-TATA⁺ and (ii) **2c**·Cl⁻-TATA⁺ as mesophases at 70 and 101°C, respectively, upon cooling from Iso, along with proposed Col_h models based on charge-by-charge and charge-segregated assemblies.

The electron-deficient nature of **3c** ↔ Cl⁻-TATA⁺ led to higher stability of electrons in the Col_h structure, resulting in relatively higher values of electron mobility.[9k] The properties and packing structures of ion-based assemblies can be modulated by further modifications of anion receptors as observed in semifluoroalkyl-substituted[9o] β-benzo- and β-coranulene-fused[9j,n] and boron-modified derivatives.[9m]

TABLE 4.1

Phase Transitions of 1c ↔ Cl⁻n_mMe$_{4-m}$N⁺

$1c \leftrightarrow Cl^-n_mMe_{4-m}N^+$		Cooling[a]	Heating[a]
$1c \cdot Cl^-n_1Me_3N^+$	$n = 8$	Cr[c] 36.3 Col$_h$ 72.1 Iso	Cr[c] 43.3 Col$_h$ 74.5 Iso
	$n = 12$	Cr[c] 37.9 Col$_h$ 72.3 Iso	Cr[c] 42.9 Col$_h$ 73.8 Iso
	$n = 16$	Cr[c] 41.9 Col$_h$ 68.1 Iso	Cr[c] 44.9 Col$_h$ 73.0 Iso
$1c \cdot Cl^-n_2Me_2N^+$	$n = 8$	Cr[c] 33.7 Col$_h$ 61.2 Iso	Cr[c] 39.2 Col$_h$ 62.6 Iso
	$n = 12$	Cr[c] 28.9 Col$_h$ 69.1 Iso	Cr[c] 35.5 Col$_h$ 70.1 Iso
	$n = 18$[b]	Cr[c] 40.2 Col$_h$ 79.8 Iso	Cr[c] 44.5 Col$_h$ 80.8 Iso
$1c \cdot Cl^-n_3Me_1N^+$	$n = 4$	Cr[c] 41.9 Col$_h$ 54.1 Iso	Cr[c] 46.2 Col$_h$ 61.9 Iso
	$n = 8$	Cr[d] 30.5 Iso	Cr[d] 35.4 Iso
	$n = 12$	Cr[d] 24.5 Iso	Cr[d] 30.8 Iso
$1c \cdot Cl^-n_4N^+$	$n = 4$	Cr[c] 40.3 Col$_h$ 84.1 Iso	Cr[c] 46.5 Col$_h$ 84.4 Iso
	$n = 8$	Cr[d] 29.2 Iso	Cr[d] 32.7 Cr'[e] 45.6 Iso
	$n = 12$	Cr[d] 23.6 Iso	Cr[d] 27.6 Cr'[e] 42.0 Iso

[a] Transition temperatures (°C, onset of peak) from DSC first cooling and second heating scans (5°C min⁻¹).

[b] Used because 16$_2$Me$_2$NCl was not readily available.

[c] Cr with Col$_h$ structures.

[d] Cr with unidentified structures.

[e] Cr′ with rectangular columnar (Col$_r$) structures.

Bulky cations can act as building blocks for the formation of ion-based assemblies in the absence of solvents even though the steric hindrance interferes with the stacking of receptor–anion complexes. In fact, Col$_h$ mesophases based on charge-by-charge assembly were obtained by complexation of **1c** with Cl⁻ as the salts of different tetraalkylammonium cations n_mMe$_{4-m}$N⁺ (C$_N$H$_{2n+1}$)$_m$Me$_{4-m}$N+: $m = 1-4$) (Table 4.1), with broken-fan-like POM textures seen upon cooling from Iso (Figure 4.7a for **1c** · Cl⁻TBA⁺ (4$_4$N⁺)).

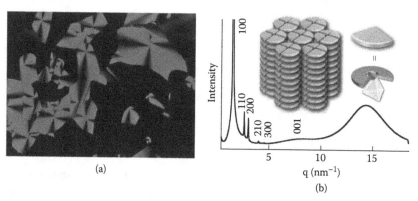

(a)

(b)

FIGURE 4.7 (a) POM texture and (b) synchrotron XRD pattern of **1c** · Cl⁻-TBA⁺ at 62°C upon cooling from Iso and proposed Col$_h$ model based on charge-by-charge assembly.

Among the various ion pairs, **1c** · Cl⁻-TBA⁺ formed an enantiotropic Col$_h$ mesophase with alternate stacking of planar **1c** · Cl⁻ and bulky TBA⁺ via charge-by-charge assembly (Figure 4.7b). Ion pairs comprising a planar anionic component along with bulky tetraalkylammonium cations showed tunable structures and properties according to the alkylammonium cations employed. In particular, as the number and length of the alkyl chains increased, the steric hindrance increased, reducing their ability to assemble with **1c** · Cl⁻, resulting in the formation of ionic liquids at fairly low temperatures.[9i]

In the preceding examples, the dimension-controlled organized structures within the ion-based materials were stabilized by van der Waals interactions between the aliphatic chains of the receptor molecules. In contrast, the anion receptors that unable alone to form soft materials could form dimension-controlled structures by combination with cationic species possessing aliphatic units. In fact, the employment of benzyltrialkylammonium chlorides (16Bn_3NCl, n = 2 and 4) and benzylpyridinium chloride (16BPyCl) (Figure 4.8a) resulted in the construction of ion-based assemblies of Cl⁻ complexes of **1a, b** and **3a, b** as 16Bn_3N⁺ (n = 2 and 4) salts. 16BPy⁺ salts showed different thermal behaviors from the individual components (Table 4.2). The delicate balance between the positively and negatively charged components (sizes of the ionic parts in the cation species and substituents on the anion receptors) enabled self-organization into mesophases.

For example, **1b** · Cl⁻-16B4_3N⁺ and **3a** · Cl⁻-16B2_3N⁺ showed flake-like and fiber-like POM textures, respectively, upon second heating (Figure 4.8b), with the formation of Col$_h$ mesophases with a = 4.79 and 3.28 nm, respectively (Figure 4.8c). In the Col$_h$ mesophase of **1b** · Cl⁻-16B4_3N⁺, the c value of 0.74 nm indicated a charge-by-charge arrangement. The c value of **3a** · Cl⁻-16B2_3N⁺ was found to be 0.38 nm ($Z \approx 2$ for $\rho = 1$), strongly suggesting local stacking of identical charged species to produce a column.

The β-substituents of the anion receptor were responsible for these charge-segregated assemblies, with the fluorine moieties withdrawing electrons by an inductive effect, making the intermolecular interactions among the anionic complexes sufficiently robust for stacking structure formation. Furthermore, flash photolysis time-resolved microwave conductivity (FP-TRMC) measurements[12] for **1b** · Cl⁻-16B4_3N⁺ and **3a** · Cl⁻-16B2_3N⁺ estimated the charge carrier mobilities Σμ to be 0.05 ± 0.01 and 0.22 ± 0.03 cm² V⁻¹ s⁻¹, respectively. The increase in mobility for the assembly with the contribution of the charge-segregated mode offers the possibility of enhancing the efficiency of charge carrier transport in organic electronic devices.[9l]

Anion receptors that lack the ability to form soft materials can form dimension-controlled organized structures by combining with anionic species possessing aliphatic units. A combination of **1a, b** with gallic carboxylates with long alkyl chains (Ar^{Cn}CO$_2$⁻, n = 16, 18, and 20, Figure 4.9a) provided mesophases, consisting mainly of lamellar structures, with transition temperatures of 42, 56, and 103; 67, 73, and 101; and 67, 81, and 96°C for **1a**·Ar^{Cn}CO$_2$⁻-TBA⁺ (n = 16, 18, and 20, respectively) and 40, 61, and 81; 53, 64, and 86; and 61, 82,

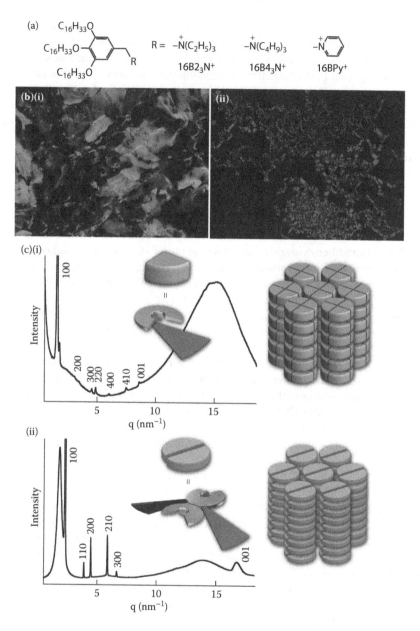

FIGURE 4.8 (a) Structures of $16Bn_3NCl$ ($n = 2$ and 4) and 16BPyCl; (b) POM textures of (i) **1b** · Cl⁻16B4₃N⁺ at 125°C and (ii) **3a** · Cl⁻16B2₃N⁺ at 85°C upon second heating; (c) synchrotron XRD patterns of (i) **1b** · Cl⁻16B4₃N⁺ at 90°C and (ii) **3a** · Cl⁻16B2₃N⁺ at 85°C upon second heating, along with proposed Col_h models. Module cations are shown as fan-like shapes based on optimized structures. Charge-by-charge and charge-segregated assemblies are represented by thick and thin disk components, respectively.

TABLE 4.2

Phase Transitions[a] of Cation Module Cl$^-$ Salts and Their Ion Pairs in Presence of 1a, b and 3a, b

	16B2$_3$NCl	16B4$_3$NCl	16BPyCl
—	Cr[b] 38.7 Cr′[b] 63.7 Col$_h$ 129.4[f] Iso	Cr[b] 42.4 Cr′[c] 54.0 Cr″[c] 76.4 Iso [Cr[b] 38.5 Col$_h$ 81.0 Iso]	Cr[b] 55.2 Cr′[b] 81.1 Col$_h$ 125.9[f] Iso
1a	Cr[b] 50.1 Col$_h$ 132.5[f] Iso	Cr[d] 43.8 Iso	Cr[b] 54.3 Col$_h$ 133.7[f] Iso
1b	Cr[c] 37.1 Cr′[c] 45.6 Iso [Cr[c] 40.4 Iso]	Cr[b] 38.6 Col$_h$ 143.1[f] Iso	Cr[c] 48.3 Cr′[c] 127.3[f] Iso
3a	Cr[c] 6.5 Cr′[b] 35.7 Col$_h$ 102.4 Iso	Cr[b] 39.1 Cr′[b] 59.6 Cr″[b] 92.1 Iso	Cr[b] 39.3 Col$_h$ 53.1[f] Iso
3b	Cr[c] 29.1 Cr′[c] 51.3 Iso [Cr[d] 36.6 M′[c] 74.2 Iso]	[Cr[c] 30.7 Cr′[c] 49.4 Cr″[c] 54.1 Iso] Cr[b] 30.2 Cr′[b] 45.7 Iso	Cr[d] 30.7 M[c] 69.8 Iso

[a] Transition temperatures (°C, onset of peak) from DSC upon second heating (5°C min^{-1}). The entries in brackets show transitions upon first cooling that exhibit phases different from those on second heating.
[b] Basically as Col$_h$ structures.
[c] Unidentified structures.
[d] Basically as lamellar structures.
[e] Basically as Col$_r$ structures.
[f] Values from POM measurements.

and 89°C for **1b**·Ar^{C_n}CO$_2^-$-TBA$^+$ ($n = 16$, 18, and 20, respectively) upon second heating (Figures 4.9b and c).

The diffractions assignable to the repeat distances of charge-by-charge assemblies could not all be observed clearly, suggesting that fairly disordered structures were produced. On consideration of the crystal-state assembled mode of **1a** · Ar^{Cl}CO$_2^-$ TPA$^+$, the anion modules may be located at distorted angles to the receptor planes and may control the assembled structures predominantly through van der Waals interactions between the aliphatic chains.

On increasing the temperature from 28 to 67°C, the electrical conductivity of **1b** · Ar^{C16}CO$_2^-$TBA$^+$ increased from 5×10^{-11} to 3×10^{-8} S m^{-1} due to the increased population of thermally activated charge carriers with equivalent mobility in the mesophases. At room temperature upon cooling from Iso, **1a** · Ar^{C16}CO$_2^-$TBA$^+$ and **1b** · Ar^{C16}CO$_2^-$TBA$^+$ showed charge carrier mobilities of 0.02 and 0.05 cm^2 V^{-1} s^{-1}, respectively.

At elevated temperatures, the values for **1a** · Ar^{C16}CO$_2^-$TBA$^+$ decreased to 0.007 and 9×10^{-4} cm^2 V^{-1} s^{-1} at 46 and 70°C, respectively, whereas the value for **1b** · Ar^{C16}CO$_2^-$TBA$^+$ first decreased to 0.003 cm^2 V^{-1} s^{-1} at 50°C and then recovered to 0.04 cm^2 V^{-1} s^{-1} at 70°C due to transitions between solid states and different mesophases.[9g]

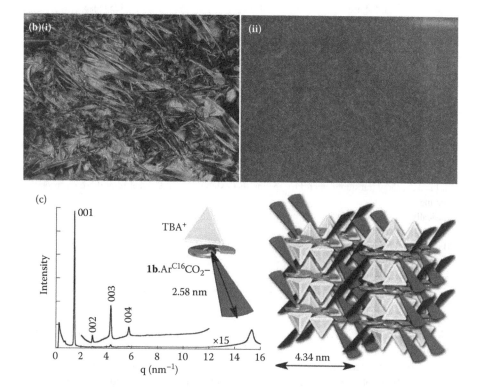

FIGURE 4.9 (a) Structures of gallic carboxylates $Ar^{Cn}CO_2^-$ (n = 1, 16, 18, 20); (b) POM images of (i) $1a \cdot Ar^{C18}CO_2^- TBA^+$ at 95°C and (ii) $1b \cdot Ar^{C16}CO_2^- TBA^+$ at 80.5°C upon cooling from Iso; (c) synchrotron XRD pattern (left) and proposed assembled model (right) of $1b \cdot Ar^{C16}CO_2^- TBA^+$ as solid state at room temperature upon cooling from Iso, suggesting that observed assembled structure was more disordered than proposed model with estimated DFT value of 2.58 nm.

4.5 SUMMARY

This chapter discusses selected examples of ion-based assemblies and materials comprising anion complexes of π-conjugated molecules and counter cations. The ordered arrangement of positively and negatively charged ionic species requires appropriate geometries and electronic properties of the charged components. Among the candidate components for achieving such materials, the pyrrole-based anion receptors synthesized by our group have been demonstrated to be efficient motifs for the formation of planar anionic complexes by binding anions, resulting in the formation of supramolecular assemblies by combination with cationic species.

Through modification of the anion receptors and the appropriate choices of anions and cations, functional ion-based materials were obtained with contributions from charge-by-charge and charge-segregated modes. As demonstrated in this chapter, a variety of ion-based assemblies have been fabricated to date; however, the exact mechanisms of assembly have not yet been elucidated fully. Innovative design and synthesis of π-electronic systems by considering their geometries, electronic properties, and stabilities would enable the achievement of fascinating ion-based materials for use in improved electronic devices in the near future.

ACKNOWLEDGMENTS

The contributions reported herein were supported by PRESTO, the Japan Science and Technology Agency (JST) Structure Control and Function 2007–2011 Grants-in-Aid for Young Scientists (B) (No. 21750155) and (A) (No. 23685032); Scientific Research in a Priority Area, Super-Hierarchical Structures (Nos. 18039038 and 19022036) from the Ministry of Education, Culture, Sports, Science, and Technology (MEXT); matching fund subsidies for private universities from MEXT, 2003–2008 and 2009–2014; and the Ritsumeikan Global Innovation Research Organization (R-GIRO) Project, 2008–2013.

The author thanks all the authors and the collaborators described in the acknowledgments in previous publications, in particular, Prof. Atsuhiro Osuka and his group for single-crystal X-ray analysis, Dr. Takashi Nakanishi of NIMS for his kind help with analyses of molecular assemblies, Prof. Shu Seki and his group at Osaka University for electrical conductivity measurements, Prof. Hitoshi Tamiaki of Ritsumeikan University for various measurements, and all the group members, especially Dr. Yohei Haketa and Dr. Bin Dong for their contributions on ion-based materials.

REFERENCES AND NOTES

1. Selected books on supramolecular assemblies: (a) G. Tsoucaris, Ed. *Current Challenges of Large Supramolecular Assemblies*, NATO Science Series (Dordrecht: Kluwer, 1999). (b) A. Ciferri, Ed. *Supramolecular Polymers* (New York: Marcel Dekker, 2000). (c) F. Würthner, Ed. *Supramolecular Dye Chemistry: Topics in Current Chemistry* (Berlin: Springer, 2005), pp. 1–324. (d) J. L. Atwood and J. W. Steed, Eds. *Organic Nanostructures* (Weinheim: Wiley-VCH, 2007). (e) P. A. Gale and J. W. Steed, Eds. *Supramolecular Chemistry: From Molecules to Nanomaterials* (Chichester: John Wiley & Sons, 2012).
2. I. W. Hamley. *Introduction to Soft Matter: Polymers, Colloids, Amphiphiles and Liquid Crystals* (Chichester: John Wiley & Sons, 2000).
3. Selected books and reviews on liquid crystals: (a) I. Dierking. *Textures of Liquid Crystals* (Weinheim: Wiley-VCH, 2003). (b) T. Kato, N. Mizoshita, and K. Kishimoto. *Angew. Chem. Int. Ed.* 45 (2006): 38–68. (c) T. Kato, Ed. *Liquid Crystalline Functional Assemblies and Their Supramolecular Structures and Bonding* (Berlin: Springer, 2008), pp. 1–237. (d) T. Kato, T. Yasuda, Y. Kamikawa et al. *Chem. Commun.* (2009): 729–739.

(e) B. R. Kaafarani. *Chem. Mater.* 23 (2011): 378–396. (f) S. Kumar. *Chemistry of Discotic Liquid Crystals: From Monomers to Polymers*, Liquid Crystals Series (Boca Raton: CRC Press, 2011).

4. Selected books, chapters, and reviews on supramolecular gels: (a) P. Terech and R. G. Weiss. *Chem. Rev.* 97 (1997): 3133–3159. (b) D. J. Abdallah and R. G. Weiss. *Adv. Mater.* 12 (2000): 1237–1247. (c) F. Fages, Ed. *Low Molecular Mass Gelators*, Topics in Current Chemistry Series (Berlin: Springer, 2005), pp. 1–283. (d) T. Ishil and S. Shinkai. In *Supramolecular Dye Chemistry*, Topics in Current Chemistry Series (Berlin: Springer, 2005), pp. 119–160. (e) R. G. Weiss and P. Terech, Eds. *Molecular Gels: Materials with Self-Assembled Fibrillar Networks* (Dordrecht: Springer, 2006). (f) D. K. Smith. In *Organic Nanostructures* (Weinheim: Wiley-VCH, 2007), pp. 111–154.

5. Selected reviews on ionic liquids: (a) T. Welton. *Chem. Rev.* 99 (1999): 2071–2084. (b) P. Wasserscheid and W. Keim. *Angew. Chem., Int. Ed.* 39 (2000): 3772–3789. (c) H. Ohno. *Bull. Chem. Soc. Jpn.* 79 (2006): 1665–1680. (d) M. A. P. Martins, C. P. Frizzo, D. N. Moreira et al. *Chem. Rev.* 108 (2008): 2015–2050. (e) P. Hapiot and C. Lagrost. *Chem. Rev.* 108 (2008): 2238–2264. (f) R. Giernoth. *Angew. Chem., Int. Ed.* 49 (2010): 2834–2839.

6. Reviews on ionic liquid crystals: (a) K. Binnemans. *Chem. Rev.* 105 (2005): 4148–4204. (b) T. L. Greaves and F. J. Drummond. *Chem. Soc. Rev.* 37 (2008): 1709–1726. (c) K. V. Axenov and S. Laschat. *Materials* 4 (2011): 206–259.

7. Selected books on anion binding: (a) A. Bianchi, K. Bowman-James, and E. García-España, Eds. *Supramolecular Chemistry of Anions* (New York: Wiley-VCH, 1997). (b) R. P. Singh and B. A. Moyer, Eds. *Fundamentals and Applications of Anion Separation* (New York: Kluwer, 2004). (c) I. Stibor, Ed. *Anion Sensing*, Topics in Current Chemistry Series (Berlin: Springer, 2005), pp., 1–238. (d) J. L. Sessler, P. A. Gale, and W.-S. Cho. *Anion Receptor Chemistry* (Cambridge: RSC, 2006). (e) R. Vilar, Ed. *Recognition of Anions: Structure and Bonding* (Berlin: Springer, 2008), pp. 1–252. (f) P. A. Gale and W. Dehaen, Eds. *Anion Recognition by Supramolecular Chemistry*, Topics in Heterocyclic Chemistry Series (Berlin: Springer, 2010), pp. 1–370. (g) K. Bowman-James, A. Bianchi, and E. García-España, Eds. *Anion Coordination Chemistry* (New York: Wiley-VCH, 2011).

8. Recent reviews: (a) B. Dong and H. Maeda. *Chem. Commun.* 49 (2013): 4085–4099. (b) H. Maeda and Y. Bando. *Chem. Commun.* 49 (2013): 4100–4113.

9. Selected reports: (a) H. Maeda and Y. Kusunose. *Chem. Eur. J.* 11 (2005): 5661–5666. (b) H. Maeda, Y. Haketa, and T. Nakanishi. *J. Am. Chem. Soc.* 129 (2007): 13661–13674. (c) H. Maeda, Y. Ito, Y. Haketa et al. *Chem. Eur. J.* 15 (2009): 3706–3719. (d) H. Maeda, Y. Terashima, Y. Haketa et al. *Chem. Commun.* 46 (2010): 4559–4561. (e) H. Maeda, Y. Bando, Y. Haketa et al. *Chem. Eur. J.* 16 (2010): 10994–11002. (f) Y. Haketa, S. Sasaki, N. Ohta et al. *Angew. Chem., Int. Ed.* 49 (2010): 10079–10083. (g) H. Maeda, K. Naritani, Y. Honsho et al. *J. Am. Chem. Soc.* 133 (2011): 8896–8243. (h) Y. Haketa, S. Sakamoto, K. Chigusa et al. *J. Org. Chem.* 76 (2011): 5177–5184. (i) B. Dong, Y. Terashima, Y. Haketa et al. *Chem. Eur. J.* 18 (2012): 3460–3463. (j) Y. Bando, S. Sakamoto, I. Yamada et al. *Chem. Commun.* 48 (2012): 2301–2303. (k) Y. Haketa, Y. Honsho, S. Seki et al. *Chem. Eur. J.* 18 (2012): 7016–7020. (l) B. Dong, T. Sakurai, Y. Honsho et al. *J. Am. Chem. Soc.* 135 (2013): 1284–1287. (m) Y. Terashima, M. Takayama, K. Isozaki et al. *Chem. Commun.* 49 (2013): 2506–2508. (n) Y. Bando, T. Sakurai, S. Seki et al. *Chem. Asian J.* 8 (2013): 2088–2095. (o) Y. Terashima, T. Sakurai, Y. Bando et al. *Chem. Mater.* 25 (2013): 2656–2662. (p) H. Maeda, W. Hane, Y. Bando et al. *Chem. Eur. J.* 19 (2013): 16263–16271 (DOI: 10.1002/chem.201301737). (q) H. Maeda, Bull. Chem. Soc. Jpn. 86 (2013): 1359–1399.

10. Reviews on anion-responsive supramolecular gels: (a) H. Maeda. *Chem. Eur. J.* 14 (2008): 11274–11282. (b) G. O. Lloyd and J. W. Steed. *Nat. Chem.* 1 (2009): 437–442. (c) M. O. M. Piepenbrock, G. O. Lloyd, N. Clarke et al. *Chem. Rev.* 110 (2010): 1960–2004. (d) J. W. Steed. *Chem. Soc. Rev.* 39 (2010): 3689–3699.

11. (a) B. W. Laursen and F. C. Krebs. *Angew. Chem., Int. Ed.* 39 (2000): 3432–3434. (b) B. W. Laursen and F. C. Krebs. *Chem. Eur. J.* 7 (2001): 1773–1783.

12. Selected reports: (a) Y. Yamamoto, T. Fukushima, Y. Suna, et al. *Science* 314 (2006): 1761–1764. (b) A. Saeki, S. Seki, T. Sunagawa et al. *Philos. Mag.* 86 (2006): 1261–1276. (c) T. Umeyama, N. Tezuka, S. Seki et al. *Adv. Mater.* 22 (2010): 1767–1770. (d) Y. Yasutani, A. Saeki, T. Fukumatsu et al. *Chem. Lett.* 42 (2013): 19–21.

5 Chemosensitive Soft Materials Based on Thermosensitive Polymers

Shogo Amemori, Kazuya Iseda, Kenta Kokado, and Kazuki Sada

CONTENTS

5.1 CHEMOSENSITIVE POLYMERS DESIGNED FROM THERMOSENSITIVE POLYMERS

In this chapter, we focus on the molecular designs of thermosensitive polymers along with their applications for sensing materials based on supramolecular interactions and molecular recognition. Thermosensitive polymers are defined as polymers that drastically change their conformations in response to thermal stimuli, i.e., small changes in temperature around the phase transition temperature (T_c).

Generally, such conformational changes directly induce changes in solubility, e.g., from insoluble to soluble or vice versa. When polymers are cross-linked chemically, they exhibit volume changes dependent on temperature that may be accompanied by swelling or collapsing of the polymer gels. These drastic macroscopic changes in thermosensitive materials are attributed to cooperative associations among the repeating units in the flexible polymer chains and prompted the development of intelligent or smart materials triggered by various stimuli.

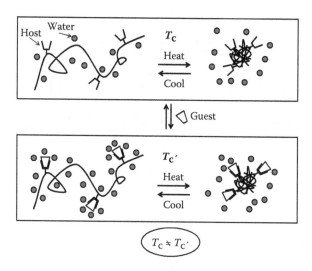

FIGURE 5.1 Concept of chemical-sensitive polymer using host–guest interactions and thermosensitive polymers (LCSF type).

In particular, the incorporation of molecular recognition sites into thermosensitive polymers should induce a change in the phase transition temperature and provide sensing materials that undergo drastic changes in their solubilities or swelling properties in response to suitable guest or host molecules as shown in Figure 5.1. The investigation of such materials should be of great interest as a synergy of supramolecular chemistry and polymer chemistry. First, we briefly introduce thermosensitive polymers, including the mechanisms for thermal responses, and then we focus on control of thermosensitive materials by guest molecules.

5.2 LOWER CRITICAL SOLUTION TEMPERATURE (LCST) POLYMERS

The critical solution (phase transition) temperature is the level at which the phase of the polymer and solution is discontinuously changed according to composition. If the polymer solution has one phase below the phase transition temperature and is phase separated above this temperature, it is generally called a lower critical solution temperature (LCST) solution. On the other hand, if the polymer solution has two phases below the phase transition temperature and becomes one phase above this temperature, it is an upper critical solution temperature (UCST) material. These phase transitions are generally monitored by a change in transmittance or turbidity using the visible absorption spectrum with respect to temperature.

Extensive studies by polymer chemists revealed that large numbers of polymers exhibit LCST or UCST behaviors in various solvents. In recent years, the LCST behaviors of amphiphilic polymers,[1] for example, poly(N-isopropylacrylamide) (PNIPAM), poly(methylvinylether) (PMVEth), and poly(N,N-diethylacrylamide) (PDEAAm) in water (Figure 5.2), have been rigorously investigated and used for many applications

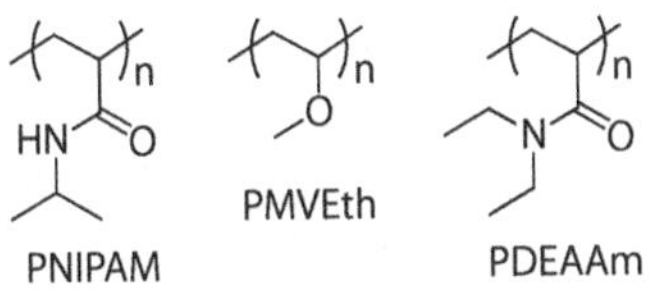

FIGURE 5.2 Typical examples of polymers having LCSTs in water.

in aqueous systems such as drug delivery,[2] gene therapy,[3] temperature-sensitive chromatography,[4] surface modification,[5] and cultivation sheets for cells[6] due to similarities of their phase transition temperatures and human body temperature.

These LCST type phase transitions in water can occur as a consequence of an appropriate balance of hydrophilic and hydrophobic moieties. Below the LCST, the polymer dissolves in water, meaning that the polymer–solvent interaction through hydrogen bonding is stronger than polymer–polymer interactions such as hydrophobic interactions and hydrogen bonding. The polymer–solvent interaction (solvation) decreases with increasing temperature. The polymer–polymer interactions increase relatively and the increase induces a collapse of the polymer. Above the LCST, the polymer–polymer interactions become stronger than the polymer–solvent interactions and aggregation and formation of compact globules occur subsequently. For these reasons, the balance of hydrophilicity and hydrophobicity of a polymer plays a key role for inducing LCST behavior in water. LCST behavior is affected by the composition[7] of the polymer and various stimuli or additives such as salts,[8] solvents,[9] surfactants,[10] pH,[11] and other molecules.[12]

Due to difficulties in directly designing LCST polymers, water-soluble LCST type thermosensitive polymers have been used to create smart polymer systems controlled by guest molecules by incorporating molecular recognition sites.[13]

Specific control of LCST behavior by guest molecules is accomplished when the balance of the hydrophilicity and hydrophobicity of a polymer is changed by molecular recognition corresponding to interactions of the polymer and small molecules. In particular, molecular recognition combinations such as (1) cyclodextrins and hydrophobic guests, (2) crown ethers and cations, and (3) boronic acids and sugars have been used.

First, we describe means of LCST control using these three interactions. Subsequently, other interactions such as cucurbit[8]uril with hydrophobic guests and single- and double-stranded DNAs will be documented. In the final section, new strategies for designing materials exhibiting LCST behaviors are presented.

5.3 THERMOSENSITIVE POLYMERS BY CYCLODEXTRIN AND HYDROPHOBIC GUESTS

Cyclodextrins (CDs) are macrocyclic oligosaccharides consisting of D-glucopyranose units. In particular, six-, seven-, and eight-membered macrocycles are, respectively, called α-cyclodextrin (α-CD), β-cyclodextrin (β-CD), and γ-cyclodextrin (γ-CD). They all exhibit different cavity diameters and chemical natures such as water solubilities and host–guest properties (Figure 5.3).

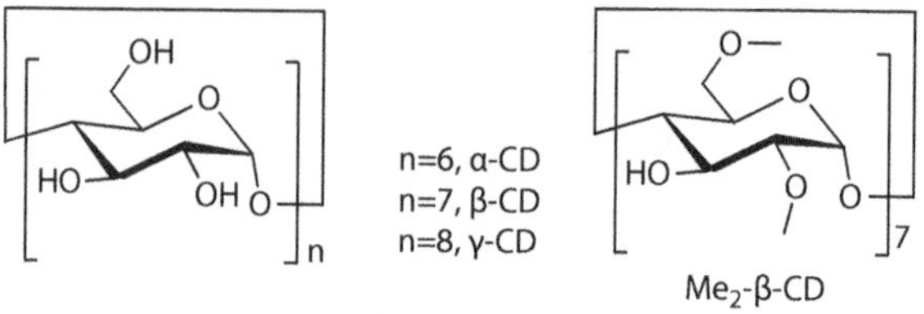

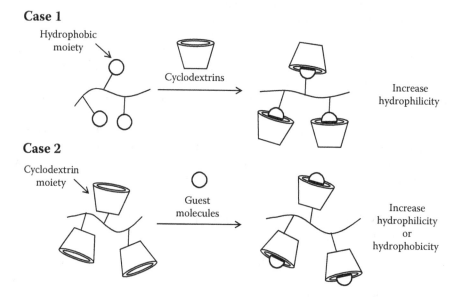

FIGURE 5.3 Structure of cyclodextrins (CDs).

Because the CDs possess hydrophobic inner surfaces and hydrophilic outer surfaces, they include hydrophobic molecules corresponding to their cavity sizes due to hydrophobic interactions.[14] Thus, the CDs are well-known host molecules that increase the solubilities of hydrophobic guest molecules in water by the formation of host–guest complexes. Since β-CD has a low solubility in water compared to the α- and γ-CDs, 2,6-dimethyl-β-cyclodextrin (Me$_2$-β-CD) and randomly methylated β-cyclodextrin (RM-β-CD) have been used in many cases (Figure 5.3).

A large number of research studies about controlling the LCST behaviors of thermosensitive polymers in water by host–guest interactions of CDs and hydrophobic molecules have been reported and fall roughly into two categories (Figure 5.4). One is the complexation between the CD and a thermosensitive polymer bearing hydrophobic moieties (Case 1). The other consists of complexation between thermosensitive polymer bearing CD moieties and guest molecules (Case 2).

In most of the former cases, the coverage of hydrophobic moieties with CDs increases the hydrophilicity of the thermosensitive polymer, leading to an increase

FIGURE 5.4 Control of thermosensitivity using molecular recognition of CDs with increase of hydrophobicity guests in water.

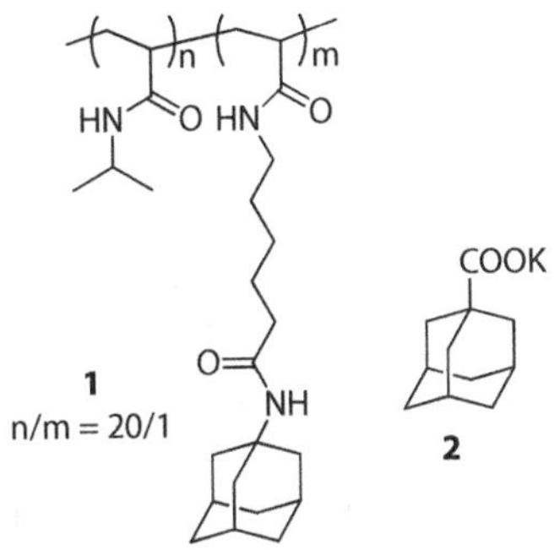

FIGURE 5.5 Structure of PNIPAM with adamantyl moiety and potassium-1-adamantylcarboxylate as competitive guest.

of the LCST. In the latter cases, both increases and decreases in LCSTs have been observed due to increases in hydrophilicity or hydrophobicity, depending on the guest molecules in the inner cavity. Hence, the molecular recognition of guest molecules makes it possible to induce control of LCST behaviors.

For controlling the thermosensitivities of LCST polymers, PNIPAM copolymers with hydrophobic moieties have been prepared and their formations of complexes with CDs as host molecules have been investigated. The LCST of unmodified PNIPAM in water is generally insensitive to CDs.[15] Ritter et al. first reported control of the LCST of a PNIPAM bearing adamantyl moieties **1** by the addition of Me$_2$-β-CD (Figure 5.5).[16] An aqueous solution of polymer **1** showed an LCST of 17°C; this temperature was significantly lower than that of typical PNIPAM copolymers due to the hydrophobicity of the incorporated adamantyl moieties.

When Me$_2$-β-CDs were added to an aqueous solution of **1**, the LCST increased to ~40°C. To investigate the thermosensitivity and host–guest interaction relationship, potassium-1-adamantylcarboxylate **2** as a competitive external guest that was expected to compete with the polymer-bound adamantyl groups was added to an aqueous solution of **1** in the presence of Me$_2$-β-CD. The LCST decreased with increasing concentration of **2**. These results show that wrapping hydrophobic adamantyl groups with Me$_2$-β-CD increased the hydrophilicity of the polymer, and thus increased the LCST.

It was also reported that the LCST behaviors of PNIPAM copolymers modified with hydrophobic moieties other than adamantyl groups could be controlled by addition of CDs.[17] In all cases, the addition of CDs increased the LCSTs.

The LCST behavior of a relatively low-molecular-weight PNIPAM was affected easily by a change in the hydrophilicity of the polymer chain ends.[18] Therefore, LCSTs can be controlled by host–guest interactions between CDs and the chain end functional groups.[19,20]

Kakuchi et al. reported the influence of the addition of β-CD on the LCST of a PNIPAM (M_n = 3000) that had a pyrenyl group at the chain end.[19] The LCST of the polymer was significantly raised by adding β-CD and saturated at a [β-CD]:[polymer] ratio of 2:1. These results suggested that β-CD formed an inclusion complex with the pyrene moiety and increased the LCST.

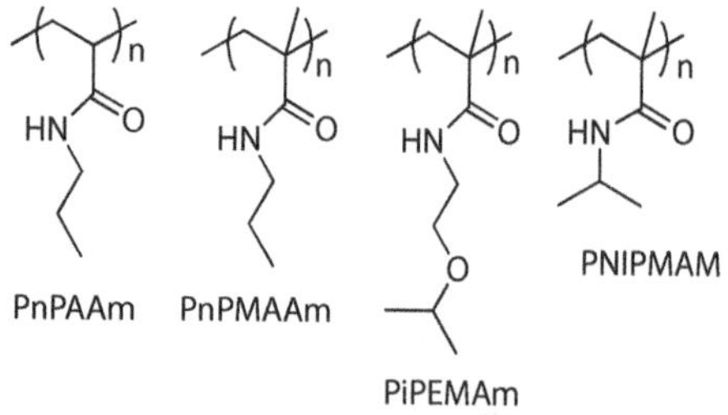

FIGURE 5.6   LCST polymers whose temperatures change in presence of α-CD.

Although the LCST of PNIPAM is affected little by the additions of CDs, control of the LCSTs of other thermosensitive polymers by the addition of CDs have been reported.[21,22] Wang et al. demonstrated that the presence of α-CD increased the LCST of poly(N-n-propylacrylamide) (PnPAAm), poly(N-n-propylmethacrylamide) (PnPMAAm), and poly(N-(2-isopropoxyethyl)methacrylamide) (PiPEMAm) in water (Figure 5.6).[21] The formation of inclusion complexes of α-CD and the side chain of these polymers was confirmed by NOESY NMR spectra.

On the other hand, the LCST values of these polymers were insensitive to the addition of β-CD, RM-β-CD and γ-CD. Also, no influence from CDs on the LCSTs of PNIPAM and poly(N-isopropylmethacrylamide) (PNIPMAM), was observed. These results clearly indicate that molecular recognition between CDs and hydrophobic groups copolymerized into the polymer chains of thermosensitive polymers plays a key role in controlling the LCST behaviors of these materials.

As noted, the addition of CDs generally induced increases in the LCSTs of the thermosensitive polymers. However, a decrease in the LCST was accomplished using dimer CD 3, which has two β-CD moieties by Ritter et al. (Figure 5.7).[23] They utilized a thermosensitive polymer with the same structure as polymer 1. The addition of 3 decreased the LCST of the polymer in contrast to the LCST increase observed from adding Me$_2$-β-CD. This can be explained by the cross-linking of the polymer chains upon complexation of the CD dimer 3.

Thermosensitive polymers bearing cyclodextrins and cyclodextrins attached to thermosensitive polymers have been extensively studied.[24–27] Researchers have shown the influence of guest molecules on the LCST of a thermosensitive polymer bearing CD.[25–27]

Yamaguchi et al. used poly(NIPAM-co-CD) 4 as a thermosensitive polymer bearing β-CD and 8-anilino-1-naphthalenesulfonic acid (ANS) as a guest molecule (Figure 5.8).[25] The LCST of poly(NIPAM-co-CD) in an aqueous solution was higher

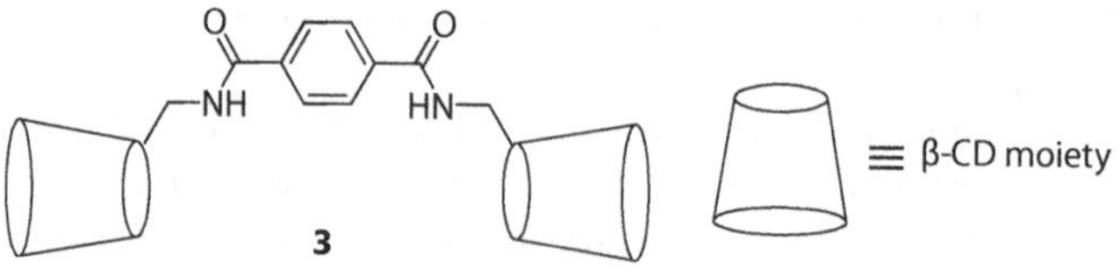

FIGURE 5.7   Dimer 3 as cross-linker using host–guest interaction.

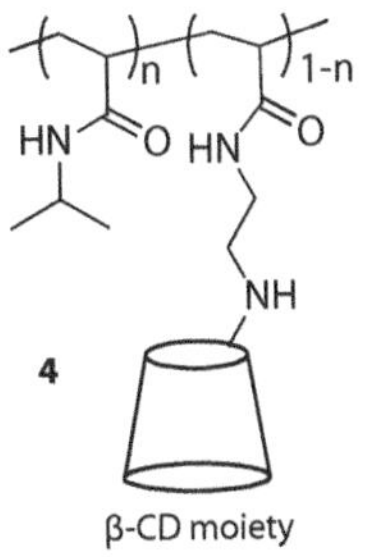

FIGURE 5.8 PNIPAM copolymer with β-CD moiety **4**.

than that of typical PNIPAMs due to the incorporation of the hydrophilic CD moiety. The LCST of poly(NIPAM-*co*-CD) decreased with the addition of ANS and the shift in the LCST was greater with higher ANS concentrations. This shows that the hydrophobicity of poly(NIPAM-*co*-CD) was increased by formation of the CD/ANS complex because the hydrophobic phenyl groups of the ANS were found to be outside the CD cavity in the formed host–guest complex.

Chu et al. evaluated the effects of guest molecules on shifts in LCSTs using ANS and 2-naphthalenesulfonic acid (NS) as guest molecules.[26] They also used a copolymer consisting of NIPAM and β-CD units. They found that the LCST of the system with the ANS guest was lower than that of the polymer without the guest in water, similar to the above case. Conversely, the addition of NS induced an increase in the LCSTs of the systems because the complexation between CD and NS could enlarge the hydrophilic moieties of the polymers due to the ionized sulfonic acid of NS in water. Also, the CD–NS complex has no hydrophobic group outside the CD cavity (Figure 5.9).

Zhang et al. demonstrated the LCST behaviors of CDs modified with oligoethylene glycols (OEGs) as thermosensitive units rather than NIPAM.[27] Their LCSTs were tunable in the range of 24 to 63°C, depending on the OEG chain lengths and CD ring

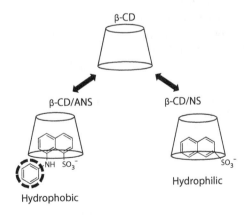

FIGURE 5.9 CD complexes with ANS or NS.

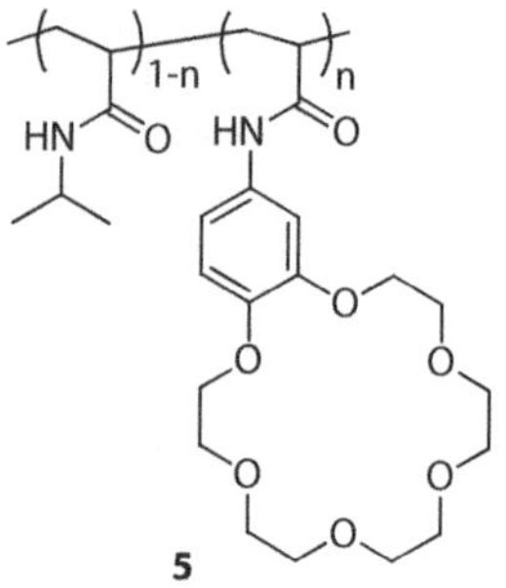

FIGURE 5.10   PNIPAM copolymer with BCAm **5**.

sizes. Moreover, α-CD bearing ethylene glycols showed much higher LCSTs after complexation with methyl orange (MO) due to the introduction of negative charges from the MO.

5.4   THERMOSENSITIVE POLYMERS UTILIZING CROWN ETHER AND METAL CATIONS

Crown ethers are heterocycles consisting of oxyethylene units, and their size depends on the number of these repeating units. They are known to capture cations into their cavities selectively.[28] Specific cation recognitions of the crown ethers led to the development of various applications such as metal cation sensors and phase transfer catalysts.[28] Since the hydrophilicities of the crown ethers increase through complexation with cations, introducing crown ether units into thermosensitive polymers should control their solubilities and LCST behaviors in the presence of cations as stimuli.

In 1992, Irie et al. synthesized the ion-sensitive poly(NIPAM-*co*-BCAm) **5** polymer consisting of *N*-isopropylacrylamide units and benzo[18]-crown-6-ether units (BCAm) for the first time (Figure 5.10).[29] The polymer **5** showed a 31.5°C LCST in an aqueous solution. The LCST value increased with increasing concentration of potassium chloride. No similar increase in the LCST was observed for only PNIPAM.

The increase in the LCST by the addition of sodium chloride was lower than in the case of potassium chloride. On the other hand, the additions of lithium chloride and cesium chloride decreased the LCST. Because the cavity size of benzo[18]-crown-6-ether fits the diameter of K^+, the binding constant of benzo[18]-crown-6-ether to K^+ is the highest among the four ions. When K^+ was captured by the crown ether moiety, the hydrophilicity of the pendent group increased. This increase enhanced the polymer solubility and shifted the LCST to higher temperatures.

Chu et al. later reported the influence of various metal cations on the LCST of poly(NIPAM-*co*-BCAm).[30] The additions of K^+, Sr^{2+}, Ba^{2+}, Hg^{2+}, and Pb^{2+} induced increases in the LCST. These ions possessed higher association constants than the others because their ionic radii fit the cavity size of the benzo[18]-crown-6-ether. However, while Cs^+ is too large to fit into the crown ether cavity of this macrocycle, it could form a stable 2:1 complex with the crown ethers.

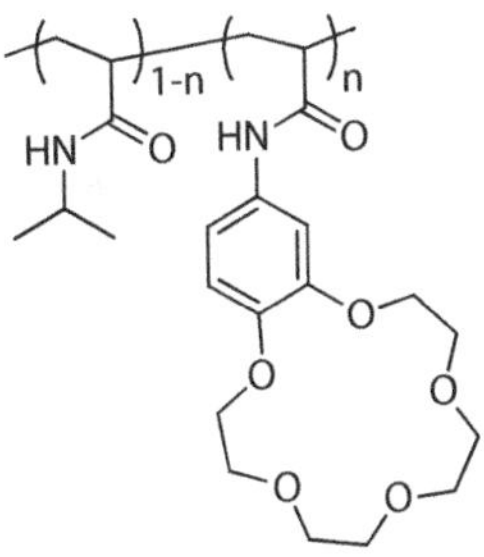

6

FIGURE 5.11   PNIPAM copolymer with benzo[15]crown-5-ether group **6**.

Cross-linking among the polymer chains through a 2:1 sandwich complex between the crown ethers and Cs^+ induced a decrease in the LCST of the polymer. Chu's group also showed that the LCST of poly(N-isopropylacrylamide-*co*-acryloyl-amidobenzo-15-crown-5) **6** decreased in the presence of K^+ because the benzo-[15]-crown-5 and K^+ can form stable 2:1 sandwich complexes (Figure 5.11).[31]

Yamaguchi et al. conducted a more detailed study of the correlation between ion recognition and the LCSTs of PNIPAM-bearing crown ethers.[32] They focused on the quantities of the complex of the crown ether receptor and ions in order to quantify the influence of the addition of ions. They used KCl, $SrCl_2$ and $BaCl_2$ as additives and PNIPAM-bearing benzo[18]-crown-6-ether **5**.

First, the complex formation constant (K) between **5** and the ions was evaluated using the ion mobility of the solution. By calculating the added ion concentration and K, they estimated the degrees of complexation as the ratios of the quantity of benzo[18]-crown-6-ether units that formed complexes in the presence of metal cations to the total quantity of the benzo[18]-crown-6-ether units. The result was about 0.02 in a 0.04 M $SrCl_2$ solution, and the change in the LCST depended on the degrees of complexation and not on the ion species. The metal ions captured by the crown ethers were restricted to contact with the water molecules by coordination bonds with the oxygen atoms of the crown ether. Thus, the hydration structure of the complex of the crown ether and the metal ion is practically the same regardless of the ion species.

The formation of a complex changes the LCSTs of thermosensitive polymers other than PNIPAM-bearing crown ether units(Figure 5.12).[33,34] For example, Moriya et al. reported systems consisting of polysilsesquioxane **7** with a bis(2-methoxyethyl) amide group and a [15]-crown-5-ether ring as the thermosensitive component and ion recognition unit, respectively. The addition of ions such as LiCl, NaCl, and KCl increased the LCST of the polymer.[33]

Yamaguchi et al. demonstrated the multi-sensitive behavior of the ternary copolymer **8** of [3-(methacryloylamino)propyl]trimethylammonium chloride, acrylic acid, and benzo[18]crown-6-acrylamide.[34] This system controlling thermosensitivity by cation recognition of the crown ether was applied to an ion gating membrane.[35]

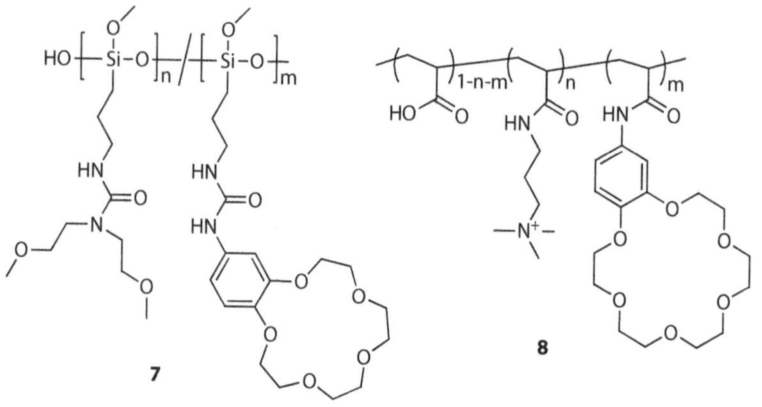

FIGURE 5.12   Some thermosensitive polymers bearing crown ether groups **7** and **8**.

5.5   THERMOSENSITIVE POLYMERS UTILIZING SUGAR AND BORONIC ACID

A characteristic chemical feature of boronic acids is the formation of reversible covalent complexes with 1,2- or 1,3-diols such as ethylene glycol, sugars, and polysaccharides. In aqueous solutions, boronic acids exist in equilibrium between an undissociated neutral trigonal form and a dissociated anionic tetrahedral form. In the presence of diols, neutral boronic acids barely form cyclic boronate esters by reaction with diols because these esters are generally hydrolytically unstable,[36] but boronate anions form stable anionic boronate esters. Thus, the formation of complexes depends on the pH of the solution and pK_a of the boronic acid.

The interactions between boronic acids and diols have been widely applied to the development of sugar sensors, insulin delivery systems, lipase inhibitors, and human immunodeficiency virus (HIV) inhibitors.[37]

As can be seen from Figure 5.13, the addition of diols such as glucose to boronic acids increases the anion forms. This means that the pK_a and hydrophilicity values increase in comparison to the starting neutral boronic acid. Moreover, if the hydrophilicity of a diol is high, the complex exhibits increased hydrophilicity. Therefore, the LCST of a thermosensitive polymer bearing a boronic acid unit or a diol unit can be controlled by the addition of a diol or boronic acid, respectively. However, because

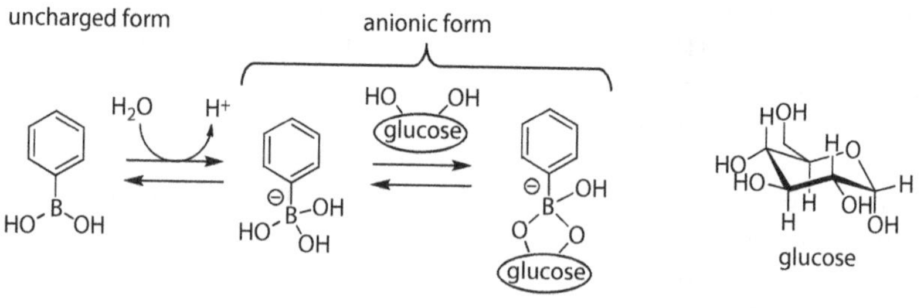

FIGURE 5.13   Equilibria of phenylboronic acid in aqueous solution in presence of glucose.

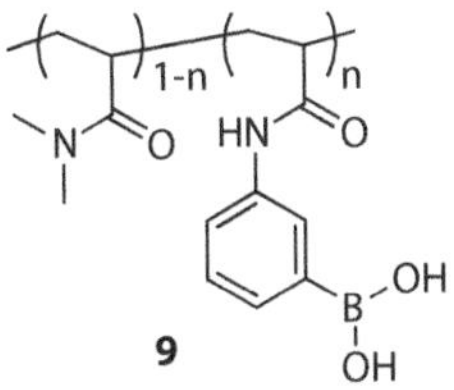

FIGURE 5.14   Copolymer 9 of *N,N*-dimethylacrylamide containing phenylboronic acid moiety.

the addition of sugars affects the LCST of PNIPAM, the change in the LCST must be monitored carefully.[12]

In 1994, Kataoka et al. demonstrated the thermosensitive behavior of a copolymer **9** of N,N-dimethylacrylamide containing 15 mol% of 3-(acrylamide)phenylboronic acid with glucose (Figure 5.14).[38] They controlled the LCST by changing the concentration of glucose. In the absence of glucose, **9** demonstrated an LCST around 27°C in HEPES-buffered saline (pH 7.4). Since the homopolymer of N,N-dimethylacrylamide did not show LCST behavior in water, the phenylboronic acid moieties in **9** played a key role in the appearance of the LCST.

In the presence of glucose, an increase in the LCST was observed and with increasing glucose concentration, the LCST increased. Based on the pK$_a$ of the boronic acid in the presence of glucose, they estimated the ratios of borate anions (degrees of complexation) in a hydroxylated boronic acid complex of glucose and boronic acid.

For example, the ratio of borate anions versus the initial phenylboronic acid fraction was about 75% at a 12.5 g/L glucose concentration. This provided a good correlation between the LCST of **9** and the degree of complexation. These results indicated that the addition of glucose increases the pK$_a$ levels of boronic acids to yield the anion forms that increased the hydrophilicity of the polymer and generated the charge for the increase in solubility. Since an anionic form of boronic acid can interact with diols, the complex is produced efficiently at a pH higher than the pK$_a$. In other words, if the control of the LCST is desired at a lower solution pH, a boronic acid unit with a lower pK$_a$ is required.

Kataoka's group also reported that adjustment of the pH at which the thermosensitivity can be controlled by the addition of glucose could be achieved by changing the substituent group of the boronic acid attached to a thermosensitive polymer (Figure 5.15).[39] They synthesized copolymers consisting of 4-carbamoylphenylboronic acid (pK$_a$ = 7.8) or 3-acrylamidophenylboronic acid (pK$_a$ = 8.2) as glucose-sensitive components and N-isopropylacrylamide acid as a thermosensitive component (**10** and **11**, respectively). The LCST changed as a function of the solution pH for **10** and **11** in the presence and absence of glucose. Moreover, the pH region at which the LCST of **10** shows a considerable change was lowered as compared to the case of **11**; this is consistent with the corresponding pK$_a$ values of boronic acid derivatives.

Although phenylboronic acid is commonly used as a diol-sensitive boronic acid unit, the copolymers of other boronic acid units and thermosensitive units are reported. For example, the LCST control of the copolymer **12** consisting of a NIPAM unit and

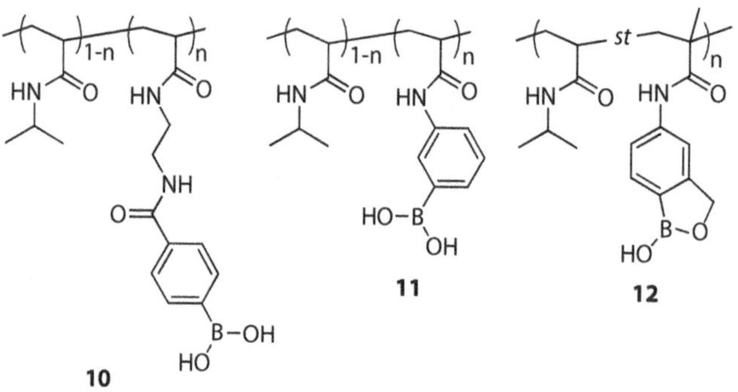

FIGURE 5.15   PNIPAM copolymer with boronic acid moieties **10**, **11**, and **12**.

1,2-benzoxaborole unit was reported by Narain et al. (Figure 5.15).[40] Benzoxaborole is known as a useful sugar-responsive moiety because it is very hydrolytically stable and exhibits high water solubility and remarkable sugar-binding properties under physiological conditions.[41] The addition of glucose induced an increase in the LCST similar to the above results.

5.6   THERMOSENSITIVE POLYMERS BY OTHER MOLECULAR RECOGNITION SYSTEMS IN WATER

Although most host–guest interactions controlling the LCSTs of thermosensitive polymers involve the three systems described earlier, several other examples with different host–guest interactions in combination with LCST polymers have been reported. Scherman et al. showed a thermosensitive system containing a PNIPAM end functionalized with a dibenzofuran moiety **13**, cucurbit[8]uril(CB[8]) and methylviologen(M_2V); see Figure 5.16.[42] Indeed, CB[8] is known to form a variety of strong, stable ternary complexes consisting of complementary pairs of an electron-deficient aromatic compound, such as M_2V and electron-rich aromatic compounds such as dibenzofuran.

The LCST of the polymer is lower than that of typical PNIPAM due to the hydrophobic nature of the dibenzofuran end functionality. However, the addition of CB[8] and M_2V increased the hydrophilicity of the polymer due to the dibenzofuran unit by CB[8], increasing the LCST. Also, pillararenes are known as hosts that incorporate hydrophobic guests in water. Ogoshi et al. demonstrated the LCST behaviors of pillar[5]arenes **14** modified with oligoethylene oxide groups and LCST control by the addition of viologen derivatives (Figure 5.16).[43]

The LCST control of a PNIPAM bearing a single-stranded DNA (ssDNA) by a single strand–double strand transformation of conjugated DNA was reported by Yamaguchi et al.[44] The ssDNA-conjugated PNIPAM shrank and aggregated with an increase in temperature because the conjugated DNA contained bound water and a small number of electric charges. In contrast, when the conjugated ssDNA was hybridized with a fully matched complementary DNA strand, the electric charge and bound water on the DNA increased. The shrinking and aggregation of the polymer chain were inhibited under the conditions in which the ssDNA-conjugated PNIPAM shrinks.

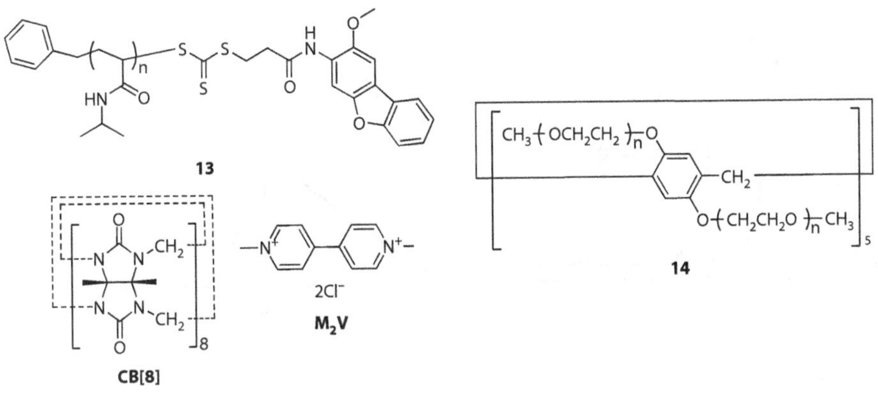

13

CB[8]

M₂V

14

FIGURE 5.16   LCST systems using host–guest complexation of cucurbituril or pillararene.

5.7   INDUCED LCST BEHAVIOR USING MOLECULAR RECOGNITION

Although controlling the LCST of thermosensitive polymers by modification of LCST polymers has been reported extensively, the de novo designs of LCST polymer systems from insoluble homopolymers in the presence of suitable guest molecules has recently attracted considerable interest with respect to their supramolecular chemistry.[45–48]

Ritter et al. reported that a polymer bearing adamantyl moieties showed LCST behavior in the presence of RM-β-CD in water due to the dissociation of RM-β-CD from the adamantyl group in the polymer chain upon heating and the subsequent association of adamantyl groups through a hydrophobic interaction.[46] Sada et al. also demonstrated readily adjustable LCST behaviors in organic solvents based on hydrogen bonds between polymer-bearing urea units and long aliphatic alcohols for solubilization of the insoluble urea polymer in nonpolar aprotic media.[47]

More recently, a charge-transfer interaction was employed for designing LCST behavior. The researchers used polymer **15**-bearing pyrene units for the insoluble homopolymers and acceptors such as pyromerite diimide **16** (Figure 5.17).[48] The

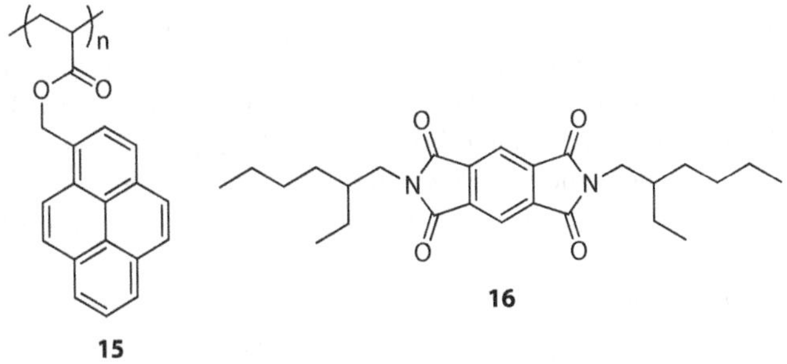

16

15

FIGURE 5.17   LCST system of polymer having pyrene unit and acceptor molecule.

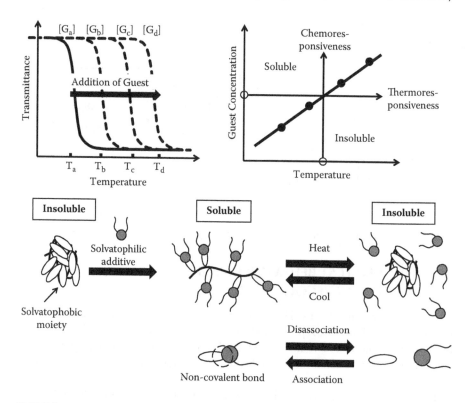

FIGURE 5.18   General concept for designing chemosensitive polymers.

compounds were all insoluble in the absence of suitable guest molecules, and in their presence, they exhibited LCST behaviors. The dissociation of supramolecular complexes at an elevated temperature triggered drastic changes in the solubilities of the polymers. Therefore, synergy in dissociation of the guest molecules and association of the polymer chains by heating plays a key role in drastic responses against temperatures.

5.8   CONCLUSION

In this chapter, we demonstrated the synergies of the thermosensitivities of LCST polymers and molecular recognition for designing chemoselective polymers. If the complexation of the recognition sites in the polymer chains with small molecules or ions as chemical stimuli increases the solubilities of polymers, the LCST increases. When complexation decreases solubility, the LCST drops. Increasing the concentration of the guest molecules increases the phase transition temperature due to enhancement of the solubility by complexation as shown in Figure 5.18. Thus, at a constant temperature, a critical concentration of guest molecules should exist for drastic changes in polymer conformation and solubility.

This should allow the de novo design of chemosensitive polymers that induce phase transitions in polymer solutions by guest molecules as chemical stimuli. Since solubility should be related closely to cooperative association of the repeating units in the flexible polymer chains against surrounding small molecules, rational design of thermosensitive and chemosensitive polymers should be of importance with respect to supramolecular chemistry between the macromolecules and small molecules.

REFERENCES

1. Aseyev, V., Tenhu, H., and Winnik, F. M. *Adv. Polym. Sci.* 2011, *242*, 29–89. Schild, H. G. *Progr. Polym. Sci.* 1992, *17*, 163–249.
2. Bajpai, A. K., Shukla, S. K., Bhanu, S. et al. *Progr. Polym. Sci.* 2008, *33*, 1088–1118. Oh, J. K., Drumright, R., Siegwart, D. J. et al. *Progr. Polym. Sci.* 2008, *33*, 448–477.
3. Jeong, J. H., Kim, S. W., Park, T. G. *Progr. Polym. Sci.* 2007, *32*, 1239–1274.
4. Kikuchi, A. and Okano, T. *Progr. Polym. Sci.* 2002, *27*, 1165–1193. Kanazawa, H. and Okano, T. *J. Chromatogr. A* 2011, *1218*, 8738–8747.
5. Lee, H. I., Pietrasik, J., Sheiko, S. S. et al. *Progr. Polym. Sci.* 2010, *35*, 24–44.
6. Brun-Graeppi, A. K., Richard, C., Bessodes, M. et al. *Progr. Polym. Sci.* 2010, *35*, 1311–1324.
7. Liu, R., Fraylich, M., and Saunders, B. R. *Colloid Polym. Sci.* 2009, *287*, 627–643.
8. Inomata, H., Goto, S., Otake, K. et al. *Langmuir* 1992, *8*, 687–690. Park, T. G. and Hoffman, A. S. *Macromolecules* 1993, *26*, 5045–5048. Shechter, I., Ramon, O., Portnaya, I. et al. *Macromolecules* 2010, *43*, 480–487. Burba, C. M., Carter, S. M., Meyer, K. J. et al. *J. Phys. Chem. B* 2008, *112*, 10399–10404.
9. Schild, H. G., Muthukumar, M., and Tirrell, D. A. *Macromolecules* 1991, *24*, 948–952. Winnik, F. M., Ottaviani, M. F., Bossmann, S. H. et al. *Macromolecules* 1992, *25*, 6007–6017. Yamauchi, H. and Maeda, Y. *J. Phys. Chem. B* 2007, *111*, 12964–12968.
10. Schild, H. G. and Tirrell, D. A. *Langmuir* 1991, *7*, 665–671. Meewes, M., Ricka, J., De Silva, M. et al. *Macromolecules* 1991, *24*, 5811–5816. Dai, S. and Tam, K. C. *Langmuir* 2004, *20*, 2177–2183.
11. Xu, F. J., Kang, E. T., and Neoh, K. G. *Biomaterials* 2006, *27*, 2787–2797.
12. Kawasaki, H., Sasaki, S., Maeda, H. et al. *J. Phys. Chem.* 1996, *100*, 16282–16284.
13. Seuring, J. and Agarwal, S. *Macromol. Rapid Commun.* 2012, *33*, 1898–1920. Seuring, J. and Agarwal, S. *ACS Macro. Lett.* 2013, *2*, 597–600.
14. Rekharsky, M. V. and Inoue, Y. *Chem. Rev.* 1998, *98*, 1875–1917. Uemura, T., Moro, T., Komiyama, J. et al. *Macromolecules* 1979, *12*, 737–739. Harada, A., Adachi, H., Kawaguchi, Y. et al. *Macromolecules* 1997, *30*, 5181–5182. Hashidzme, A. and Harada, A. *Polym. Chem.* 2011, *2*, 2146–2154.
15. Han, S. J., Yoo, M. K., Sung, Y. K. et al. *Macromol. Rapid Commun.* 1998, *19*, 403–407.
16. Ritter, H., Sadowski, O., and Tepper, E. *Angew. Chem. Int. Ed.* 2003, *42*, 3171–3173. Ritter, H., Sadowski, O., and Tepper, E. *Angew. Chem. Int. Ed.* 2005, *44*, 6090–6099.
17. Gingter, S., van Sloun, C., and Ritter, H. *Polym. Int.* 2012, *61*, 1234–1237. Ogoshi, T., Masuda, K., Yamagishi, T. et al. *Macromolecules* 2009, *42*, 8003–8005.
18. Chung, J. E., Yokoyama, M., Suzuki, K. et al. *Colloids Surf. B* 1997, *9*, 37–48. Duan, Q., Narumi, A., Miura, Y. et al. *Polym. J.* 2006, *38*, 306–310.
19. Duan, Q., Miura, Y., Narumi, A. et al. *J. Polym. Sci. A* 2006, *44*, 1117–1124.
20. Chem, F., Zhang, X., Li, W. et al. *Soft Matter.* 2012, *8*, 4869–4872.
21. Yuan, G., Yin, X., Sun, L. et al. *ACS Appl. Mater. Interfaces* 2012, *4*, 950–954.
22. Zhao, Y., Guo, K., Wang, C. et al. *Langmuir* 2010, *26*, 8966-8970. Schmitz, S. and Ritter, H. *Macromol. Rapid Commun.* 2007, *28*, 2080–2083.
23. Kretschmann, O., Choi, S. W., Miyauchi, M. et al. *Angew. Chem. Int. Ed.* 2006, *45*, 4361–4365.

24. Nozaki, T., Maeda, Y., Ito, K. et al. *Macromolecules* 1995, *28*, 522–524. Hirasawa, T., Maeda, Y., and Kitano, H. *Macromolecules* 1998, *31*, 4480–4485.
25. Ohashi, H., Hiraoka, Y., and Yamaguchi, T. *Macromolecules* 2006, *39*, 2614–2620.
26. Yang, M., Chu, L. Y., Xie, R. et al. *Macromol. Chem. Phys.* 2008, *209*, 204–211.
27. Yan, J., Li, W., Zhang, X. et al. *J. Mater. Chem.* 2012, *22*, 17424–17428.
28. Gokel, G. W., Leevy, W. M., and Weber, M. E. *Chem. Rev.* 2004, *104*, 2723–2750.
29. Irie, M., Misumi, Y., and Tanaka, T. *Polymer* 1993, *34*, 4531–4535.
30. Zhang, B., Ju, X. J., Xie, R. et al. *J. Phys. Chem. B* 2012, *116*, 5527–5536.
31. Liu, Z., Luo, F., Ju, X. J. et al. *Adv. Funct. Mater.* 2012, *22*, 4742–4750.
32. Ito, T., Sato, Y., Yamaguchi, T. et al. *Macromolecules* 2004, *37*, 3407–3414.
33. Matsuoka, T., Yamamoto, S., Moriya, O. et al. *Polym. J.* 2010, *42*, 313–318.
34. Hara, N., Ohashi, H., Ito, T. et al. *Macromolecules* 2009, *42*, 980–986.
35. Yamaguchi, T., Ito, T., Sato, T. et al. *J. Am. Chem. Soc.* 1999, *121*, 4078–4079. Ito, T. and Yamaguchi, T. *J. Am. Chem. Soc.* 2004, *126*, 6202–6203.
36. Lorand, J. P. and Edwards, J. D. *J. Org. Chem.* 1959, *24*, 769–774.
37. James, T. D., Sandanayake, K. R. A. S., and Shinkai, S. *Nature* 1995, *374*, 345–347. Ravaine, V., Ancla, C., and Catargi, B. *J. Controlled Release* 2008, *132*, 2–11. Cambre, J. N. and Sumerlin B. S. *Polymer* 2011, *52*, 4631–4643.
38. Kataoka, K., Miyazaki, H., Okano, T. et al. *Macrolecules* 1994, *27*, 1061–1062.
39. Matsumoto, A., Ikeda, S., Harada, A. et al. *Biomacromolecules* 2003, *4*, 1410–1416.
40. Kotsuchibashi, Y., Agustin, R. V. C., Lu, J. Y. et al. *ACS Macro Lett.* 2013, *2*, 260–264.
41. Dowlut, M. and Hall, D. G. *J. Am. Chem. Soc.* 2006, *128*, 4226–4227. Tomsho, J. W., Pal, A., Hall, D. G. et al. *ACS Med. Chem. Lett.* 2012, *3*, 48–52.
42. Rauwald, U., Barrio, J. D., Loh, X. J. et al. *Chem. Commun.* 2011, *47*, 6000–6002.
43. Ogoshi, T., Shiga, R., and Yamagishi, T. *J. Am. Chem. Soc.* 2012, *134*, 4577–4580.
44. Sugawara, Y., Tamaki, T., Ohashi, H. et al. *Soft Matter* 2013, *9*, 3331–3340.
45. Schmitz, S. and Ritter, H. *Angew. Chem. Int. Ed.* 2005, *44*, 5658–5661.
46. Kretschmann, O., Steffens, C., and Ritter, H. *Angew. Chem. Int. Ed.* 2007, *46*, 2708–2711.
47. Amemori, S., Kokado, K., and Sada, K. *J. Am. Chem. Soc.* 2012, *134*, 8344–8347.
48. Amemori, S., Kokado, K., and Sada, K. *Angew. Chem. Int. Ed.* 2013, *52*, 4174–4178.

6 Synergistic Effect on Superlattice Formation in Self-Assembled Monolayers at Liquid– Solid Interfaces

Kazukuni Tahara, Steven De Feyter, and Yoshito Tobe

CONTENTS

6.1 INTRODUCTION

Synergy in molecular recognition and self-assembly occurs in various phases in both solutions and also in solid states and liquid crystals. On-surface self-assembly possesses a special advantage in that molecules are visualized by scanning tunneling microscopy (STM). This allows us to gain detailed insight into how molecules act cooperatively with each other.

During the past decade, we engaged in forming porous networks constituted by self-assemblies of alkoxy-substituted dehydro[12]annulenes (DBAs) at liquid–solid interfaces and observed many interesting events arising from the high level of adaptability of these building blocks. DBAs are adaptable for (i) modification of pore size, (ii) pore functionalization for selective binding of guest molecules, (iii) installation of reaction sites for covalent bonds between themselves leading to two-dimensional (2D) polymers and with substrates such as graphene, (iv) generation of chirality on surfaces at single molecular and supramolecular levels, and (v) formation of superlattice structures on surfaces.

In this chapter, we describe two examples of superlattice formation through synergetic interactions between building blocks and between a molecular network and a guest molecule at the liquid–solid interface. In both instances, various levels of adaptability of the DBA building block play central roles.

6.2 SELF-ASSEMBLED MONOLAYERS FORMED ON SOLID SURFACES OR AT LIQUID–SOLID INTERFACES

Self-assembly in 2D spaces on solid surfaces has been a topic of intense interest in recent decades because of the perspectives in numerous applications in the fields of nanoscience and nanotechnology.[1] Among many instrumental methods used in these fields, scanning tunneling microscopy (STM) is extremely useful for structural evaluation with submolecular resolution of self-assembled monolayers formed on the surfaces of conducting solid materials.[2]

STM operates both under ultrahigh vacuum (UHV) conditions and at the interface between liquid and solid phases. At the liquid–solid interface, the experimental approach is more straightforward than at UHV and a wide range of molecular building blocks can be used.[3] Design and control of structures and functions of molecular networks require understanding of both the structural and electronic features of the building blocks and also the topology due to surface confinement. This strategy, known as *crystal engineering* in 3D systems, has rapidly developed into 2D systems.[4]

6.3 POROUS SELF-ASSEMBLED MONOLAYERS OF ALKOXY-SUBSTITUTED TRIANGLE BUILDING BLOCKS DBA-OCN VIA VAN DER WAALS INTERACTIONS AMONG ALKYL CHAINS

Recently we showed the possibility of triangular DBA substituted with six alkoxy groups, here denoted as **DBA-OCn**, where n represents the number of carbon atoms in the alkoxy chains (Figure 6.1a), for use as a model system to study molecular self-assembly on surfaces.[5]

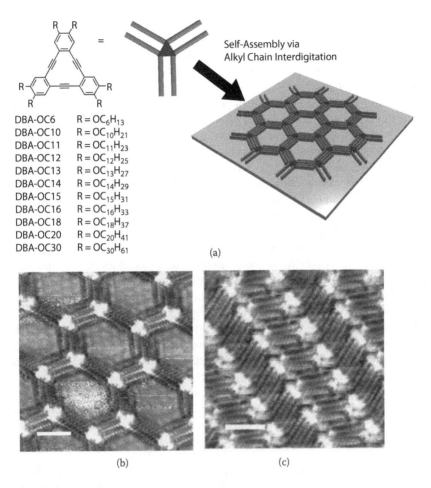

DBA-OC6 $R = OC_6H_{13}$
DBA-OC10 $R = OC_{10}H_{21}$
DBA-OC11 $R = OC_{11}H_{23}$
DBA-OC12 $R = OC_{12}H_{25}$
DBA-OC13 $R = OC_{13}H_{27}$
DBA-OC14 $R = OC_{14}H_{29}$
DBA-OC15 $R = OC_{15}H_{31}$
DBA-OC16 $R = OC_{16}H_{33}$
DBA-OC18 $R = OC_{18}H_{37}$
DBA-OC20 $R = OC_{20}H_{41}$
DBA-OC30 $R = OC_{30}H_{61}$

(a)

(b)   (c)

FIGURE 6.1 (a) Chemical structures of **DBA-OCn**, and their self assembly to form a honey-comb network. STM images of porous network (b) and nonporous network (c) of **DBA-OC16** at 1,2,4-trichlorobenzene–HOPG interface. (*Source:* Adapted from *J. Am. Chem. Soc.* 2013, 135, 12068–12075. Copyright (2013), American Chemical Society. With permission.)

At the interface of an organic solvent (typically 1,2,4-trichlorobenzene, 1-pheny-loctane, or octanoic acid) and a solid substrate (highly oriented pyrolytic graphite (HOPG) or Au(111)), depending on experimental conditions, **DBA-OCn** forms either porous honeycomb-type molecular networks or nonporous close-packed assemblies by van der Waals interactions between interdigitated alkyl chains of neighboring molecules (Figure 6.1a).[6] Typical STM images for these phases are shown in Figures 6.1b and c, respectively. The size of the DBA core plays a crucial role in the formation of the porous networks linked by interdigitated alkoxy chains, since triangular building blocks with smaller or larger cores do not form regular porous networks.[7]

In view of their potential applications, porous molecular networks have attracted a great deal of interest.[8] Fortunately, through the investigation of concentration effects on network morphology, we found that porous networks with pore diameters of 2.7 to 7.5 nm were formed selectively when the solute concentration was reduced as low as about 10^{-6} M for **DBA-OCn** (n ≤ 20).[9] Moreover, examination of both concentration and temperature effects led us to propose a thermodynamic model for the transition between the two phases.[10]

Depending on the interdigitation directions of the alkoxy chains, supramolecular chirality of a porous structure is generated on surfaces. Based on the interdigitation pattern we tentatively call (+)-type at the dimer level, a chiral C_6-symmetric hexamer in which the inner alkyl chains run in a counterclockwise direction is formed. (See Figure 6.3 in Section 6.5.2.) Conversely, (–)-type interdigitation results in a C_6-hexamer of clockwise chirality. Experimentally, the domains of chiral networks co-exist in equal numbers for the achiral **DBA-OCn** molecules. However, by simple modification of the alkoxy chain, we were able to create homochiral porous networks at the liquid–HOPG interface.[11]

The porous networks formed by self-assembly of **DBA-OCn** at the liquid–solid interface offer the possibility of immobilizing guest molecules by recognizing the sizes and shapes of the pore spaces. Although matrix rigidity favors guest selectivity, flexible host networks such as those formed by **DBA-OCn** may exhibit induce-fit mechanisms similar to those of bioenzymes.[12]

Another advantage of the two-dimensional (2D) networks formed by **DBA-OCn** is the ability to fine tune pore sizes by changing the alkyl chain length by increments of 1.25 Å (one methylene unit). We found that in the honeycomb networks of **DBA-OCn**, various guest molecules were co-adsorbed with reasonable recognition of the size and shape of the pore space.

The guest molecules include coronene (**COR**),[6d,13] a large triangular aromatic hydrocarbon (Figure 6.2),[14] a giant wheel-like molecule with a diameter of 5.7 nm,[6c] and a heterocluster formed by coronene and isophthalic acid.[6d,15]

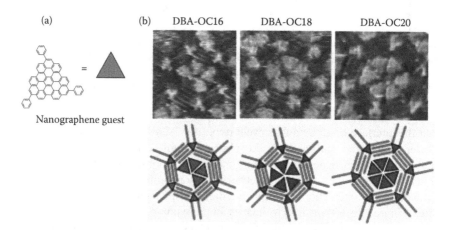

FIGURE 6.2 (a) Chemical structure of nanographene molecule and (b) STM images and corresponding models of porous networks **DBA-OCn** (n = 16, 18, and 20) co-adsorbed with 4, 5, and 6 molecules of nanographene at the 1,2,4-trichlorobenzene/HOPG interface. (*Source: Cryst. Eng. Commun.* 2010, 12: 3369–3381. Reproduced by permission of Royal Society of Chemistry.)

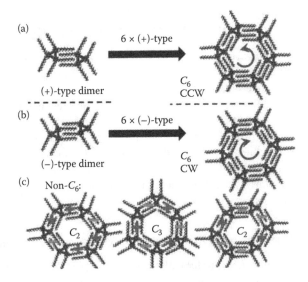

FIGURE 6.3 (a) (+)-Type interdigitation pattern forming counterclockwise (CCW) C_6-hexamer of **DBA-OCn**. (b) (−)-Type interdigitation pattern forming clockwise (CW) C_6-hexamer. (c) Lower symmetry non-C_6 hexamers arising from +/− mixed interdigitation. (*Source: ACS Nano* 2013, 7: 8031–8042. Copyright (2013), American Chemical Society. With permission.)

6.4 SUPERLATTICE FORMATION IN SELF-ASSEMBLED MONOLAYERS

Although hierarchical self-assembly is ubiquitous in biological systems, it is still difficult to create such sophisticated architectures in artificial systems. Construction of complex artificial systems of high level of periodicity, typically called superlattices or superstructures, represents a challenge in on-surface 2D supramolecular chemistry because of interest in understanding the principles governing transitions to highly organized architectures, potential applications in nanopatterning and molecular electronic devices, and aesthetically pleasant appearances.[16]

Many superlattice structures are formed by synergetic interactions of various kinds of molecules co-adsorbed on surfaces. In general, when more than two molecular components are co-adsorbed on solid surfaces, the three possible phase behaviors are phase segregation,[17] random mixing,[18] and co-crystallization, in addition to preferential adsorption by which certain components adsorb preferentially over others.[19]

If the components do not mix, phase separation representing the structure of the pure system occurs. When one component is incorporated in the lattice of another or both components act cooperatively to form a new structure, mixing on the surface occurs. Depending on whether this mixing is random or orderly, random mixing or co-crystallization is seen on the surface.[20] To create a superlattice structure by co-crystallization, each component should participate in favorable inter-component interactions.[21]

6.5 SUPERLATTICE FORMATION BY SYNERGISTIC INTERACTIONS OF MOLECULAR BUILDING BLOCKS

Understanding and predicting the assembling behaviors of multi-component mixtures on solid surfaces is very challenging because of the formation of complex assemblies such as superlattice structures and quasi-crystals on surfaces.[22] In this section, we describe the formation of superlattice structures at liquid–solid interfaces by co-adsorption of two structurally similar molecules, i.e., **DBA-OCn** bearing alkoxy chains that differ by only one methylene unit, via synergetic interactions between mutual components.[23] Since **DBA-OCn** at the monocomponent level exhibits an odd–even effect related to molecular chirality that is an origin of superlattice formation, we will start with a discussion on the odd–even effect on molecular self-assemblies on surfaces.

6.5.1 ODD–EVEN EFFECTS IN SELF-ASSEMBLED MONOLAYERS ON SURFACES

The phase behaviors of straight-chain alkanes and the corresponding alcohols demonstrate how subtle difference in structures and functional groups affect the outcomes of mixing. Straight-chain alkanes that differ only by one carbon atom usually phase separate in 3D due to the differences in the crystal structures of alkanes with odd- (*pgg* space group) and even-numbered (*cm* arrangement) carbon atoms.[24] The difference in space groups is mainly due to the different orientations of the terminal methyl groups that lead to differences in intermolecular steric repulsion.

The densely packed organic thin films of alkanes tend to exhibit alterations of the network structures, depending on the parity of the number of the methylene units—a result called the *odd–even effect*.[25] Binary mixtures of primary linear alcohols that differ by one methylene group show ideal mixing if the shorter alcohol possesses an odd number of carbon atoms.[26] In contrast, if the shorter component is even numbered, co-crystals are formed. The appearance of mixing (random or ordered) on the surface rather than phase separation is ascribed to the similar symmetries of the unit cells of the pure components.[27]

6.5.2 ODD–EVEN EFFECTS IN SELF-ASSEMBLED MONOLAYERS OF DBA-OCN

In Section 6.3, we described the formation of porous self-assembled monolayers of triangle-shaped molecular building blocks—**DBA-OCn** substituted with six alkoxy groups at the periphery, with a simplified model in which the direction of alkoxy chains with respect to the edge of the DBA cores is assumed to be orthogonal (Figure 6.1). However, close inspection of the registry of the monolayers and that of the underlying graphite lattice reveal that the orientations are different for **DBA-OCn$_{odd}$** and **DBA-OCn$_{even}$** bearing alkoxy chains with odd or even numbers of carbon atoms, respectively.

Moreover, regular hexagonal pores of C_6-symmetry and distorted hexagonal pores of lower symmetry arising from the combination of (+) and (−) interdigitation patterns are observed, particularly when a system evolves toward thermodynamic

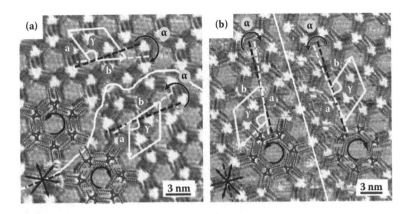

FIGURE 6.4 High-resolution STM images of mirror image domains of (a) **DBA-OC11** and (b) **DBA-OC12**. Unit cells are indicated by rhombi. HOPG reference axis used to measure angle α is shown as dark dotted line. As indicated by short curved arrow, CCW and CW domains have +α (unit cell vector tilted clockwise to HOPG axis) and −α (unit cell vector tilted counterclockwise to HOPG axis), respectively, with the absolute value of α being characteristic to **DBA-OC11** and **DBA-OC12**. Long curved arrow indicates CW/CCW supramolecular chirality of C_6-symmetric pores in each domain. (*Source: ACS Nano.* 2013, 7: 8031–8042. Copyright (2013), American Chemical Society. With permission.)

equilibrium. Of the 12 possible combinations, C_3- and C_2-symmetric structures are most common (Figure 6.3).

DBA-OCn derivatives with different alkyl chain lengths (n = 11 to 16) were used to investigate mixing behaviors. As a single component, they formed at the 1-phenyloctane–HOPG interface C_6-symmetric structures as observed by scanning tunneling microscopy (STM). Inspection of the unit cell alignment angle α with respect to the HOPG lattice revealed the difference between **DBA-OCn$_{odd}$** and **DBA-OCn$_{even}$** as shown in Figure 6.4 for **DBA-OC11** and **DBA-OC12**. The mean average angle α of the domains of **DBA-OCn$_{odd}$** (n = 11, 13, 15) was 13 degrees, whereas it was 4.5 degrees for **DBA-OCn$_{even}$** (n = 12, 14, 16).

To understand these different orientation angles, we performed molecular modeling experiments for a dimer model. Starting from a D_{3h}-symmetric DBA molecule optimized in vacuo, the structural change induced by adsorption on graphite in combination with maximizing the van der Waals interactions with the alkyl chains of the neighboring molecule resulted in the C_3-symmetric structure of **DBA-OCn** when adsorbed on graphite. By deforming from D_{3h}- to C_3-symmetric structure, adsorbed DBA molecules became chiral.

Therefore, in addition to supramolecular chirality arising from the (+) and (−) directions for the interdigitation of the alkyl chains on the surface (Section 6.3), each DBA molecule possesses specific molecular chirality, here denoted as R-**DBA** or S-**DBA**, depending on whether the angle of the alkyl chains denoted θ with respect to the DBA edge is larger or smaller than 90 degrees (Figure 6.5a).

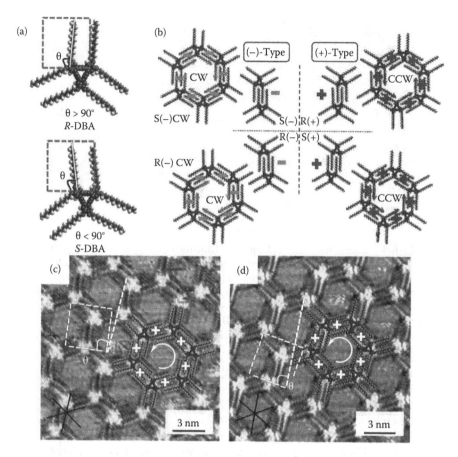

FIGURE 6.5 (a) Molecular chirality of **DBA-OCn**, R-DBA ($\theta > 90$ degrees) or S-DBA ($\theta < 90$ degrees) generated upon adsorption. (b) Combination of molecular chirality of **DBA-OCn** with supramolecular chirality due to alkyl chain interdigitation pattern yielding four types of dimers and C_6-symmetric hexamers. (c) High resolution STM images of CCW domain of **DBA-OC12**. (d) CCW domain of **DBA-OC13**. The white dotted square represents a 90-degree angle with respect to core edge of **DBA-OCn**. The additional dotted line indicates alkyl chain direction that coincides with HOPG reference axis. The angle between them defines tilt angle θ that is characteristic of **DBA-OC12** and **DBA-OC13**. The additional curved arrow indicates CW/CCW supramolecular chirality of C_6-symmetric pores in each domain. (*Source: ACS Nano* 2013, 7: 8031–8042. Copyright (2013), American Chemical Society. With permission.)

An analysis of high-resolution STM images confirmed that the directions of the alkyl chains were not orthogonal with respect to the edge of the DBA core (Figures 6.5c and d). The experimentally measured values of angle for **DBA-OC12** are 85 ± 3 degrees and 95 ± 3 degrees. Therefore, at the level of DBA dimmers, four combinations are possible (Figure 6.5b).

While $R(+)/S(-)$ and $R(-)/S(+)$ pairs are enantiomeric and have same stability, $R(+)/S(+)$ and $R(-)/S(-)$ are diastereomeric pairs and therefore should have different stability. Intriguingly, the existence of a specific combination depends on the parity of carbons in the alkoxy chains: $R(+)$ and $S(-)$ for **DBA-OCn$_{even}$**, while $R(-)$ and $S(+)$ are most favorable for **DBA-OCn$_{odd}$** (Figures 6.5c and d).

The observed odd–even effect in porous 2D assemblies originates from the asymmetry in van der Waals interactions of the interdigitated alkyl chains and DBA core. Since alkyl chains are known to align along main symmetry axes of graphite, the overall alignment of DBA assemblies with respect to the graphite lattice is also structure specific. By using this model, it was possible to reproduce the experimentally determined unit cell alignment angle α with respect to the HOPG lattice in good agreement, thus supporting the validity of the simulations.

6.5.3 ODD–EVEN EFFECTS IN MIXING BEHAVIORS OF DBA-OCN BEARING ALKYL CHAINS DIFFERING BY ONE METHYLENE UNIT

We first examined the mixing behaviors of bi-component mixtures of **DBA-OCn** (n = 10, 14, 16, 18, 20) bearing alkoxy chains with an even number of carbon atoms at the 1,2,4-trichlorobenzene–HOPG interface.[28] For a binary mixture of **DBA-OC14/DBA-OC16** that differed in alkoxy chain length by two methylene units, phase separation with pure domains of the porous pattern of **DBA-OC14** and the nonporous pattern of **DBAOC16** were observed. However, a considerable fraction of mixed domains in which hybrid porous networks consisted of both components was also observed.

Hybrid networks were identified by distorted hexagonal structures, irregular sizes, and periodicities. For a mixture of **DBA-OC16** and **DBA-OC18**, however, it was difficult to evaluate phase behaviors because of the smaller proportional difference between the alkoxy chain lengths of the two components.

As the alkoxy chain length difference increased to four methylene units, the phase separation became more pronounced due to the increased differences in unit cell parameters. For a binary mixture of **DBA-OC14** and **DBA-OC18**, pure domains of each component could be distinguished clearly. Only at domain boundaries was some amount of mixing observed. For mixtures **DBA-OC14/DBA-OC20** and **DBA-OC10/DBA-OC20** with even larger differences in alkoxy chain lengths (6 and 10, respectively), complete phase separation was observed without mixing even at domain boundaries.

Next, we investigated the phase behaviors of bi-component mixtures of **DBA-OCn** bearing alkoxy chains that differed by one methylene unit to determine the mixing behavior and odd–even effect. Premixed solutions of **DBA-OC11/DBA-OC12**, **DBA-OC12/DBA-OC13**, **DBA-OC13/DBA-OC14**, **DBA-OC14/DBA-OC15**, and **DBA-OC15/DBA-OC16** in a 1:1 molar ratio in 1-phenyloctane (total concentrations: ~10^{-6} M) were probed. As shown in Figure 6.6, for mixtures of **DBA-OC12/DBA-OC13** and **DBA-OC16** for all mixtures porous networks were formed.

Surprisingly, however, detailed analysis of the relative abundances of different types of hexagonal pores suggested that their distributions on the surface are classified into two groups. Namely, when the alkoxy chains of **DBA-OCn$_{even}$** are smaller than that of **DBA-OCn$_{odd}$** (i.e., **DBA-OC12/DBA-OC13** and **DBA-OC14/DBA-OC15** mixtures), the surfaces are covered mainly with domains of porous patterns of C_6-symmetry (Figure 6.6a).

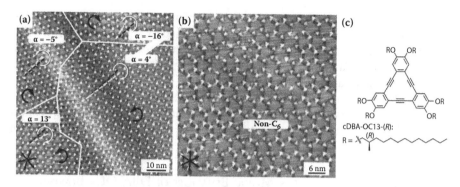

FIGURE 6.6 (a) STM image of **DBA-OC12/DBA-OC13** with domains of CW or CCW hexagons. Clockwise (+) or counterclockwise (−) direction of angle (α) between unit cell vector (white dotted line) and HOPG reference axis (dark dotted line) is indicated by white curved arrow. Dark curved arrow indicates CW/CCW supramolecular chirality of C_6-symmetric pores in each domain. Domain borders are highlighted by white lines. (b) **DBA-OC15/DBA-OC16** whose surface is covered mainly with non-C_6-symmetric hexagons. (c) Chemical structure of **cDBA-OC13(R)**. (*Source: ACS Nano* 2013, 7: 8031–8042. Copyright (2013), American Chemical Society. With permission.)

On the other hand, when the alkoxy chains of **DBA-OCn$_{odd}$** are smaller than that of **DBA-OCn$_{even}$** (i.e., **DBA-OC11/DBA-OC12, DBA-OC13/DBA-OC14**, and **DBA-OC15/DBA-OC16** mixtures), the surfaces are covered mainly with non-C_6-symmetric hexagons of lower symmetry (Figure 6.6b).

The above results indicate a clear difference in the phase behaviors of the respective mixtures. However, the composition of each network, i.e., whether it consisted of a single component via preferential adsorption or phase separation or two, it consisted of components via co-crystallization or random mixing on the surface, is not certain. Although direct assignment of the adsorbed DBA molecules by STM is not possible because of small differences in alkoxy chain lengths for regular hexagons of C_6-symmetry, it is possible to identify each DBA molecule on the basis of the unit cell alignment angle α with respect to the HOPG lattice discussed for single component networks in the previous section.

The alignment angles α in the domains of C_6-symmetric hexagons formed from **DBA-OC12/DBA-OC13** and **DBA-OC14/DBA-OC15** mixtures show clear bimodal distribution with maxima at 5.1 and 12.9, and 4.9 and 13.5 degrees, respectively. The observed angles of ~5 and 13 degrees correspond to those of the single component networks of **DBA-OCn$_{odd}$** and **DBA-OCn$_{even}$**, respectively. It is clear that self-assembly of the **DBA-OCn** mixtures in which n_{even} is smaller than n_{oddn} leads to phase separation on the surface.

For the **DBA-OCn** mixtures in which n_{odd} is smaller than n_{even} that form mainly disordered domains, similar analysis cannot be applied. The disorder indicates random mixing of both components on the surface since mono-component systems have already proven to form highly ordered hexagonal networks. However, to gain further structural insights into phase behaviors of bi-component DBA assemblies at the molecular level, we used as a marker molecule **cDBA-OC13(*R*)** (Figure 6.6c) that has a stereocenter in each alkoxy chain because it can be readily distinguished from DBAs without stereocenters by its appearance with a dark contour.[11]

STM observations were performed for a binary mixture of **DBA-OC12** with **cDBA-OC13(*R*)** to observe phase separation wherein domains consisting of hexagons with dark contours and those without markers were observed. In the case of **cDBA-OC13(*R*)/DBA-OC14**, dark contour molecules are distributed rather randomly in the domains, confirming random mixing of two components in the disordered domains.

6.5.4 SUPERLATTICE FORMATION BY SYNERGISTIC INTERACTIONS OF DBA-OCN MIXTURES

To investigate whether the observed phase behaviors of the mixtures of **DBA-OCn** resulted from thermodynamically equilibrated or kinetically trapped meta-stable assemblies, annealing experiments were performed because the observed structures resulting from self-assembly on surfaces do not always represent thermodynamic equilibrium. For the **DBA-OCn** mixtures in which n_{even} is smaller than n_{odd}, all non-C_6-symmetric pores present when the monolayers are prepared at room temperature disappear by annealing at high temperature, furnishing only large separated phases of pure components. Interestingly, significant preferential adsorption (>90%) of shorter DBAs was noted.

Unexpectedly, in the case of **DBA-OCn** mixtures in which n_{odd} was smaller than n_{even}, thermal annealing of the disordered self-assemblies led to the formation of a new superlattice structure comprising both C_6- and C_2-symmetric pores (Figure 6.7). This structure consists of two distinct sites, one for DBAs forming

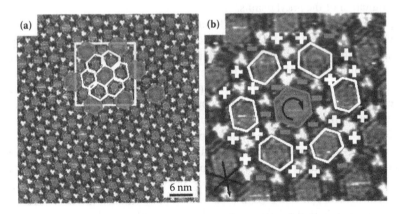

FIGURE 6.7 STM images of superlattice structure formed by **DBA-OC11/DBA-OC12** mixture after thermal annealing at 80°C. (a) Large-scale image containing ordered combination of chiral C_6- (solid hexagons) and achiral C_2-symmetric (open hexagons surrounding C_6-hexagon). (b) Close-up image of white square of (a). Bright plus and gray minus signs indicate interdigitation patterns. (*Source: ACS Nano* 2013, *7*: 8031–8042. Copyright (2013), American Chemical Society. With permission.)

C_6-symmetric rings and the other for single DBAs joining such rings into the perfectly hexagonal superstructure.

Note that the alkyl chain interdigitation patterns of DBAs forming the C_6-symmetric ring are mixed, i.e., either +/+/− or +/−/−. The patterns of DBAs at the junctions are uniform. Such superlattice structures were never observed in the self-assembly of single component systems and thus require both components. Again using the marker molecule **cDBA-OC13(R)** in a binary mixture with **DBA-OC14**, we found that **DBA-OCn$_{even}$** forms C_6-symmetric rings and **DBA-OCn$_{odd}$** occupies the joint sites in the superlattice structure.

Modeling experiments for heterodimers composed of two DBAs (**DBA-OCn$_{odd}$** and **DBA-OCn$_{even}$**) provided important insights about the reasons for the odd–even effect and the formations of superlattice structures. Depending on whether the longer alkoxy chain of **DBA-OCn** has an odd or even number of alkoxy chains, two different structures are predicted by comparing the intermolecular interaction energy (ΔE) of the heterodimer with the average of the intermolecular binding energies of the homodimers built from its components. When $n_{odd} > n_{even}$, ΔE is negative, indicating that the interaction energy of the heterodimer is lower than the average of the homodimer binding energies.

Unlike mono-component systems, the overall stability of these hetero-assemblies is a compromise of several factors such as mismatched interdigitation of the alkyl chains and ineffective packing with a gap between some of the methyl groups of **DBA-OCn$_{even}$** and the DBA cores of **DBA-OCn$_{odd}$**. These factors make the mixing of DBAs unfavorable, thereby leading to phase separation as observed experimentally. On the other hand, when $n_{odd} < n_{even}$, ΔE is positive; the interaction energy of the heterodimer is larger than the average of the homodimer binding energies. For these mixtures, in contrast to the case $n_{odd} > n_{even}$, molecular modeling predicts matched interdigitation of the alkyl chains.

This discussion is still limited to heterodimer formation. For network formation, molecular chirality of **DBA-OCn** induced by adsorption of the surface and supramolecular chirality arising from interdigitation pattern plays a key role in the formation of a superlattice structure (Figure 6.8). In Figure 6.8a, all DBA molecules have well-defined spatial positions, chemical structures (**DBA-OC11** or **DBA-OC12**), and molecular (R- or S-enantiomer) and supramolecular ((−) or (+)) chiralities in good agreement with experimental data. Alternative arrangements shown in panels (b) and (c) reveal incorrect molecular and supramolecular chiralities, respectively. These arrangements show inefficient packing (arrows in panel (b)) or absence of additional van der Waals contacts (open arrows in panel (c)).

The superlattice structure originates from synergy of several factors, including chemical structure of **DBA-OCn** (odd–even alkoxy chains), favorable geometry on the surface (angle θ of alkoxy chain with respect to DBA edge), and molecular and supramolecular chiralities. From an opposite view, in the superlattice structure, a high level of complexity involving composition and spatial order and also molecular and supramolecular chiralities of the components is defined. We should emphasize that the adaptability of **ODA-OCn** at various levels plays a crucial role in superlattice formation.

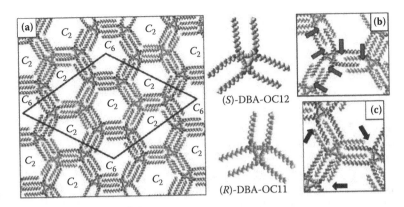

FIGURE 6.8 (a) Model for superlattice structure formed from **DBA-OC11** (light and marked with asterisks) and **DBA-OC12** (dark). (b) Structure with wrong molecular chirality and inefficient packing (arrows). (c) Structure with wrong supramolecular chirality and missing some additional van der Waals contacts (arrows). (*Source: ACS Nano.* 2013, 7: 8031–8042. Copyright (2013), American Chemical Society. With permission.)

6.6 SUPERLATTICE FORMATION BY SYNERGISTIC INTERACTIONS WITH GUEST MOLECULES

As described in Section 6.3, in the nanoscale pores of the self-assembled monolayers of **DBA-OCn**, many different kinds of guest molecules are co-adsorbed in a size- and shape-complementary manner. Here we describe an example in which co-adsorption of a guest induces a periodic deformation of the network structure from C_6- to C_3-symmetric hexagons, leading to the formation of a superlattice structure at the 1-phenyloctane–HOPG interface.

6.6.1 FUNCTIONALIZATION OF HEXAGONAL PORES IN SELF-ASSEMBLED MONOLAYERS OF DBA-OCN

Modifying the interior spaces of porous networks to construct tailored functional pores is an important challenge in the field of on-surface self-assembly. A few studies reported the influence of chemical modification of pore structures on guest co-adsorption properties.[29] However, no work focused on guest-specific binding that exhibits a tight stoichiometric selectivity toward a guest molecule until we succeeded in constructing tailored 2D pores by modifying network pores using a specially designed DBA molecule.[30]

Our general molecular design is to introduce functional groups linked by phenylene units at the ends of three of the six alkoxy chains of DBA in an alternating fashion. When thus modified DBA forms self-assembled monolayers on surfaces via interdigitation of the alkoxy chains, the three alkyl chains bearing one functional group each should be located outside the interdigitation pairs, whereas the remaining unfunctionalized chains occupy the inner position, thereby furnishing a hexagonal pore containing six functional groups (Figure 6.9).

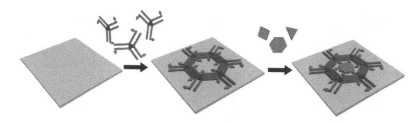

FIGURE 6.9 Pore modification using functionalized DBA.

As a first example, **DBA-AB** bearing three photoresponsive azobenzene groups was designed. To each azobenzene group, two carboxy groups are introduced at the meta positions to create a confined space for a guest molecule by forming a hydrogen-bonded hexamer because a cyclic hexamer of isophthalic acid is known to immobilize one coronene (**COR**) molecule on surfaces via size and shape recognition.[31]

The dicarboxyazobenzene units are connected via *meta*-phenylene linkers at the ends of C_{12} chains. The length of the unfunctionalized chains is set to C_{14} based on molecular modeling to fit with the functionalized chains for making the interdigitated alkyl chain linkage. Indeed, STM observation showed that by self-assembly of **DBA-AB** at the 1-octanoic acid/HOPG interface, a honeycomb network was formed wherein six azobenzene units were visualized as spoke-like features. This indicates the formation of hydrogen-bonded hexamers of the isophthalic acid parts of the molecules.

Reversible photoisomerization of the azobenzene between planar *trans*- and kinked *cis*-configurations took place to change the pore size and shape upon irradiation of UV and visible light (Figure 6.10). This change in pore geometry led to a change in

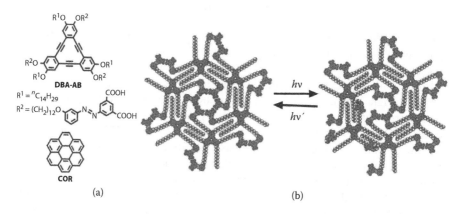

FIGURE 6.10 (a) Chemical structures of **DBA-AB** and coronene (COR). (b) Pore size changes induced by reversible *trans–cis* isomerization of azobenzene unit in **DBA-AZO** resulting from irradiation of UV and visible light.

the number of adsorbed **COR** molecules. This is, to our knowledge, the first example of the construction of 2D pores that respond to external stimuli in a specific manner.[32]

6.6.2 Guest Binding in Porous Network of Modified DBA: Superlattice Formation via Induced-Fit Mechanism

Since the known examples of co-adsorption of guest molecules in host networks are based solely on van der Waals interactions, only sizes and shapes of guests can be recognized. We designed DBA molecules that can bind guests via non-covalent interactions such as hydrogen bonding, ion–dipole interactions, and fluorophilic interactions to recognize electrostatic information of a guest. As we performed these studies, we unexpectedly found a superlattice structure created by an induced-fit mechanism.[33]

As a reference compound, we prepared **DBA-OC10,10** in which a C_{10} chain was linked by a *para*-phenylene unit at the ends of three of the six alkoxy chains of DBA in an alternating fashion. The remaining alkoxy chain length was set to C_{14}. As shown in Figure 6.11, at the 1-phenyloctane/HOPG interface, this compound formed a porous network. In the STM image, the phenylene units appear as dim dots. The terminal alkyl chains running along the edge of the hexagonal pore are also visible.

When a premixed solution of **DBA-OC10,10** with hexakis(phenylethynyl)benzene **HPEB** (about 2.5 equivalent excess) whose molecular size would fit the pore size formed on the surface by **DBA-OC10,10** was subjected to STM observations, two types of pores were visualized: one containing a clear star-shaped feature ascribed to immobilized **HPEB** and the other with a fuzzy feature, most likely due to mobile **HPEB** (Figure 6.12). In addition, a small number of vacant pores were also observed.

A statistical analysis revealed their proportions to be 59, 32, and 9%. Most unexpectedly, the pores with clear HPEB distributed in an ordered manner over the

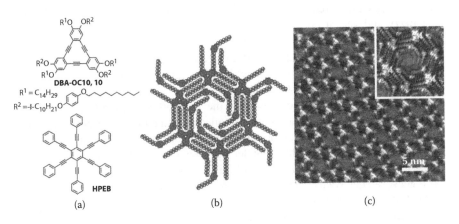

FIGURE 6.11 (a) Chemical structures of **DBA-OC10,10** and HPEB. (b) Molecular model of honeycomb structure formed by **DBA-OC10,10**. (c) STM image of honeycomb structure of **DBA-OC10,10** formed at 1-phenyloctane–HOPG interface.

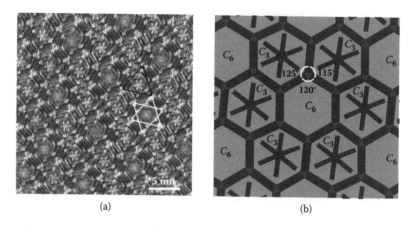

(a)   (b)

FIGURE 6.12 (a) STM image of superlattice formed by **DBA-H** and HPEB at 1-phenyloc-tane–HOPG interface. Arrows indicate distances to next nearest neighbors in hexagonal net-works. (b) Representation of superlattice. Regular hexagon at center is surrounded by slightly distorted hexagons.

domain, indicating the formation of a superlattice structure. The long-range period-icity was confirmed by analysis of the distribution of the nearest-neighbor distance between the pores assigned as fuzzy and vacant.

Moreover, careful inspection of the molecular networks revealed that the sizes and shapes of the two types of the pores were different. The corner-to-corner distances between the phenylene linkers was 2.73 ± 0.24 nm for the pores containing immobi-lized guests. The corresponding distances were longer in the fuzzy and vacant pores (2.97 ± 0.16 and 3.09 ± 0.22 nm, respectively), indicating that the size of the former pores is smaller than pore sizes of other types.

To further evaluate the deformation of the hexagonal shape constituted by six DBA cores, we measured the distances between the centers of every other DBA core. The fuzzy and vacant pores adopted C_6-symmetry with the same inter-DBA dis-tances of 5.07 ± 0.11 nm and 5.03 ± 0.15 nm, respectively. In contrast, the pores with immobilized guests exhibited two different inter-DBA distances in an alternating fashion (5.00 ± 0.11 and 4.74 ± 0.09 nm; Figure 6.12a), indicating that the hexagon is distorted to a C_3-symmetric structure.

As shown in the geometric drawing of the superlattice (Figure 6.12b), one of the angles in the C_3-symmetric hexagon is smaller than that of regular hexagon and the other is larger. Due to pore deformation, the average area of the pores containing immobilized guests was slightly smaller than the areas of the other types of pores. These results indicate that by inclusion of **HPEB**, the pore size is reduced to maximize host–guest van der Waals interactions at least at three corners of the hexagonal pore.

Such deformation can occur only periodically to furnish the observed superlat-tice structure because the remaining C_6-symmetric pores cannot be deformed in the same way any longer. The formation of a superlattice structure arises from synergetic interactions between the host network and the guest via an induced-fit mechanism. It originates from the adaptability of surface-confined DBA molecules to adjust their conformations to maximize intermolecular interactions while keeping the hexagonal network symmetry.

6.7 CONCLUSIONS AND OUTLOOK

The formation of superlattice structures in molecular self-assembly on-surfaces provides detailed insight in how molecules act cooperatively by means of STM. By studying monolayers formed by the self-assembly of alkoxy-substituted dehydro[12] annulenes (DBAs) at the liquid–solid interface, we realized that the high adaptability of this building block led to the generation of chirality on surfaces at single molecular and supramolecular levels and formation of superlattice structures on surfaces.

The two examples described in this chapter demonstrate how the superlattice structures are formed via synergetic interactions between the building blocks and between a molecular network and a guest molecule at the liquid–solid interface. These results suggest that the balance of rigidity and flexibility of building blocks plays a crucial role in the formation of complex structures. Acquiring experimental examples coupled with understanding the controlling principles with aid from simulations would make it possible to build predictably even more complex 2D patterns on surfaces—one of major goals of 2D crystal engineering.[4c,d]

ACKNOWLEDGMENTS

This work was supported by the Fund of Scientific Research (FWO), Flanders, Academy of Innovation by Science and Technology (IWT). Flanders, Katholieke Universiteit Leuven (GOA), Belgian Federal Science Policy Office (IAP-7/05), Hercules Foundation, and a grant in aid for scientific research from the Ministry of Education, Culture, Sports, Science, and Technology of Japan.

REFERENCES

1. (a) Barth, J.V., G. Costantini, and K. Kern. 2005. *Nature* 437: 671–679. (b) Bartels, L. 2010. *Nat. Chem.* 2: 87–95.
2. (a) Samorì, P. 2004. *J. Mater. Chem.* 14: 1353–1366. (b) Samorì, P. 2005. *Chem. Soc. Rev.* 34: 551–561.
3. Rabe, J.P. and S. Buchholz. 1991. *Science* 253: 424–427.
4. (a) Plass, K.E., A.L. Grzesiak, and A.J. Matzger. 2007. *Acc. Chem. Res.* 40: 287–293. (b) Furukawa, S. and S. De Feyter. 2009. *Top. Curr. Chem.* 287: 87–133. (c) Elemans, J.A.A.W., S. Lei, and S. De Feyter. 2010. *Angew. Chem. Intl. Ed.* 48: 7298–7332. (d) Mali, K.S., J. Adisoejoso, E. Ghijens et al. 2012. *Acc. Chem. Res.* 45: 1309–1320.
5. (a) Lei, S., K. Tahara, J. Adisoejoso, T. Balandina et al. 2010. *Cryst. Eng. Commun.* 12: 3369–3381. (b) Tahara, K., S. Lei, J. Adisoejoso et al. 2010. *Chem. Commun.* 8507–8525.
6. (a) Furukawa, S., H. Uji-i, K. Tahara et al. 2006. *J. Am. Chem. Soc.* 128: 3502–3503. (b) Tahara, K., S. Furukawa, H. Uji-i et al. 2006. *J. Am. Chem. Soc.* 128: 16613–166 25. (c) Tahara, K., S. Lei, D. Mössinger et al. 2008. *Chem. Commun.* 3897–3899. (d) Balandina, T., K. Tahara, N. Sändig et al. 2012. *ACS Nano* 6: 8381–8389.

7. (a) Wu, P., Q. Zeng, C. Xu. et al. 2001. *Chem. Phys. Chem.* 2: 750–754. (b) Tahara, K., C. A. Johnson, T. Fujita et al. 2007. *Langmuir* 23: 10190–10197.

8. (a) Bonifazi, D., S. Mohnani, and A. W. Llanes-Pallas. 2009. *Chem. Eur. J.* 15: 7004–7025. (b) Kudernac, T., S. Lei, J. A. Elemans, and S. De Feyter. 2009. *Chem. Soc. Rev.* 38: 402–421. (c) Xuemei, Z., Z. QingDao, and W. Chen. 2014. *Sci. China Chem.* 57:13–25.

9. Lei, S., K. Tahara, F.C. De Schryver et al. 2008. *Angew. Chem. Int. Ed.* 47: 2964–2968.

10. Blunt, M.O., J. Adisoejoso, K. Tahara et al. 2013. *J. Am. Chem. Soc.* 135: 12068–12075.

11. (a) Tahara, K., H. Yamaga, E. Ghijsens et al. 2011. *Nat. Chem.* 3: 714–719. (b) Destoop, I., E. Ghijsens, K. Katayama et al. 2012. *J. Am. Chem. Soc.* 134: 19568–19571.

12. Davis, A.M. and S.J. Teague. 1999. *Angew. Chem. Int. Ed.* 38: 736–749.

13. Furukawa, S., K. Tahara, F.C. De Schryver et al. 2007. *Angew. Chem. Intl. Ed.* 46: 2831–2834.

14. Lei, S., K. Tahara, X. Feng et al. 2008. *J. Am. Chem. Soc. 130*: 7119–7129.

15. Lei, S., M. Surin, K. Tahara et al. 2008. *Nano Lett.* 8: 2541–2546.

16. (a) Spillmann, H., A. Dmitriev, N. Lin et al. 2003. *J. Am. Chem. Soc.* 125: 10725–10728. (b) Xiao, W., X. Feng, P. Ruffieux et al. 2008. *J. Am. Chem. Soc.* 130: 8910–8912. (c) Liu, J., T. Chen, X. Deng et al. 2011. *J. Am. Chem. Soc.* 133: 21010–21015. (d) Jester, S.S., E. Sigmund, L.M. Röck et al. 2012. *Angew. Chem. Intl. Ed.* 51: 1–6.

17. (a) Baker, R.T., J.D. Mougous, A. Brackley et al. 1999. *Langmuir* 15: 4884–4891. (b) Kim, K., K.E. Plass, and A.J. Matzger. 2005. *J. Am. Chem. Soc.* 127: 4879–4887.

18. (a) Padowitz, D. F., D.M. Sada, E.L. Kemer et al. 2002. *J. Phys. Chem. B* 106: 593–598. (b) Tahara, K., E. Ghijsens, M. Matsushita et al. 2011. *Chem. Commun.* 47: 11459–11461.

19. (a) Xia, T.K. and U. Landman. 1993. *Science* 261: 1310–1312. (b) Venkataraman, B. J.J. Breen, and G.W. Flynn. 1995. *J. Phys. Chem.* 99: 6608–6619.

20. (a) Xue, Y. and M.B. Zimmt. 2012. *J. Am. Chem. Soc.* 134: 4513–4516. (b) Silly, F., A.Q. Shaw, M.R. Castell et al. 2008. *Chem. Commun.* 1907–1909.

21. (a) Eichhorst-Gerner, K., A. Stabel, G. Moessner et al. 1996. *Angew. Chem. Intl. Ed.* 35: 1492–1495. (b) Kampschulte, L., S. Griessl, W.M. Heckl et al. 2005. *J. Phys. Chem. B* 109: 14074–14078. (c) Nath, K.G., O. Ivasenko, J.A. Miwa et al. 2006. *J. Am. Chem. Soc.* 128: 4212–4213. (d) Theobald, J.A., N.S. Oxtoby, M.A. Phillips et al. 2003. *Nature* 424: 1029–1031.

22. Clarke, S.M., L. Messe, J. Adams et al. 2003. *Chem. Phys. Lett.* 373: 480–485.

23. Ghijsens, E., O. Ivasenko, K. Tahara et al. 2013. *ACS Nano* 7: 8031–8042.

24. (a) Castro, M.A., S.M. Clarke, A. Inaba et al. 1999. *Phys. Chem. Chem. Phys.* 1: 5017–5023. (b) Castro, M., S.M. Clarke, A. Inaba et al. 2001. *Phys. Chem. Chem. Phys.* 3: 3774–3777.

25. (a) Ramin, L. and A. Jabbarzadeh. 2011. *Langmuir* 27: 9748–9759. (b) Liu, X.F., T.Y. Wang, and M.H. Liu 2012. *Langmuir* 28: 3474–3482. (c) Nerngchamnong, N., L. Yuan, D.C. Qi et al. 2013. *Nat. Nanotechnol.* 8: 113–118. (d) Tao, F. and S.L. Bernasek. 2007. *Chem. Rev.* 107: 1408–1453. (e) Xu, L., X.R. Miao, B. Zha et al. 2013. *Chem. Asian J.* 8: 926–933.

26. (a) Messe, L., S.M. Clarke, C.C. Dong et al. 2002. *Langmuir* 18: 9429–9433. (b) Wang, G.J., S.B. Lei, S. De Feyter et al. 2008. *Langmuir* 24: 2501–2508.

27. Clarke, S.M., L. Messe, J. Adams et al. 2003. *Chem. Phys. Lett.* 373: 480–485.

28. Lei, S., K. Tahara, K. Müllen et al. 2011. *ACS Nano* 5: 4145–4157.

29. (a) Phillips, A.G., L.M.A. Perdigão, P.H. Beton et al. 2010. *Chem. Commun.* 46, 2775–2777. (b) Stepanow, S., M. Lingenfelder, A. Dmitriev et al. 2004. *Nat. Mater.* 3: 229–233; (c) Räisänen, M.T., A.G. Slater, N.R. Champness et al. 2012. *Chem. Sci.* 3: 84–92.

30. Tahara, K., K. Inukai, J. Adisoejoso et al. 2013. *Angew. Chem. Intl. Ed. 52:* 8373–8376.
31. (a) Lei, S., M. Surin, K. Tahara et al. 2008. *Nano Lett.* 8: 2541–2546. (b) Lackinger, M. and W.M. Heckl. 2009. *Langmuir* 25: 11307–11321. (c) Shen, Y. T., M. Li, Y. Y. Guo et al. 2010. *Chem. Asian. J.* 5: 787–790.
32. Browne, W.R. and B.L. Feringa. 2009. *Annu. Rev. Phys. Chem.* 60: 407–28.
33. Tahara, K., K. Katayama, M.O. Blunt et al. 2014. *Submitted for publication.*

7 Polymeric Architectures Formed by Supramolecular Interactions[*]

Takeharu Haino

CONTENTS

7.1 INTRODUCTION

Supramolecular chemistry focuses on self-assembled molecular architectures that are held together by supramolecular interactions such as hydrogen bonding, van der Waals forces, and coulombic and solvophobic interactions.[1,2] The association and dissociation of molecular constituents occur simultaneously in a molecular assembly process. Subsequently, a self-assembled supramolecular architecture emerges as a thermodynamic minimum.

Supramolecular assembly offers quick access to structurally large and elaborate molecular architectures ranging from the nano- to micrometer scales. At present, *synergistic collaboration* between supramolecular chemistry and material science has become one of the most effective bottom-up strategies to generate functional nano-size objects.[3–7]

A *supramolecular polymer*[8] is any type of supramolecular assembly formed from one or more molecular components; therefore, monomeric and polymeric states are in equilibrium over the relevant experimental timescale.[9,10] Liquid crystals, micelles,

---

[*] This chapter is dedicated to the memory of Prof. Hiroshi Tsukube.

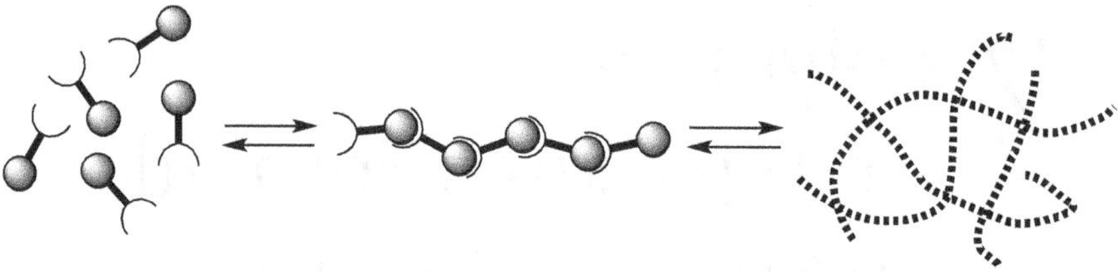

FIGURE 7.1 Supramolecular polymerization driven by molecular recognition.

vesicles, and lipid bilayers may be considered classes of supramolecular polymers; however, these molecular assemblies are not commonly considered as such, although they are in equilibrium and composed of repeating monomers. Supramolecular polymers may also be defined as polymeric arrays of monomeric units that are driven by reversible and directional supramolecular interactions and exhibit chain-like behaviors in dilute solutions, concentrated solutions, and in bulk.[11]

A supramolecular polymerization is illustrated in Figure 7.1. Initially, a monomeric unit has two or more interaction sites that find their complements on other units, generating strong supramolecular interactions through molecular recognition and giving rise to polymeric aggregates. This process is thermodynamically reversible. An average degree of polymerization can be estimated based on the association constant of each connection site. Thus, *predetermined supramolecular interactions* are responsible for the size, direction, and dimensions of the polymeric aggregates.[12] Extensive growth of the polymeric aggregates leads to sizable polymers that produce three-dimensional networks. The interchain interactions among supramolecular polymeric chains may become prominent. Supramolecular polymers may also display physical properties similar to those of conventional polymers.[13]

7.2 HYDROGEN-BONDED SUPRAMOLECULAR POLYMERS

Hydrogen bonding is one of the most common non-covalent interactions. These types of interactions exhibit high directionality and versatility and are thus popular for the development of supramolecular polymers. A complementary donor–acceptor arrangement is able to construct multiple hydrogen bonds with high stability and directionality. Lehn et al. reported the pioneering concept of using linear-chain non-covalent polymers driven by multiple hydrogen bonding interactions as a starting point for functional supramolecular polymer chemistry (Figure 7.2).[14]

A complementary hydrogen-bonded pair between N,N′-2,6-pyridinediylbisacetoamide and uracil has the potential to develop hydrogen-bonded supramolecular polymers. Two homoditopic monomers **1** and **2** were synthesized: **2** was composed of two diaminopyridines connected by a tartrate linker, and two uracil derivatives were used in **1**. The complementary triple hydrogen bonds facilitated the polymerization between **1** and **2** to produce liquid crystalline hydrogen-bonded supramolecular polymers.

The liquid crystallinity was observed over a wide range of temperatures. Additionally, the molecular recognition-directed polymerization and subsequent

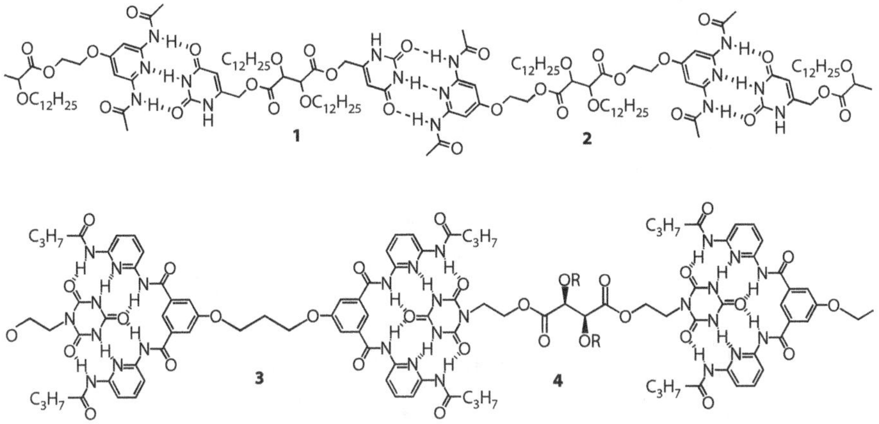

FIGURE 7.2 Supramolecular polymers formed by complementary hydrogen bonding.

self-organization resulted in helical supramolecular entities through the translation of the molecular chirality to the supramolecular level on the nanoscale.[15]

The multiple hydrogen bonds between diaminopyridine-substituted isophthalimide and cyanuric acid moieties produced a large association constant ($K_a > 10^4$ L mol^{-1}) in chloroform (Figure 7.2).[16] The complementary multiple hydrogen bonds between **3** and **4** led to effective supramolecular polymerization, resulting in a highly viscous solution due to the formation of long, entangled fibers in the organic solution. Observation of the fibers with an electron microscope provided evidence for the formation of helical fibers.

Ureidopyrimidone (UPy) formed a self-complementary dimer via quadruple hydrogen bonding (DDAA-AADD) with an unusually strong dimerization constant ($K_a > 10^6$ L mol^{-1}) in chloroform. Bifunctional monomer **5**, bearing two UPy moieties, iteratively generated hydrogen-bonded pairs in chloroform (Figure 7.3).[17–19] The supramolecular polymers produced a highly viscous solution. The solution properties of the supramolecular polymers were dependent on the concentration and temperature because the polymers were constitutionally dynamic due to the reversible nature of the hydrogen-bonding interactions.

A calix[4]arene-based supramolecular polymer is a remarkable example (Figure 7.3).[20] Tetra-urea-substituted calix[4]arene dimerizes to form a capsule capable of binding small molecules by inclusion. Biscalix[4]arene **6** self-assembled to generate the hydrogen-bonded polymeric calixarenes known as "polycaps." A small guest complexation into the capsule greatly stabilized the dimeric form that facilitated the supramolecular polymerization. The supramolecular polymeric chains maintained measurable and meaningful mechanical integrity even in solution.[21]

A strong dipolar molecule usually prefers an antiparallel orientation in supramolecular ensembles due to the strong dipole–dipole interactions. A unique supramolecular approach was taken to drive the head-to-tail parallel orientation of an anisotropic

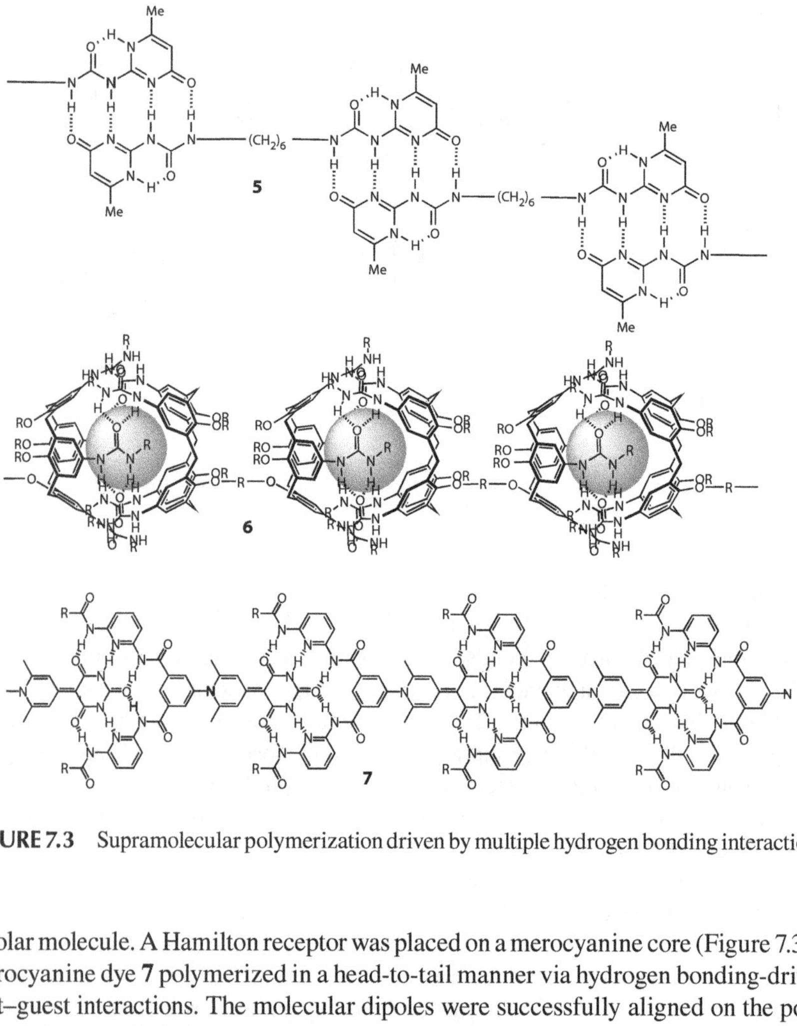

FIGURE 7.3   Supramolecular polymerization driven by multiple hydrogen bonding interactions.

dipolar molecule. A Hamilton receptor was placed on a merocyanine core (Figure 7.3).[22] Merocyanine dye **7** polymerized in a head-to-tail manner via hydrogen bonding-driven host–guest interactions. The molecular dipoles were successfully aligned on the polymer chain. In addition, the supramolecular polymer gelled in a chloroform–hexane solution. The gel phase displayed a remarkable bluish-white fluorescence.

Resorcinarene is a well-known cyclic tetramer prepared by the condensation of resorcinol and alkyl aldehydes.[23,24] Four of the eight phenolic hydroxyl groups generate interaromatic hydrogen bonds that maintain a bowl-shaped cavity. The other four phenolic hydroxyl groups point upward and remain available for intermolecular hydrogen bonding. A resorcinarene was assembled indirectly into a homodimeric capsule held together by multiple hydrogen-bonding networks with eight protic solvent molecules oriented rim-to-rim (Figure 7.4).[25–28] A small guest could be encapsulated within the cavity.

Rim-to-rim connected bisresorcinarenes **8a** through **d** are potential platforms for the synthesis of supramolecular polymers because their homoditopic nature promotes the iterative dimerization of the resorcinarene moieties via multiple hydrogen bonds and the complexation of small guests (Figure 7.5). Kudo and Nishikubo first discovered rim-to-rim connected bisresorcinarene **8a** during the synthesis of a

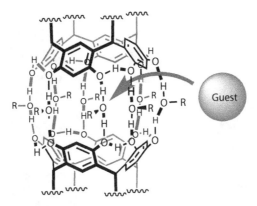

FIGURE 7.4   Dimeric structure of resorcinarene with eight alcohol molecules.

double-cyclic ladder-type oligomer.[29] The reported conditions with resorcinol and 1,6-hexandial gave **8a** in poor yield. Employing dialdehydes masked with dimethyl-acetals greatly improved the yield of bisresorcinarenes **8a** through **d**.[30]

The supramolecular polymerization of **8a** through **d** was investigated in the solid state. Morphological insights of the supramolecular polymers were obtained using scanning electron microscopy (SEM) and atomic force microscopy (AFM). The supramolecular polymers were grown using only toluene as a co-solvent (Figures 7.6a and b).

Polymeric fibrillar networks of **8b** in the presence of toluene were found, suggesting anisotropic growth of the supramolecular assembly. The 12-nm height profile was represented in the topological image (Figure 7.6b). Judging from the capsule diameter of ~1.6 nm, the supramolecular polymers were bundled to form fibrils 12 nm in diameter. A smaller chloroform guest did not enhance the polymerization (Figure 7.6c).

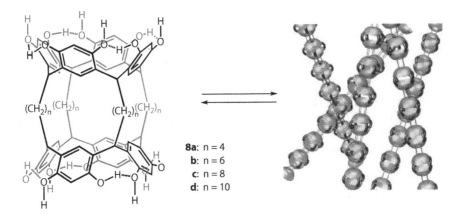

8a: n = 4
b: n = 6
c: n = 8
d: n = 10

FIGURE 7.5   Supramolecular polymerization of bisresorcinarenes 8a through d connected rim to rim.

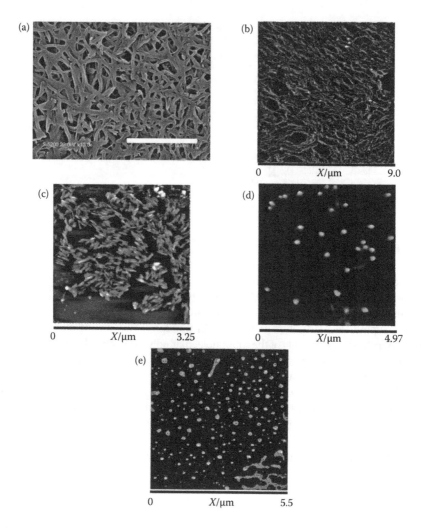

FIGURE 7.6 Scanning electron microscopy (a) and atomic force microscopy (b) through (d) images of cast films prepared from ethanolic solutions of 8b with toluene (a) and (b), chloroform (c), no co-solvent (d), and DMF (e). White scale bar in (a) represents 5 μm.

The supramolecular polymerization did not proceed without a co-solvent (Figure 7.6d). The polymeric morphologies were completely disrupted in the presence of DMF (Figure 7.6e). The competitive hydrogen bonding between the DMF molecules and the phenolic hydroxyl groups most likely collapsed the dimeric structures.

The co-solvents can be encapsulated within the cavity of the capsular structure. The volume of the inner cavity was calculated to be ~237 Å.[3] Rebek reported that a self-assembled dimeric capsule prefers to be filled by a guest with a volume occupying approximately 55% of the binding space.[31] Accordingly, toluene is a good guest for the dimeric capsule, while the molecular volume of chloroform is too small to occupy 55% of the binding space. The guest encapsulation generates enough stabilization to facilitate supramolecular polymerization.

7.3   SUPRAMOLECULAR POLYMERS VIA MOLECULAR RECOGNITION

Hydrogen bonding and coordination interactions are the most commonly used for the development of supramolecular polymers. By contrast, non-hydrogen bonding host–guest interactions have been employed less often for supramolecular polymerization that requires host–guest interactions with very high affinities to achieve a substantial degree of polymerization. This requirement limits the number of possible variations of host–guest interactions for supramolecular polymerizations. Crown ethers,[32] cyclodextrins,[33–36] cucurbit[n]urils,[37–40] calix[n]arenes,[41,42] and other macrocycles[43] are well-recognized as capable host molecules that show very high affinities for certain guest molecules and provide potential supramolecular structures for polymerization.[44]

Cram et al. originally suggested the formation of polymeric species via host–guest complexation of a bis(crown ether) and an α,ω-diammonium salt in 1977.[45] Supramolecular polymerization of heteroditopic monomer **9** via ion-dipole interactions was reported by Gibson et al.[46–49] The diphenylene-32-crown-10 unit selectively captured the paraquat moiety in a head-to-tail manner (Figure 7.7). Mass spectrometry of **9** confirmed the formation of supramolecular polymers.

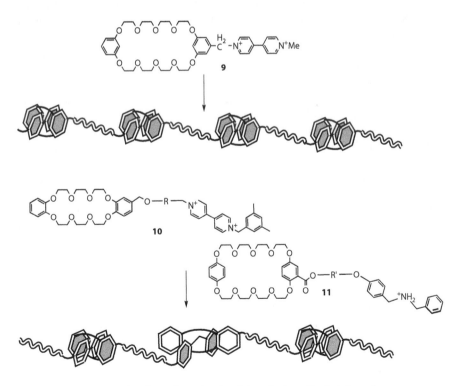

FIGURE 7.7   Supramolecular polymerization driven by ion–dipole interactions.

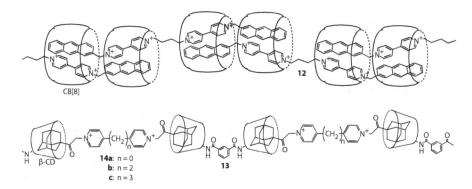

FIGURE 7.8 Polymeric assemblies via host-stabilized charge transfer (HSCT) and host–guest interactions.

The high viscosity of the polymer solutions in organic solvents was characteristic of linear polymeric aggregates. Fibers of the supramolecular polymers were drawn from concentrated solutions. The supramolecular polymers of **9** acted like a conventional polymer, although the repeating units were associated by non-covalent bonds. This strategy was extended to enable the self-sorting organization of supramolecular alternating copolymers.[50]

Two host–guest pairs (**10** and **11**) were synthesized, with each monomer composed of a guest moiety that was complementary to the host moiety of the other. The orthogonal host–guest interactions achieved perfect self-sorting of the host–guest pairs. Consequently, an alternating copolymerization was established by the two heteroditopic monomers.

Charge transfer complexes of aromatic donor–-acceptor pairs are greatly stabilized within the hollow cavities of cucurbit[n]uril (CB[n]) in aqueous environments.[51,52] This host-stabilized charge transfer (HSCT) interaction has been recently applied to developing supramolecular polymers.[53,54] The supramolecular polymerization of multifunctional monomer **12** via HSCT interactions was demonstrated (Figure 7.8).[55]

The supramolecular polymerization of **12** occurred with the aid of CB[8] to form stable, deep-purple hydrogels. Complexation of methylviologen within the hollow cavity inhibited supramolecular polymerization. Consequently, the supramolecular gel collapsed. Accordingly, the supramolecular polymer networks were crucial for gelation.

Cyclodextrins (CDs) encapsulate a variety of guests into their cavities through hydrophobic interactions. This molecular recognition event is valuable for the synthesis of supramolecular polymers.[56] Harada et al. reported that the adamantyl groups of homoditopic guests **14** were captured in the hydrophobic cavity of bis-β-CD host **13** in water (Figure 7.8).[57,58] Extended linker lengths between the two adamantyl groups unfavorably influenced the growth of the supramolecular polymers in an aqueous environment.

The host–guest complexation between **13** and **14a** resulted in supramolecular polymers with molecular weights greater than 90,000 g mol⁻¹, as obtained by

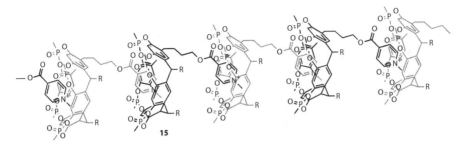

15

FIGURE 7.9 Cation–π interaction-directed supramolecular polymerization.

vapor pressure osmometry analysis. The smaller supramolecular assemblies formed between **13** and **14b** and **c** were independent of the concentrations, indicating that **14b** and **c** formed cyclic oligomers with **13** due to the flexible linker units.

The calix[n]arene family is probably the most widely used platform in the field of supramolecular chemistry.[59] For the small calix[n]arenes (n = 4 or 5), the cyclic hydrogen bonds of the phenolic hydroxyl groups on the lower rim produce a highly symmetric cone cavity in which a variety of complementary guest molecules can be encapsulated through weak non-covalent interactions such as cation–π, CH–π, aromatic stacking, and van der Waals.

The cation–π interaction-directed supramolecular polymerization of heteroditopic monomer **15** was demonstrated by Dalcanale et al. (Figure 7.9).[58,59] Phosphonate cavitand **15** encapsulated the positively charged pyridinium tail with a very high association constant, leading to the head-to-tail supramolecular polymers in chloroform. The weight-average molecular weight (M_w) of the polymer reached 26,300 g mol^{-1}, which corresponds to an average degree of polymerization of 18.

Calix[5]arene-based supramolecular polymers have also been developed. Parisi and Cohen et al. used a heteroditopic calix[5]arene bearing an amino group on the lower rim.[62–64] The amino group was protonated readily by contact with acidic solution. The ammonium terminus was encapsulated iteratively within the calix[5]arene cavity to produce the calix[5]arene-based supramolecular polymers. DLS experiments supported the formation of sizable supramolecular polymers in solution. SEM provided clear evidence for the formation of fibrillar networks of the supramolecular polymers.

A pillar[5]arene was recently discovered by Ogoshi et al.[63] The pillarcalix[5]arene cavity entrapped a variety of small guests. A pillar[5]arene bearing an octyl tail was polymerized via CH–π interactions to form a supramolecular polymer in organic solutions.[66] The crystal structure of the polymer indicated that the alkyl tail was inside the cavity, producing an attractive CH–π interaction. The solution viscosity of the supramolecular polymer was concentration-dependent, indicating that physical contacts of the supramolecular polymers caused entanglement of the polymer main chains.

7.4 SUPRAMOLECULAR FULLERENE POLYMERS

Fullerene has attracted significant attention due to its unique three-dimensional geometry and outstanding electrochemical and photophysical properties.[67,68] Fullerene chemistry has been incorporated into macromolecular science. This

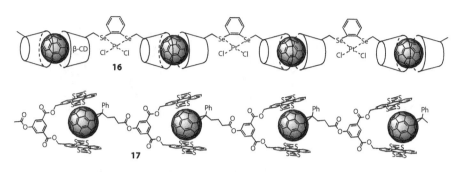

FIGURE 7.10 Supramolecular fullerene-containing polymers.

synergetic collaboration offers a great opportunity to generate new fullerene-based polymer materials.[67]

The synthesis of fullerene-containing polymeric materials was first attempted via direct polymerization achieved through cycloaddition reactions of C_{60}.[68–70] The equal chemical reactivities of the 30 double bonds resulted in one-, two-, and three-dimensionally extended polymeric architectures and a loss of stereo- and region-regularity. Supramolecular chemistry can be helpful in developing structurally regulated fullerene polymers.

C_{60} is extremely hydrophobic; thus, some of the initial attempts to capture C_{60} molecules were carried out in water. C_{60} forms a 1:2 complex with a cyclodextrin molecule. Liu et al. synthesized water-soluble double-CD **16** (Figure 7.10).[73] End-to-end intermolecular inclusion complexation of C_{60} with double-CD **16** led to the formation of a supramolecular polymer in water. An STM image of the polymer displayed the regular linear arrangement of the C_{60} nano-array. TEM observation of the supramolecular polymer confirmed the presence of a linear structure with a length in the range of 150 to 250 nm, suggesting that the polymer consists of 60 to 80 units of the minimum component.

Martín et al. demonstrated that a double-armed fullerene host possessing a π-extended TTF analogue, 2-[9-(1,3-dithiol-2-ylidene)anthracen-10(9H)-ylidene]-1,3-dithiole (exTTF), displayed a good affinity for fullerene (Figure 7.10).[72,73] Supramolecular association of head-to-tail donor-acceptor hybrid **17** resulted in supramolecular fullerene-containing polymers. Dynamic light scattering (DLS) confirmed the presence of sizable aggregates larger than 400 repeating units in solution. AFM measurements revealed the formation of supramolecular polymer networks in the solid state.

Haino et al. actively investigated the development of calixarene-based host molecules via upper rim functionalization.[74,77] The motivation of the research was to create a much larger cavity capable of encapsulating sizable guest molecules. A calix[5]arene served as a potential host for encapsulating a large molecular guest, C_{60}, within the cone cavity.[78,79] The covalent connection of two calix[5]arene molecules generated a larger cavity capable of encapsulating C_{60} as well as fullerenes that resulted in a large increase in association constants (Ka > $10^{4–5}$ L mol^{-1}).[80,89]

This strong ball-and-socket interaction was applied in the construction of supramolecular fullerene polymers. Homoditopic monomer **18** bears two binding pockets

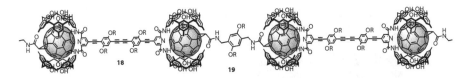

FIGURE 7.11 Supramolecular polymerization directed by molecular recognition between calix[5]arene and C_{60}.

(each pocket consists of two calix[5]arenes) to encapsulate a C_{60} molecule in each binding site (Figure 7.11).[90] The supramolecular polymerization arises through a complexation of the C_{60} and calix[5]arene moieties.

The supramolecular interactions between **18** and **19** were detected with 1H NMR and absorption spectroscopy in organic solvents. The resonance of the bridge methylene on a calix[5]arene was broadened because of the interconversion of the ring between the cone conformation and its mirror image that occurred on the NMR time scale. The supramolecular complexation of **18** and **19** resulted in the AB quartet-like splitting of the bridging methylene protons, indicating that the interconversion process became slow on the NMR timescale. Apparently, the energetic barrier of the inversion process was increased due to the host–guest interaction between the C60 moiety and the cavity.

Diffusion-ordered NMR spectroscopy (DOSY) is a powerful tool for structural studies of molecular aggregates in solution. Translational diffusion coefficients (D) determined by DOSY correlate directly with the sizes and dimensions of aggregates. The diffusion coefficients of **18** and a 1:1 mixture of **18** and **19** at a low concentration in chloroform-d_1 are 4.09×10^{-10} and 2.57×10^{-10} m^2 s^{-1}, respectively. The ratio of $D_{18 \cdot 19}/D_{19}$ of 0.63 was reasonably close to the theoretical range of 0.59 to 0.60 expected for a linear trimer. These results confirmed the formation of oligomeric assemblies even at low concentrations.

The polymer-like morphologies were observed by scanning electron microscopy (Figure 7.12a). The thicker entwined fibers had lengths exceeding 100 μm and widths between 250 and 500 nm. The addition of C_{60} completely disrupted the network because C_{60} molecules competitively occupied the cavities of the hosts, thus interfering with the complexation of **19**. Consequently, further growth of the supramolecular polymers was not permitted in the solid state.

Figure 7.12b shows the nano networks of the fibrillar assemblies that resulted from the aggregation of the supramolecular linear polymers. Judging from the calculated structure of the oligomers (Figure 7.12c), the alkyl chains adopted a parallel arrangement on the mica surface and the nano assemblies were most likely composed of bundles of 40 to 60 polymer chains generated by the interdigitation of the alkyl side chains through van der Waals interactions.

Cross-linking linear polymers produced three-dimensional networks with reduced structural flexibility, resulting in dramatic changes in the macroscopic morphology and properties. The introduction of a reversible polymer cross-linkage is the predominant challenge leading to the generation of stimuli-responsive polymeric materials. Reversible cross-linkages can be realized by employing non-covalent bonds.[93–96] When supramolecular entities are grafted onto conventional polymeric

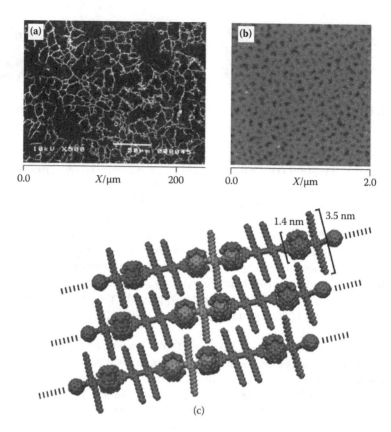

FIGURE 7.12 Images of cast films prepared from solution of **18** and **19** obtained by scanning electron microscopy (a) and atomic force microscopy (b). (c) Plausible supramolecular polymeric structures.

scaffolds, the reversible cross-linking of polymer chains can be established with supramolecular methodologies.

Ditopic host **18** selectively encapsulated C_{60} moieties grafted on a large polymer backbone (Figure 7.13).[95] This molecular recognition event created a remarkably stable cross-linkage. The complexation of **18** to poly-**1** increased the molecular weight to more than twice the weight of free poly-**1** (Figure 7.14a). Remarkably, excess **18** did not break the pre-formed cross-linkages in toluene even though the host–guest complexation of C_{60} with a calix[5]arene was driven by non-covalent interactions.

The cross-linking behavior was influenced by the solvent properties. The competitive solvation of o-dichlorobenzene (relative to toluene) weakened the host–guest interaction, causing host **18** to be unable to create stable cross-linkages with poly-**1**.

The macroscopic solid-state morphology of poly-**1** was highly influenced by supramolecular cross-linking. The agglomerated nanostructures of poly-**1** consisted of particle-like aggregates that originated from the cohesion of C_{60} moieties through π–π interactions and the immiscible nature of C_{60} (Figure 7.14b).

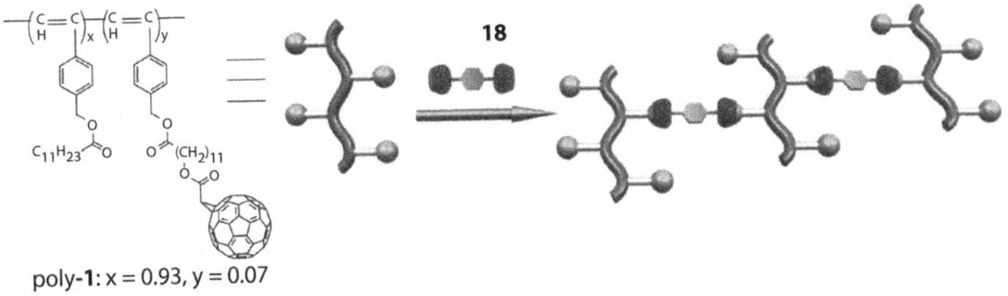

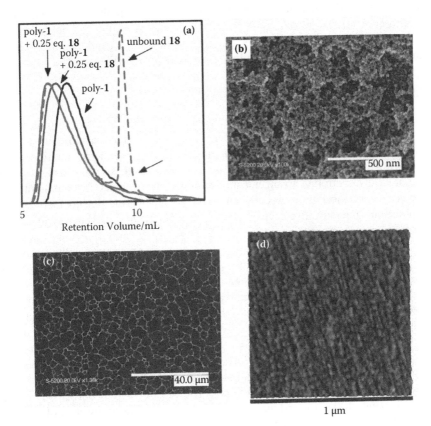

poly-1: x = 0.93, y = 0.07

FIGURE 7.13 Supramolecular cross-linking.

The uptake of C_{60} moieties by **18** disrupted the interfullerene interactions, and the polymer chain became longer via supramolecular cross-linking. This elongation led to a dramatic morphological change in which the particle-like nanostructures completely disappeared and a widespread fibrous network was formed (Figure 7.14c). The cross-linked fibrils aligned into a well-oriented 2D array on a HOPG surface

FIGURE 7.14 (a) Size-exclusion chromatograms of poly-**1** in the presence of **18** in toluene. (b) and (c) SEM images and (d) AFM images of thin films of poly-**1** (b) and poly-**1** with **18** (c) and (d).

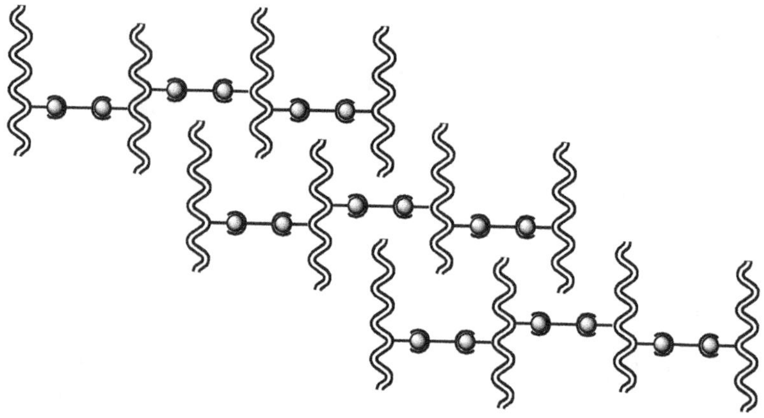

FIGURE 7.15   Plausible surface orientations of cross-linked polymer chains.

(Figure 7.14d). The peak-to-peak profile showed a uniform pitch of 19.6 nm, which was significantly larger than the predicted diameter (4 nm) of the polymer chain. Interdigitation of the highly aligned polymer chains may expand the interchain distance more than the diameter of individual polymer chain (Figure 7.15).

7.5   SUPRAMOLECULAR PORPHYRIN POLYMERS

A variety of porphyrin-based model compounds have been created to mimic natural light-harvesting complexes, with the goal of applying the materials to artificial light-harvesting systems and molecular photonic devices. Nanometric multiporphyrin arrays are the most challenging targets; their covalent synthesis offers a highly stable and precisely controlled arrangement of multiporphyrin arrays.[98,99] Supramolecular porphyrin polymers have recently attracted great attention due to their creative applications in photoactive devices. Because certain metal–ligand pairs can produce appreciably strong coordination bonds, coordination-driven self-assembly is one of the most useful approaches for building large, elaborate porphyrin architectures.[100]

The coordination-driven supramolecular polymerization of metalloporphyrins provides quick access to huge nanometric porphyrin arrays (Figure 7.16). Kobuke's porphyrin **20** generated supramolecular polymers in a self-complementary manner.[101] Size-exclusion chromatographic analysis of the supramolecular polymers indicated that approximately 400 monomers assembled to form the supramolecular polymers with lengths that reached 550 nm. Hunter's cobalt porphyrin **21** also polymerized in a self-complementary manner.[102] The size-exclusion chromatographic trace was dependent on the concentration of the monomer, indicating that the coordinated polymers are in equilibrium in solution. The degree of polymerization reached approximately 100 units, and the polymer had a mean molecular weight of 136 kDa.

Hydrogen bonding-mediated supramolecular porphyrin polymerization was developed by Aida et al.[101] Carboxylic acid functionalities dimerized through complementary hydrogen bonding. A porphyrin bearing two carboxylic acid groups in opposite *meso* positions assembled into supramolecular hydrogen-bonded porphyrin

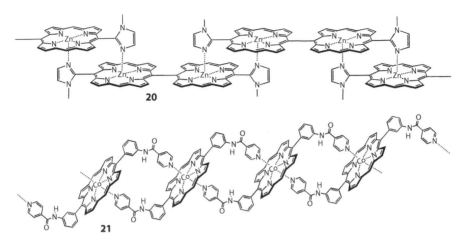

FIGURE 7.16 Coordination-driven supramolecular polymerization of multiporphyrins.

polymers. Strikingly, spin coating of the supramolecular polymer produced optically active films in which either of the two enantiomeric forms was dictated by the spinning direction.

Haino et al. recently discovered that a bisporphyrin molecule connected by a pyridine dicarboxamide linker assembled to form a unique complementary dimer in organic media.[104,105] The competitive complexation of electron-deficient planar aromatic guests (trinitrofluorenone, trinitrobenzene, tetracyanobenzene, and pyromellitic dianhydride) into the bisporphyrin cleft led to a π donor–acceptor-type host–guest complex.

Based on these findings, a π donor–acceptor-type host–guest interaction was incorporated into heteroditopic monomer **22** (Figure 7.17).[104] The electron-deficient guest moiety, 4,5,7-trinitrofluorenone-2-carboxylate (TNF), binds within the bisporphyrin cleft via a charge transfer interaction. The iterative head-to-tail host–guest complexation produces a new supramolecular polymer.

The sizes of the polymeric aggregates at different concentrations were investigated using DOSY experiments (Figure 7.18a). The diffusion coefficients of **22** were highly dependent on the concentration, whereas those of its analogue bearing the acetyl group instead of the TNF moiety were not influenced by the concentration. At low concentrations, **22** existed in its monomeric form. At high concentrations, the diffusion coefficient dropped to 10% of its initial value. Accordingly, **22** formed large polymeric aggregates in response to increased concentrations. The average size of the supramolecular polymer was calculated to have an approximate DP of 660 by simplistically assuming that all the aggregates were spherical.

A concentration-dependent viscosity change is indicative of a transition from a dilute to a semidilute concentration regime in which overlaps of the polymer chains provide viscous drag (Figure 7.18b). The dilute concentration regime implied that the oligomeric assemblies were not entangled. In the semidilute regime, a scale exponent of 3.07 in chloroform suggested that sizable supramolecular polymers were

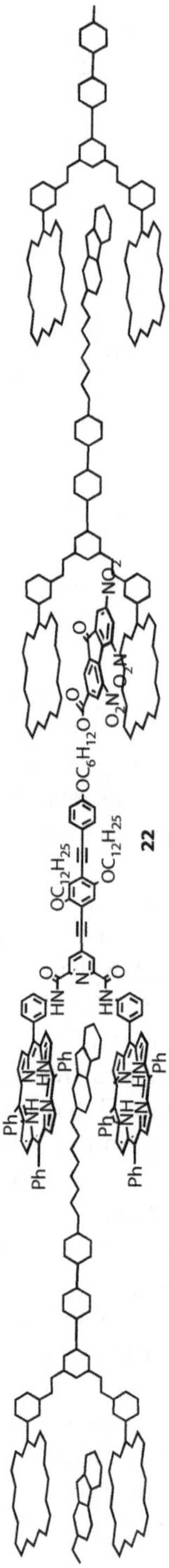

FIGURE 7.17 Head-to-tail supramolecular porphyrin polymerization via donor–acceptor interactions.

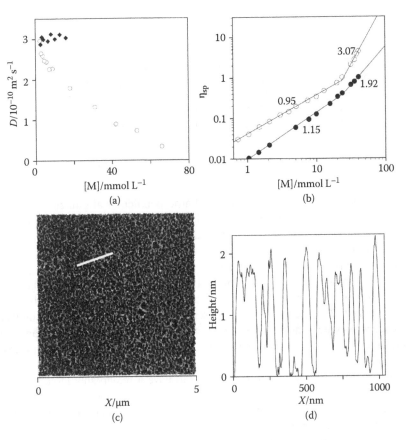

FIGURE 7.18 (a) Diffusion coefficients (D) of **22** (open circles) and its acetyl analogue (filled squares) at 298 K in chloroform-d_1. (b) Specific viscosity (η_{sp}) of **22** in chloroform (open circles) and toluene (filled circles) at 293 K. (c) Atomic force microscopy images of drop-cast films of **22** (5.4×10^{-5} mol L^{-1}). (d) Height profile of white line in image (c).

produced, and the chains became entangled when the polymers came into contact, which is commonly observed in polymer solutions. The relaxation of the supramolecular polymers occurred via a repetition mechanism even though the supramolecular polymer was constitutionally dynamic.

The well-developed supramolecular polymer networks of **22** were visualized by AFM measurements (Figure 7.18c). The fibrillar morphology disappeared after the addition of 2,4,7-trinitrofluorenone, which interrupted the host–guest interaction. These substantial morphological differences clearly indicated that the head-to-tail host–guest complexation of the bisporphyrin cleft and the TNF moiety caused the directional growth of the supramolecular polymer.

The fibers had a uniform height of 1.8 ± 0.1 nm (Figure 7.18d). The width (1.7 nm) of the tetraphenylporphyrin moiety closely matched the observed height;

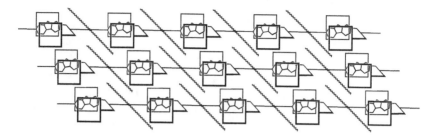

FIGURE 7.19 Plausible orientation of supramolecular polymers on surface.

therefore, the porphyrin moieties adopted a perpendicular alignment on the mica surface that stabilized the oriented fibrillar nano network via the stacking of the porphyrin moieties and the van der Waals interactions between the alkyl side chains (Figure 7.19).

Another type of stimulus-responsive supramolecular polymer was produced using the complementary affinity of a bisporphyrin (Figure 7.20).[107] The bisporphyrin units were connected with 1,3-butadiyne units to produce homoditopic tetrakisporphyrin **23**. The iterative self-assembly of **23** gave rise to supramolecular nanometric polymeric assemblies. An electron-deficient aromatic guest can aggressively bind within the bisporphyrin cleft, causing the supramolecular polymeric assemblies to dissociate.

The self-assembly of **23** was determined to have a remarkably large association constant of 1,500,000 L mol^{-1}, which implies that sizable supramolecular polymeric aggregates were formed at millimolar concentrations. The stimuli-responsive behavior of the supramolecular polymers was observed in solution (Figure 7.21a and b). The resonance peaks in the ^{1}H NMR spectrum of the supramolecular polymers were highly broadened, most likely due to the numerous conformational and structural possibilities. Upon addition of an electron-deficient aromatic guest (e.g., 2,4,7-trinitrofluorenone) to the solution, the broad signals became sharpened, indicating that 2,4,7-TNF$_2 \subset$ **23** was formed with the simultaneous dissociation of the supramolecular polymers.

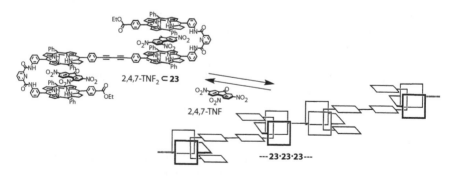

FIGURE 7.20 Guest-induced regulation of supramolecular polymerization of **23**.

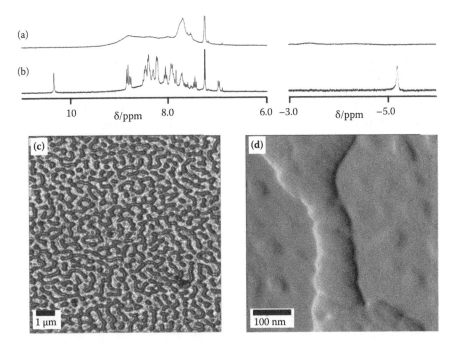

FIGURE 7.21 ^1H NMR spectra of **23** (a) and **23** (b) with 2,4,7-TNF in chloroform-d_1. Atomic force microscopy images of topography (c) and phase (d).

In the solid state, AFM images of the long winding, fibrillar fragments demonstrated that 23 iteratively created complementary connections with each terminal functionality, resulting in supramolecular polymers (Figure 7.21c). Coiled structures with pitches of 64 ± 6 nm were observed in the magnified images (Figure 7.21d). The bundled supramolecular polymeric assemblies most likely adopted coiled superstructures. Based on the pitch length, the coiled tape was composed of a bundle of approximately 30 chains of polymeric assemblies (Figure 7.22).

FIGURE 7.22   Formation of supramolecular polymer and its highly organized helical morphology.

7.6 CONCLUSION AND OUTLOOK

Supramolecular interactions are useful for constructing supramolecular polymeric aggregates with high orthogonality. Unique hosts including calixarene, cyclodextrin, crown ether, and porphyrin bind certain guests, displaying large association constants with high guest selectivities. Therefore, supramolecular complexes offer numerous methods for the development of supramolecular polymers. However, a lack of robustness may restrict the practical use of these polymers in real-world applications. To overcome this downside, further research is needed. Supramolecular polymer chemistry may be coming into its own, even though its inception occurred over 20 years ago. The development of fascinating supramolecular materials is eagerly anticipated.

ACKNOWLEDGMENTS

The author would like to thank Grants-in-Aid for Scientific Research (B) (No. 21350066) and Challenging Exploratory Research (No. 23655105) of JSPS, as well as Grants-in-Aid for Scientific Research on Innovative Areas ("Stimuli-Responsive Chemical Species for the Creation of Functional Molecules" and "New Polymeric Materials Based on Element Blocks," Nos. 25109529, 25102532) for financial support.

REFERENCES

1. D. J. Cram, *Angew. Chem. Int. Ed. Engl.* 27: 1009–1020, 1988.
2. J-M. Lehn, *Angew. Chem. Int. Ed. Engl.* 27: 89–112, 1988.
3. M. Fujita, *Chem. Soc. Rev.* 27: 417–425, 1998.
4. G. M. Whitesides, J. P. Mathias, and C. T. Seto, *Science* 254: 1312–1319, 1991.
5. D. L. Caulder and K. N. Raymond, *Acc. Chem. Res.* 32: 975–982, 1999.
6. T. Aida, E. W. Meijer, and S. I. Stupp, *Science* 335: 813–817, 2012.
7. K. Ariga, H. Ito, J. P. Hill et al., *Chem. Soc. Rev.* 41: 5800–5835, 2012.
8. A. Ciferri, Ed., *Supramolecular Polymers,* 2nd ed. CRC Press, Boca Raton, 2005.
9. J-M. Lehn, *Macromol. Symp.* 174: 5–6, 2001.
10. J-M. Lehn, *Polym. Int.* 51: 825–839, 2002.
11. L. Brunsveld, B. J. B. Folmer, E. W. Meijer et al., *Chem. Rev.* 101: 4071–4097, 2001.
12. M. J. Serpe and S. L. Craig, *Langmuir.* 23: 1626–1634, 2007.
13. T. Haino, *Polym. J.* 45: 363–383, 2013.
14. C. Fouquey, J-M. Lehn and A. M. Levelut, *Adv. Mater.* 2: 254–257, 1990.
15. T. Gulikkrzywicki, C. Fouquey, and J-M. Lehn, *Proc. Natl. Acad. Sci. USA* 90: 163–167, 1993.
16. V. Berl, M. Schmutz, M. J. Krische et al., *Chem. Eur. J.* 8: 1227–1244, 2002.
17. R. P. Sijbesma, F. H. Beijer, L. Brunsveld et al., *Science* 278: 1601–1604, 1997.
18. S. H. M. Söntjens, R. P. Sijbesma, M. H. P. van Genderen et al., *Macromolecules* 34: 3815–3818, 2001.
19. E. W. Meijer, J. H. K. K. Hirschberg, L. Brunsveld et al., *Nature* 407: 167–170, 2000.
20. R. K. Castellano, D. M. Rudkevich, and J. Rebek, Jr., *Proc. Natl. Acad. Sci. USA* 94: 7132–7137, 1997.
21. R. K. Castellano, R. Clark, S. L. Craig et al., *Proc. Natl. Acad. Sci. USA* 97: 12418–12421, 2000.

22. R. Schmidt, M. Stolte, M. Grüene et al., *Macromolecules* 44: 3766–3776, 2011.
23. P. Timmerman, W. Verboom, and D. N. Reinhoudt, *Tetrahedron* 52: 2663–2704, 1996.
24. D. J. Cram and J. M. Cram. *Container Molecules and Their Guests*; Royal Society of Chemistry, Cambridge, 1994.
25. N. K. Beyeh and K. Rissanen, *Isr. J. Chem.* 51: 769–780, 2011.
26. A. Shivanyuk, J. C. Friese, S. Döring et al., *J. Org. Chem.* 68: 6489–6496, 2003.
27. A. Shivanyuk and J. Rebek, Jr., *Chem. Commun.* 2374–2375, 2001.
28. K. Murayama and K. Aoki, *Chem. Commun.* 607–608, 1998.
29. H. Kudo, R. Hayashi, K. Mitani et al., *Angew. Chem. Int. Ed.* 45: 7948–7952, 2006.
30. H. Yamada, T. Ikeda, T. Mizuta et al., *Org. Lett.* 14: 4510–4513, 2012.
31. S. Mecozzi and J. Rebek, *Chem. Eur. J.* 4: 1016–1022, 1998.
32. G. Gokel, *Crown Ethers and Cryptands*; Royal Society of Chemistry: Cambridge, 1991.
33. M. J. W. Ludden, D. N. Reinhoudt, and J. Huskens, *Chem. Soc. Rev.* 35: 1122–1134, 2006.
34. A. Harada, Y. Takashima, and H. Yamaguchi, *Chem. Soc. Rev.* 38: 875–882, 2009.
35. H. J. Schneider, F. Hacket, V. Rüdiger et al., *Chem. Rev.* 98: 1755–1786, 1998.
36. S. A. Nepogodiev and J. F. Stoddart, *Chem. Rev.* 98: 1959–1976, 1998.
37. J. W. Lee, S. Samal, N. Selvapalam et al., *Acc. Chem. Res.* 36: 621–630, 2003.
38. J. Lagona, P. Mukhopadhyay, S. Chakrabarti et al., *Angew. Chem. Intl. Ed.* 44: 4844–4870, 2005.
39. K. Kim, N. Selvapalam, Y. H. Ko et al., *Chem. Soc. Rev.* 36: 267–279, 2007.
40. L. Isaacs, *Chem. Commun.* 619–629, 2009.
41. C. D. Gutsche, *Calixarenes: An Introduction,* 2nd ed. Royal Society of Chemistry: Cambridge, 2008.
42. C. D. Gutsche, *Calixarenes Revisited*; Royal Society of Chemistry: Cambridge, 1998.
43. F. Diederich, *Cyclophanes*; Royal Society of Chemistry: Cambridge, 1991.
44. J. W. Steed and J. L. Atwood, *Supramolecular Chemistry*; John Wiley & Sons: Chichester, 2009.
45. R. C. Helgeson, T. L. Tarnowski, J. M. Timko et al., *J. Am. Chem. Soc.* 99: 6411–6418, 1977.
46. N. Yamaguchi, D. S. Nagvekar, and H. W. Gibson, *Angew. Chem. Int. Ed.* 37: 2361–2364, 1998.
47. H. W. Gibson, N. Yamaguchi, and J. W. Jones, *J. Am. Chem. Soc.* 125: 3522–3533, 2003.
48. J. W. Jones, W. S. Bryant, A. W. Bosman et al., *J. Org. Chem.* 68: 2385–2389, 2003.
49. Z. Niu, F. Huang, and H. W. Gibson, *J. Am. Chem. Soc.* 133: 2836–2839, 2011.
50. F. Wang, C. Y. Han, C. L. He et al., *J. Am. Chem. Soc.* 130: 11254–11255, 2008.
51. H-J. Kim, J. Heo, W. S. Jeon et al., *Angew. Chem. Intl. Ed.* 40: 1526–1529, 2001.
52. Y. Xu, M. Guo, X. Li et al., *Chem. Commun.* 47: 8883–8885, 2011.
53. K. Kim, D. Kim, J. W. Lee et al., *Chem. Commun.* 848–849, 2004.
54. Y. L. Liu, Y. Yu, J. A. Gao et al., *Angew. Chem. Intl. Ed.* 49: 6576–6579, 2010.
55. Y. Hasegawa, M. Miyauchi, Y. Takashima et al., *Macromolecules.* 38: 3724–3730, 2005.
56. K. Ohga, Y. Takashima, H. Takahashi et al., *Macromolecules.* 38: 5897–5904, 2005.
57. C. D. Gutsche, *Calixarenes*; Royal Society of Chemistry: Cambridge, 1989.
58. E. Dalcanale, R. M. Yebeutchou, F. Tancini et al., *Angew. Chem. Intl. Ed.* 47: 4504–4508, 2008.
59. F. Tancini, R. M. Yebeutchou, L. Pirondini et al., *Chem. Eur. J.* 16: 14313–14321, 2010.
60. M. F. Parisi, D. Garozzo, G. Gattuso et al., *Org. Lett.* 5: 4025–4028, 2003.
61. Y. Cohen, S. Pappalardo, V. Villari et al., *Chem. Eur. J.* 13: 8164–8173, 2007.
62. M. F. Parisi, G. Gattuso, A. Notti et al., *J. Org. Chem.* 73: 7280–7289, 2008.
63. T. Ogoshi, S. Kanai, S. Fujinami et al., *J. Am. Chem. Soc.* 130: 5022–5023, 2008.
64. Z. Zhang, Y. Luo, J. Chen et al., *Angew. Chem. Intl. Ed.* 50: 1397–1401, 2011.
65. H. W. Kroto, J. R. Heath, S. C. O'Brien et al., *Nature.* 318: 162–163, 1985.

66. W. Krätschmer, L. D. Lamb, K. Fostiropoulos et al., *Nature*. 347: 354–358, 1990.
67. F. Giacalone and N. Martín, *Chem. Rev.* 106: 5136–5190, 2006.
68. M. N. Regueiro, P. Monceau, and J-L. Hodeau, *Nature*. 355: 237–239, 1992.
69. P. Zhou, Z-H. Dong, A. M. Rao et al., *Chem. Phys. Lett.* 211: 337–340, 1993.
70. Y. Iwasa, T. Arima, R. M. Fleming et al., *Science*. 264: 1570–1572, 1994.
71. Y. Liu, H. Wang, P. Liang et al., *Angew. Chem. Intl. Ed.* 43: 2690–2694, 2004.
72. G. Fernandez, E. M. Pérez, L. Sánchez et al., *J. Am. Chem. Soc.* 130: 2410–2411, 2008.
73. G. Fernández, E. M. Pérez, L. Sánchez et al., *Angew. Chem. Intl. Ed.* 47: 1094–1097, 2008.
74. T. Haino, T. Harano, K. Matsumura et al., *Tetrahedron Lett.* 36: 5793–5796, 1995.
75. T. Haino, K. Matsumura, T. Harano et al., *Tetrahedron*. 54: 12185–12196, 1998.
76. T. Haino, K. Yamada, and Y. Fukuzawa, *Synlett*. 673–674, 1997.
77. T. Haino, Y. Katsutani, H. Akii et al., *Tetrahedron Lett.* 39: 8133–8136, 1998.
78. T. Haino, M. Yanase, and Y. Fukazawa, *Tetrahedron Lett.* 38: 3739–3742, 1997.
79. T. Haino, M. Yanase, and Y. Fukazawa, *Angew. Chem. Int. Ed. Engl.* 36: 259–260, 1997.
80. T. Haino, M. Yanase, and Y. Fukazawa, *Angew. Chem. Int. Ed.* 37: 997–998, 1998.
81. M. Yanase, T. Haino, and Y. Fukazawa, *Tetrahedron Lett.* 40: 2781–2784, 1999.
82. M. Yanase, M. Matsuoka, Y. Tatsumi et al., *Tetrahedron Lett.* 41: 493–497, 2000.
83. T. Haino, H. Araki, Y. Yamanaka et al., *Tetrahedron Lett.* 42: 3203–3206, 2001.
84. T. Haino, H. Araki, Y. Fujiwara et al., *Chem. Commun.* 2148–2149, 2002.
85. T. Haino, Y. Yamanaka, H. Araki et al., *Chem. Commun.* 402–403, 2002.
86. T. Haino, J. Seyama, C. Fukunaga et al., *Bull. Chem. Soc. Jpn.* 78: 768–770, 2005.
87. T. Haino, C. Fukunaga, and Y. Fukazawa, *Org. Lett.* 8: 3545–3548, 2006.
88. T. Haino, M. Yanase, C. Fukunaga et al., *Tetrahedron*. 62: 2025–2035, 2006.
89. T. Haino, C. Fukunaga, and Y. Fukazawa, *J. Nanosci. Nanotechnol.* 7: 1386–1388, 2007.
90. T. Haino, Y. Matsumoto, and Y. Fukazawa, *J. Am. Chem. Soc.* 127: 8936–8937, 2005.
91. T. Oku, Y. Furusho, and T. Takata, *Angew. Chem. Intl. Ed.* 43: 966–969, 2004.
92. O. Kretschman, S. W. Choi, M. Miyauchi et al., *Angew. Chem. Intl. Ed.* 45: 4361–4365, 2006.
93. L. R. Rieth, R. F. Eaton, and G. W. Coates, *Angew. Chem. Intl. Ed.* 40: 2153–2156, 2001.
94. O. Bistri-Aslanoff, Y. Blériot, R. Auzély-Velty et al., *Org. Biomol. Chem.* 8: 3437–3443, 2010.
95. T. Haino, E. Hirai, Y. Fujiwara et al., *Angew. Chem. Int. Ed.* 49: 7899–7903, 2010.
96. Y. Nakamura, N. Aratani, and A. Osuka, *Chem. Soc. Rev.* 36: 831–845, 2007.
97. E. Iengo, E. Zangrando, and E. Alessio, *Acc. Chem. Res.* 39: 841–851, 2006.
98. K. Toyofuku, M. A. Alam, A. Tsuda et al., *Angew. Chem. Intl. Ed.* 46: 6476–6480, 2007.
99. K. Ogawa and Y. Kobuke, *Angew. Chem. Intl. Ed.* 39: 4070–4073, 2000.
100. U. Michelsen and C. A. Hunter, *Angew. Chem. Intl. Ed.* 39: 764–767, 2000.
101. T. Yamaguchi, T. Kimura, H. Matsuda et al., *Angew. Chem. Intl. Ed.* 43: 6350–6355, 2004.
102. T. Haino, T. Fujii, and Y. Fukazawa, *Tetrahedron Lett.* 46: 257–260, 2005.
103. T. Haino, T. Fujii, and Y. Fukazawa, *J. Org. Chem.* 71: 2572–2580, 2006.
104. T. Haino, A. Watanabe, T. Hirao et al., *Angew. Chem. Intl. Ed.* 1473–1476, 2012.
105. T. Haino, T. Fujii, A. Watanabe et al., *Proc. Natl. Acad. Sci. USA* 106: 10477–10481, 2009.

8 Self-Assembled Molecular Capsule

Dariush Ajami

CONTENTS

8.1 INTRODUCTION

The well-known discovery of crown ethers[1] by Charles Pedersen in the 1960s led to the manifestation of a new field of molecular recognition, now fundamental to physical organic chemistry. These macrocyclic polyethers (hosts) encircle alkali metal cations (guests) through ion–dipole interactions between the metal ion and oxygen atoms (Figure 8.1).

In this classic example, a host with a *convergent* binding site forms a complex with a guest with *divergent* binding sites via non-covalent interactions. The complex is held in a *synergistic* manner through a complementary combination of satisfactory chemical attractions and size rather than sheer forces. Subsequently, the outer surfaces of more complex target structures demonstrated the development of new receptors with diverse concave shapes such as clefts,[2] tweezers,[3] and bowls.[4]

As the field matured, hosts with the capabilities to surround guests completely and isolate them from outside media were developed. The cryptophanes[5] and carcerands[6] are pioneer cases in which guests are held within hosts via a web of covalent bonds. Those systems, however, lack reversibility—a vital component of a dynamic system. A creative solution to this challenge led to the formation of self-assembled hosts.[7] These are complexes containing inner cavities in which two or more copies of a same molecule or different molecules associate through non-covalent interactions.

The aggregates formed should have well-defined structures in solution and be capable of binding behaviors that none of their individual components displays alone.

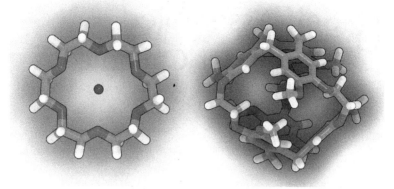

FIGURE 8.1 Modeled structures of 18-crown-6 and potassium cation complex (left) and crytophane (right).

Self-assembled molecular capsules come in different shapes and cavity sizes and hold together through a variety of non-covalent forces; they are capable of reversible encapsulation. Small molecule guests are entirely surrounded by larger molecular assemblies and the steric barriers imposed by a host retain the encapsulated molecules.

To prevent build-up of vacuum conditions in their cores, molecular capsules are assembled only if suitable guests are present. The selection of an appropriate solvent for an encapsulation study is critical. Solvents that complement cavities are reluctant to be displaced by solutes since the solutes are present in much higher concentrations.

These molecular capsules self-assemble through a range of forces such as metal–ligand interactions,[8] hydrophobic interactions,[9] hydrogen bonding,[10] and dynamic covalent chemistry[11] (Figure 8.2). Their lifetimes vary from milliseconds to days,

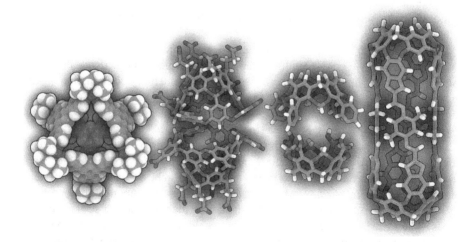

FIGURE 8.2 Modeled structures of molecular capsules assembled with (left to right) metal–ligand coordination, hydrophobic interaction, hydrogen bond interaction, and dynamic covalent chemistry. Peripheral groups have been removed for viewing clarity.

depending on the natures of solvents, temperatures, types of non-covalent interactions of capsule fragments, and the sizes and shapes of encapsulated molecules.

The behaviors of molecules in dilute solutions are very different when isolated from bulk media in the confined spaces inside molecular capsules. The molecular capsules are novel tools of modern physical organic chemistry and have been used as reaction chambers,[12] chiral receptors,[13] stabilizers of reactive intermediates,[14] transition states,[15] and modifiers of reactions in limited quarters.[16]

In these systems, synergy is at play. The interplay of capsule components, solvent molecules, and guest molecules is governed by attraction forces and the interplay is reversible. System structure and behavior have been reviewed extensively.[17] A brief review of hydrogen-bonded capsules follows.

8.2 EARLY EXAMPLES OF HYDROGEN-BONDED CAPSULES

The first example of a self-assembled molecular capsule was developed by properly positioning hydrogen bonding donor and acceptor sites on a concave molecular platform.[18] The platform consisted of a planar spacer fused to glycoluril units to provide self-complementary hydrogen bonding sites.

The glycoluril structure is unique; in addition to the hydrogen bonding donor and acceptor possibilities, it has a curvature formed by the *cis* fusion of the five-membered rings and the solubilizing groups are presented on its convex face. The structure can self-assemble to a semispherical dimeric structure held together by a seam of eight hydrogen bonds resembling two halves of a tennis ball **1** (Figure 8.3).

This capsule possesses rather small cavity with a flattened sphere shapes and inner volumes of ~70 Å^3 that can be occupied by small guest molecules such as methane or ethane. The assembly was characterized in solution by NMR spectroscopy. The signal for the encapsulated guest molecule appears far upfield, a result

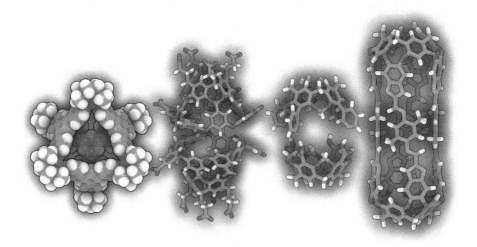

FIGURE 8.3 Structural drawings and models of tennis ball 1 (top) and soft ball 2 (bottom). Peripheral groups have been removed for viewing clarity.

of magnetic anisotropy exposure, caused by aromatic subunits of the capsule that surround the guest molecule. The exchange of guests into and from the capsules is slow (typical lifetime ~1 s). The separate signals for free and bound guests allows integration and the direct determination of equilibrium constants.

Modification of the spacer between two glycolurils gives rise to larger molecular containers that can encapsulate increasingly larger guests or multiple copies of the same molecule.[19] In a laborious effort, a monomer consisting of 13 fused rings in its extended and rigid structure dimerized to a closed-shell capsule with spherical shape held together with eight hydrogen bonds along its seams (Figure 8.3).[20] The formation of this assembly was shown to be reversible but an entropy-driven process, due to its unusual temperature-dependent guest binding process.

In general, molecular assembly formations are believed to be enthalpically driven, where favorable binding forces (enthalpy) compete with energy loss due to the decreased freedom of the individual subunits (entropy). The increased guest binding character of **2** at elevated temperature is due to the fact that more than one solvent molecule is involved in the initial complex and their substitution by a single large molecule is entropically favorable.

The ability of this host to encapsulate more than one guest suggests its use as a "reaction chamber" for bimolecular activities such as the Diels-Alder reaction. Indeed, when both p-benzoquinone and cyclohexadiene were present in solution, an initial 200-fold rate acceleration was observed, but the optimism faded upon discovery of a product inhibition effect. Since the product is a better fit for the cavity, no turnover could be observed.[21] However, the catalytic turnover was observed for a pair of p-benzoquinone and thiophene dioxide derivatives since the adduct is ejected from the capsule in favor of two reactant molecules.[22]

A tetrameric capsule **3** was devised from the preferential hydrogen bonding of the cyclic sulfamide donor and glycoluril acceptor that results in a head-to-tail arrangement of four monomers and encapsulates a variety of cyclic guest molecules (Figure 8.4).[23] Again, the entropic penalty of bringing together four monomers and

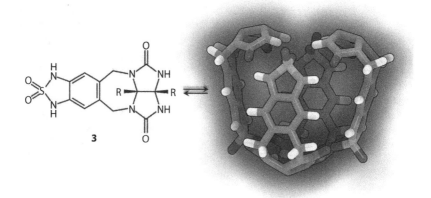

FIGURE 8.4 Structural drawing (left) and model (right) of tetrameric capsule **3**. Peripheral groups have been removed for viewing clarity.

one guest in a single discrete complex is forfeited by the enthalpic gains provided by the formation of 16 hydrogen bonds and whatever host–guest interactions are offered. For guest exchange to occur with a tetrameric glycoluril capsule, partial dissociation of the assembly must take place.

8.3 RESORCIN[4]ARENE-BASED CAPSULES

The concave shape of resorcin[4]arene is stabilized by intramolecular hydrogen bonds of eight phenols on its edge, and its synthetic availability makes it an attractive scaffold for molecular recognition.[24] Its capsular formation was first determined by a spectacular solid-state structure in which a spherical hexameric molecular capsule assembled together by 60 hydrogen bonds between 6 resorcin[4]arene and 8 water molecules created a spherical cavity of ~1375 Å^3, but the enclosed guest molecules were too disordered for assignment (Figure 8.5).

NMR spectroscopy was used to study its behavior in wet organic solvents and showed six chloroforms or eight benzene molecules could be encapsulated in its cavity.[25] Due to its relatively large and spherical inner cavity, small guests such as heptane or benzene can freely tumble and exchange positions within the cavity without any translational and rotational limitations imposed by the host.

Solvent molecules within the hexameric capsule can be displaced by sizable quaternary ammonium salts, and dissociation of at least one resorcinarene molecule from the hexamer is required for guest exchange. More details about the dynamic process of this assembly at nano- and micromolar concentrations were acquired using

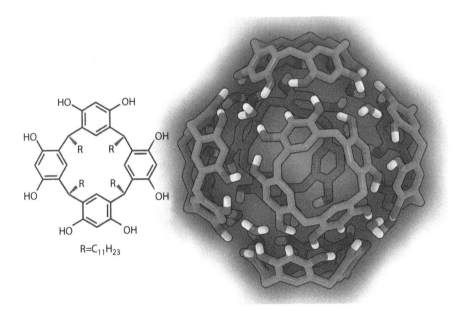

FIGURE 8.5 Structural drawing of resorcin[4]arene (left) and model of hexameric capsule (right). Peripheral groups have been removed for viewing clarity.

fluorescence resonance energy transfer (FRET) techniques with the attachment of suitable fluorophores to the assembly subunits.[26]

8.4 CYLINDRICAL CAPSULES

Utilizing the concave surface of resorcin[4]arene and appending imide walls to its four corners resulted in a cavitand that, in the presence of suitable complementary guests, dimerizes through a seam of eight bifurcated hydrogen bonds (Figure 8.6). The resulting 425 Å^3 cavity can encapsulate up to three different guest molecules.[27] Mesitylene-d$_{12}$ is the preferred solvent for encapsulation studies, since it is most commercially available deuterated solvent. The lifetime of the capsule is around 1 to 2 s and the exchange of guests in and out is slow at NMR timescale at room temperature.

The hydrogen bond donor and acceptor-rich polar belt at the mid-line of the capsule control the exchange of guests in and out of cylindrical capsules **4**. A wide variety of guests, both rigid and flexible, fit inside the confines of this cylindrical capsule.

The aromatic surface of the capsule forms favorable CH–π interactions with included guests and provides anisotropic shielding of guest molecules. A map of induced magnetic shielding for the inner space was obtained through nucleus independent chemical shift (NICS) calculations.[28] Comparison of these values with the experimentally measured ^{1}H NMR upfield shifts provides a means to pinpoint the exact position of guest inside the capsule.

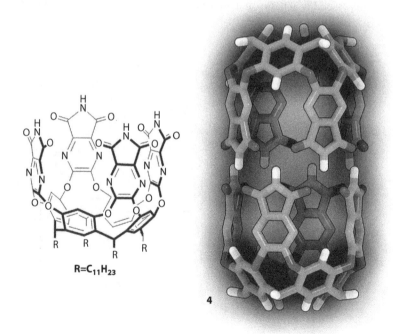

FIGURE 8.6 Structural drawing (left) and model (right) of cylinderical capsule **4**. Peripheral groups have been removed for viewing clarity.

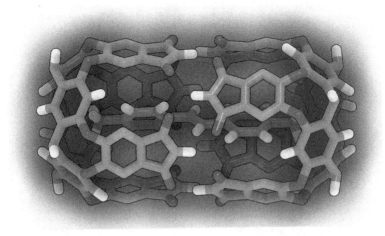

FIGURE 8.7   Modeled structures of 4,4′-dimethylstilbene encapsulated within **4**. Peripheral groups have been removed for viewing clarity.

Hydrocarbon chains such as n-tetradecane (C14) that are longer than the internal length of the capsule cavity form encapsulation complexes by assuming helical conformations as confirmed by through-space nuclear Overhauser effect (NOE) cross peaks along their chains (Figure 8.10). Apparently, the unfavorable gauche interactions are compensated by attractive CH– π interactions.

Although these helical structures are chiral, experiments suggest that the helical structures can interconvert between enantiomers within the capsule.

The narrow cylindrical shape of the capsule aligns two encapsulated molecules in a way that limits the interactions between the molecules. This property of the capsule has been exploited to investigate the reaction mechanisms of carboxylic acids and isonitriles, explore the possibility of funneling reactants toward a specific reaction pathway, and highlight the importance of effective molarity.[29]

The effects of three-dimensional geometrical restriction imposed by a capsule on the physical properties of a guest were also studied.[30] For example, the fluorescence of encapsulated 4,4′-dimethylstilbene was quenched, since this fluorophore adopts a twisted conformation to avoid unfavorable π–π interactions with the capsule walls and produces the twisting of the two halves of the capsule by 45 degrees (Figure 8.7). Self-assembly of the external host can also affect the excited state of its guest through geometrical control of the guest's surroundings in which the irradiation of encapsulated benzil derivatives results in blue fluorescence from a singlet excited state for encapsulated benzil and green phosphorescence from the triplet excited state for encapsulated 4,4′-dimethylbenzil.[31]

8.5   HYBRID ASSEMBLIES

The structures of molecular capsules described to this point are composed of identical subunits that form homomeric assemblies and achieve high symmetry. Demonstration of disproportional capsular formation of two modules capable of

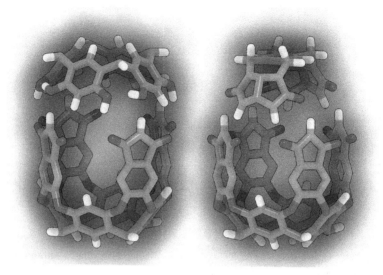

FIGURE 8.8 Model of two hybrid assemblies. Left: Complex of resorcin[4]arene with **4**. Right: Complex of tennis ball subunit with **4**. Peripheral groups have been removed for viewing clarity.

forming their own homomeric capsules through hydrogen bonding may indicate some flexibility in the rules of self-sorting. As a matter of fact, a hybrid assembly formed in response to the presence of appropriate guests when the cylindrical capsule **4** was mixed with resorcin[4]arene, itself a starting material for the synthesis of **4** (Figure 8.8). This new hybrid assembly exhibited host properties that differed noticeably from those of the parent hexameric or cylindrical capsules.[32]

Half of the tennis ball **1** capsule can also form a hybrid assembly with **4**, suggesting that other capsular assemblies may emerge from apparently distinct components that have concave surfaces and are rich in hydrogen bonding capabilities.[33]

8.6 MULTICOMPONENT ASSEMBLIES

De novo design and synthesis of molecular capsules with appropriate curvatures and complementary binding sites is not an easy task. Covalent modifications on the structure of an already known module are laborious and may encounter unwelcome solubility problems or collapses of concave surfaces.

A highly favorable and indirect approach is the development of external spacer elements that can reach and bind to both edges of the parent assembly and create a new multicomponent assembly with a distinct extended cavity. The first successful example was the extension of dimension and cavity size of cylindrical capsule **4** by non-covalent insertion of four glycolurils into the polar middle of the capsule.[34] The patterns of the donors and acceptors on glycolurils and wall of the cavitand are complementary. The only disconnect is the slight difference between the angle of curvature (113 degrees) of the glycoluril and the corner of the cavitand (90 degrees).

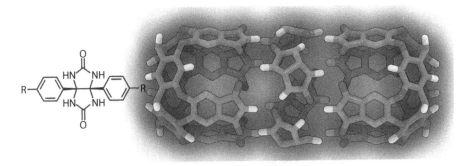

FIGURE 8.9   Left: Chemical structure of glycoluril spacer. Right: Model of mono-extended capsule. Peripheral groups have been removed for viewing clarity.

Spacers can correct this by adopting a twisted arrangement to maximize hydrogen bonding interactions. Four glycoluril spacers are needed for extension of the assembly. Each glycoluril forms four hydrogen bonds with the cavitands and four hydrogen bonds with its adjacent spacers. In this arrangement, superior hydrogen bond donors and acceptors are paired. The acidic N-Hs of the imides interact to the ureido carbonyls of glycolurils. Overall, 40 hydrogen bonds exist in the new assembly and a belt of glycolurils extends the length of the capsule by 7Å and its cavity by 200 Å^3 (Figure 8.9).

Only one carbonyl group of each imide wall is left without a hydrogen bond donor, and this asymmetric arrangement causes a twist in the array of the imide walls that reduces the symmetry of the cavitands from C_{4v} to C_4 and creates a chiral space at two ends of the new assembly. As evidenced by 2D NMR spectroscopy, the two enantiomers of the new assembly interconvert at room temperature. The racemization process occurs in a concerted fashion through rotation of all spacers in one direction, transitioned through an achiral structural intermediate.[35]

A number of other guests that cannot be accommodated in **4** were taken up in the new assembly, including longer alkanes up to n-nonadecane, alkylated arenes, rigid quadro-phenyl, and four molecules of cyclopropane.[36] The increased dimensions of the extended capsule permitted encapsulation of two long molecules or even several short ones. Some guests have restricted motion inside the cavity; they are too thick to slip past each other and too long to tumble rapidly. This results in observation of multiple capsular arrangements. These systems have been called social isomers in which the orientation of one guest is dependent on the presence and nature of the other co-encapsulated molecules.[37]

The magnetic anisotropy character of the two ends of the extended capsule induces magnetic shielding, similar to what we observed for the original assembly. Surprisingly, the middle of the capsule has an opposite effect and provides moderate deshielding to the guest protons in the space defined by the four glycolurils. The nuclei in this area are exposed to the edges of the aromatic units on the convex surface of the glycolurils of the capsule.[38]

Compression and expansion of encapsulated C_{14} can be controlled reversibly by an external stimulus. The encapsulated coiled alkane in **4** with eight gauche interactions

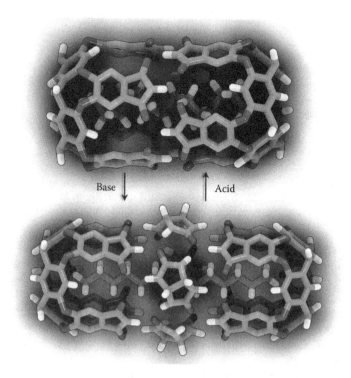

FIGURE 8.10 Coiling and uncoiling of n-tetradecane (C14) is controlled reversibly by addition of acid or base.

in its backbone acts as loaded spring. In the expanded capsule, C_{14} relaxes into an extended conformation as a result of extra accessible inner space. A glycoluril bearing a weakly basic site that can be protonated by strong acids was prepared.[39] Reversible spring loaded devices can be realized by the addition of acids and bases in which glycoluril inserts into the assembly under basic conditions; acidic conditions precipitate the glycoluril and regenerate the capsule **4** with coiled C_{14} inside (Figure 8.10).

The extension of capsule **4** is not limited to the addition of four glycolurils and one belt of spacer element. Alkanes longer than C_{19} require incorporation of an additional belt of glycolurils into the assembly. Eight glycolurils form 40 hydrogen bonds with the cavitands and themselves and form a double-extended capsule with an inner cavity around 800 Å³—almost twice the size of the original cylinder. Biologically active molecules such as omega-6 and omega-3 fatty acids and anandamide which is an endogenous ligand for the cannabinoid receptor form inclusion complexes.[40]

Encapsulation of multiple guests inside this hyperextended capsule has been studied as well. Two γ-picolines and two trifluoroacetic acid molecules present as acid–base pairs form a discrete 14-component supramolecular structure (Figure 8.11). The ¹H NMR spectrum of the γ- picoline–trifluoroacetic acid ion pair shows a signal at δ = 18.7 ppm, indicating the acidic proton is in contact with both the picoline nitrogen and the trifluoroacetate oxygen.[41] Further extension of the length of the capsule

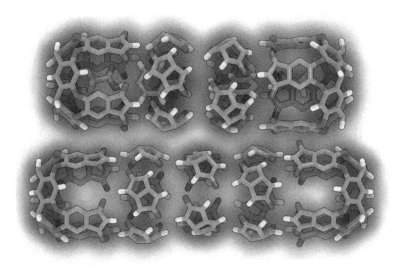

FIGURE 8.11 Modeled structures of hyperextended molecular capsule. Top: Double extended with 8 spacer molecules. Bottom: Triple extended with 12 spacer molecules. Peripheral groups have been removed for viewing clarity.

was observed by encapsulation of even longer hydrocarbons such as C_{24} to C_{29} in a complex involving three glycoluril belts. This 15-component assembly held together through a network of 56 hydrogen bonds (Figure 8.11).

As demonstrated, for each three to four methylenes longer than the alkane, another belt of glycolurils is recruited into the assembly structure. The energetics involved in formation of these multicomponent assemblies are very complex, but some of the entropically and enthalpically favorable processes are recognized. At first glance, assembly of 15 subunits seems to be entropically disfavored, but some of this entropical penalty was paid during synthesis of the cavitand, where four hydrogen-bonding modules are placed in a concave surface. These subunits are all solvated and involved in an unsymmetrical network of hydrogen bonds in solution.

The final assembly has an entropic advantage since a minimum molecules are involved in its structure. For example, a long chain alkane guest needs to be surrounded by many solvent molecules, and these solvent molecules are released to the bulk medium when the alkane enters the capsule. The closed capsule assembly maximizes the number of possible hydrogen bonds formed and also matches the best donors with the best acceptors, so an enthalpic contribution is present.

8.7 PROPANDIUREA DERIVATIVES AS SPACERS

The curvature of propandiurea derivatives is a better match for the corners of tetraimide cavitands **4**. The angle of plane of the two ureido functionalities is 99 degrees—much better than the 113 degrees of glycoluril. The twisted belt arrangement of glycolurils in the extended capsules is linked to this geometric mismatch. The addition of propandiurea to a complex of **4** with C_{14} resulted in formation of two new assemblies, one chiral and the other achiral (Figure 8.12).[41]

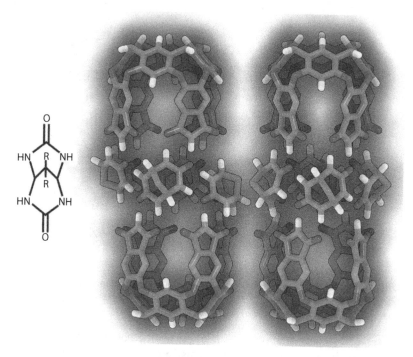

FIGURE 8.12 Chemical structure of propandiurea (left) and models of achiral (middle) and chiral (right) extended capsules. Peripheral groups have been removed for viewing clarity.

The chiral structure has been shown to hold the twisted belt structure, similar to glycoluril-based assemblies. The dynamic motion in the spacer belt is fast and enantiomers are interconverted (racemized) very rapidly even at lower temperatures. This is evidenced by coalescence of the diastereotopic proton signals of the geminal hydrogen atoms of encapsulated alkanes even at reduced temperature and 2D NOE analysis. Presumably, the achiral intermediate requires a lower activation barrier.

In the second observed assembly with achiral character, two propandiureas are placed at each corner of the cavitands and the other two are twisted to form close assembly with a maximum number of hydrogen bonds. The achiral structure has slightly longer dimensions and is the only assembly formed when longer alkanes such as heptadecane were offered for encapsulation.

The encapsulation of n-nonadecane required formation of yet another new chiral assembly since the diastereotopic proton signals of the guest can be observed (Figure 8.13).

Integration of the relevant signals revealed that this new assembly contains six units of propandiurea—an unprecedented number of spacer units since all other extensions involve multiples of four units. Detailed NMR studies and molecular modeling revealed an unusual banana-shaped structure for this new assembly. The unusual shape suggested the potential of the structure for shape-selective encapsulation of rigid guests and this proved to be the case. Structures with kinks at the middle

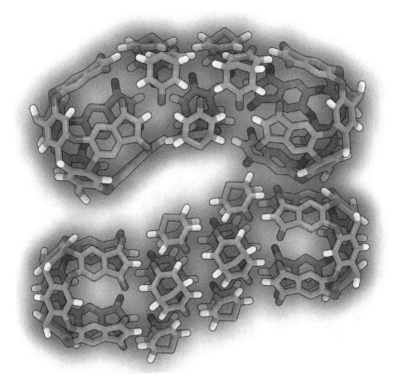

FIGURE 8.13 Hyperextended capsules with 6 (top) and 8 (bottom) propandiureas incorporated in their structures. Peripheral groups have been removed for viewing clarity.

were preferred to the rigid rectilinear polycyclic aromatic types. Further extension of the capsule with a double belt of propandiureas was also observed and its inner space accommodated alkanes such as C_{21} to C_{23}.

8.8 DEEP CAVITANDS

Prevention of closed capsular formation through disruption of hydrogen bonding along one of the edges of the spacer module in the previous systems resulted in formation of a deeper cavitand. Covalent deepening of these open-ended systems proved to be challenging and required multistep synthesis. Furthermore, solution studies of covalent systems with elongated walls resulted in aggregates with no internal spaces.

The use of a N-monosubstituted glycoluril was considered to "short circuit" the network of hydrogen bonds in the extended assembly.[42] In fact, the addition of N-methylated glycoluril resulted in formation of a chiral deep cavitand in the presence of alkanes shorter than C_{14} and longer than C_{19} (Figure 8.14).

The system is open-ended and alkanes are partially encapsulated; only four of the terminal methyls and methylenes of the alkane are exposed to the magnetic shielding of the aromatic ends of the cavitand and experienced upfield shifting of their

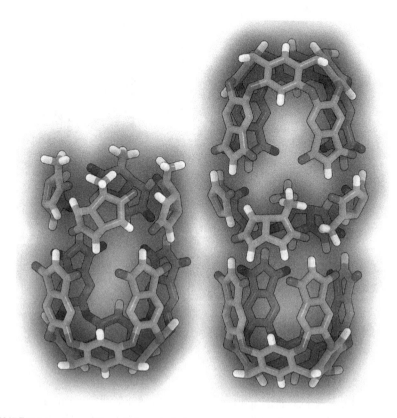

FIGURE 8.14   Assemblies of N-methyl glycoluril with **4**. Modeled structures of deep cavitand (left) and extended capsule (right). Peripheral groups have been removed for viewing clarity.

chemical shifts. The four methyl groups on x are oriented in the same direction and the array of glycolurils is highly symmetric and diastereoselective, providing an asymmetric nanoenvironment near its rim.

The encapsulation of alkanes from C_{15} to C_{19} resulted in two unexpected new assemblies in which hydrocarbons were coiled into the cavity, reminiscent of capsular formation. The N-methylated glycolurils positioned themselves in the middle of assembly in an effort to maximize their superior hydrogen bonds, where the methyl groups of the spacer adopt an up–down–up–down orientation.

8.9   SUMMARY

In conclusion, self-assembled molecular capsules are held together by weak intermolecular forces such as hydrogen bonds, CH–π interactions, van der Waals forces, strong metal–ligand binding, and even dynamic covalent chemistry. Capsule formation creates an inner cavity with a unique shape that can accommodate guests with complementary shapes and binding sites. The behaviors of molecules in dilute solution are different from their behaviors in confined spaces inside molecular capsules. The resourcefulness of these self-assembled molecular capsules points to a range of conceivable applications.

REFERENCES

1. Pedersen, C. J. (1967). *J. Am. Chem. Soc.* 89, 2495–2496.
2. Rebek, J. (1987). *Science* 235, 4795, 1478–1484.
3. Zimmerman, S. and Wu, W. (1989). *J. Am. Chem. Soc.* 111, 8054–8055.
4. Sanderson, P. G. E., Kilburn, J. D., and Still, W. C. (1989). *J. Am. Chem. Soc.* 111, 8314–8315.
5. Canceill, J., Lacombe, L., and Collet, A. (1986). *J. Am. Chem. Soc.* 108, 4230–4232.
6. Cram, D. J., Choi, H. J., Bryant, J. A. et al. (1992). *J. Am. Chem. Soc.* 114, 7748–7765.
7. Lehn, J. M. (1990). *Angew. Chem.* 102, 1347; *Angew. Chem. Intl. Ed. Engl.* 29, 1304–1319.
8. Fujita, M., Oguro, D., Miyazawa, M. et al. (1995). *Nature* 378, 469–471.
9. Gibb, C. L. D. and Gibb, B. C. (2004). *J. Am. Chem. Soc.* 126, 11408–11409.
10. Kobayashi, K., Ishii, K., Sakamoto, S. et al. (2003). *J. Am. Chem. Soc.* 125, 10615–10624.
11. Asadi, A., Ajami, D., and Rebek, J. (2013). *Chem. Sci.* 4, 1212–1215.
12. Chen, J. and Rebek, J. (2002). *Org. Lett.* 4, 327–329.
13. Scarso, A., Shivanyuk, A., Hayashida, O. et al. (2003). *J. Am. Chem. Soc.* 25, 6239–6243.
14. Dong, V. M., Fiedler, D., Carl, B. et al. (2006). *J. Am. Chem. Soc.* 128, 14464–14465.
15. Yoshizawa, M., Tamura, M., and Fujita, M. (2006), *Science* 312, 251–254.
16. Kaanumalle, L. S., Gibb, C.L.D., Gibb, B.C. et al. (2004). *J. Am. Chem. Soc.* 126, 14366–14367.
17. Fujita, M., Tominaga, M., Hori, A. et al. (2005). *Acc. Chem. Res.* 38, 369–378.
18. Branda, N., Wyler, R., and Rebek, J. (1994). *Science* 263, 1267–1268.
19. Meissner, R. S., Rebek, J., and Demendoza, J. (1995). *Science* 270, 1485–1488.
20. Kang, J. M. and Rebek, J. (1996). *Nature* 382, 239–241
21. Kang, J. M. et al. (1998). *J. Am. Chem. Soc.* 120, 3650–3656.
22. Kang, J. M., Santamaria, J., Hilmersson, S. et al. (1998). *J. Am. Chem. Soc.* 120, 7389–7390.
23. Martin, T., Obst, U., and Rebek, J. (1998). *Science* 281, 1842–1845.
24. MacGillivray, L. R. and Atwood, J. L. (1997). *Nature* 389, 469–471.
25. Shivanyuk, A. and Rebek, J. (2001). *Proc. Natl. Acad. Sci. USA* 98, 7662–7665.
26. Barrett, E. S., Dale, T., and Rebek, J. (2007). *J. Am. Chem. Soc.* 129, 3818–3819.
27. Heinz, T., Rudkevich, D. M., and Rebek, J. (1998), *Nature* 394, 764–766.
28. Ajami, D., Iwasawa, T., and Rebek, J. (2006), *Proc. Natl. Acad. Sci. USA* 103, 8934–8936.
29. Hou, J. L., Ajami, D., and Rebek, J. (2008). *J. Am. Chem. Soc.* 130, 7810–7811.
30. Ams, M. R., Ajami, D., and Rebek, J. (2009). *Beilstein J. Org. Chem.* 5.
31. Ams, M. R., Ajami, D., and Rebek, J. (2009). *J. Am. Chem. Soc.* 131, 13190–13191.
32. Ajami, D., Schramm, M. P., Volonterio, A. et al. (2007). *Angew. Chem. Intl. Ed.* 46, 242–244.
33. Ajami, D., Hou, J. L., Dale, T. J. et al. (2009). *Proc. Natl. Acad. Sci. USA* 106, 10430–10434.
34. Ajami, D. and Rebek, J. (2006). *J. Am. Chem. Soc.* 128, 5314–5315.
35. Ajami, D. and Rebek, J. (2009). *Nat. Chem.* 1, 87–90.
36. Ajami, D. and Rebek, J. (2008). *Angew. Chem. Intl. Ed.* 47, 6059–6061.
37. Ajami, D. and Rebek, J. (2007). *Proc. Natl. Acad. Sci. USA* 104, 16000–16003.

38. Ajami, D. and Rebek, J. (2009). *J. Org. Chem.* 74, 6584–6591.
39. Ajami, D. and Rebek, J. (2006). *J. Am. Chem. Soc.* 128, 15038–15039.
40. Ajami, D. and Rebek, J. (2007). *Angew. Chem. Intl. Ed.* 46, 9283–9286.
41. Tiefenbacher, K., Ajami, D., and Rebek, J. (2011). *Angew. Chem. Intl. Ed.* 50, 12003–12007.
42. Yamauchi, Y., Ajami, D., Lee, J. Y. et al. (2011). *Angew. Chem. Intl. Ed.* 50, 9150–9153.

9 Synergistic Effects in Double Helical Foldamers

Yun Liu and Amar H. Flood

CONTENTS

9.1 INTRODUCTION

Double helices (duplexes) are self-assembled foldamers that can equilibrate across two levels of organization (Figure 9.1). The first level is between a random coil and a folded state akin to α-helices and the second level involves the generation of intertwined double helices from singles akin to the structure of double-stranded DNA.

Due to their hierarchical ordering and attendant structural complexity, duplexes have become attractive supramolecular systems to study. Early examples employed metal–ligand interactions to assemble multidentate organic ligands into double helix configurations.[1] More recently, metal-free duplexes make use of direct interstrand interactions, e.g., hydrogen bonds,[2] π–π interactions,[3] and salt bridges,[4] as a means to collapse two elegantly designed foldamers into well-organized structures.

In this chapter, we focus on the metal-free duplexes and discuss the role of cooperativity as it emerges in both their formation and function.

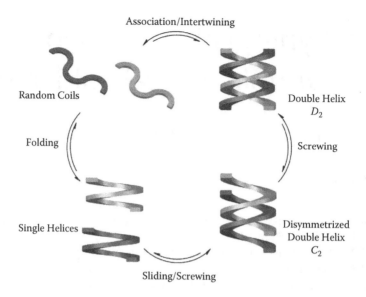

FIGURE 9.1   Principal equilibria and proposed mechanistic pathways for formation of double helices.

Cooperativity is an essential element in designing duplexes because the final folded structure reflects the coming together of many interaction sites that were originally separated in the random coils. As soon as the first interstrand interaction is established, two preorganized helices have the opportunity to slip into each other in a sliding and/or screwing motion.[5] Alternatively, two random coils can directly wrap around each other to form a duplex.[6] Along both pathways, the stability (ΔG) of the folded intertwined state becomes substantially enhanced as more and more interactions contribute to the structure.

The equilibrium representing duplex formation can also be expressed by the dimerization constant, K_{dim}:

$$2 \text{ random coils} \rightleftharpoons \text{duplex} \qquad K_{dim}$$

This reaction can be treated as the forward association rate (k_f) ratioed against the backward dissociation rate (k_b). In a positively cooperative system, multiple contact points join to form completely folded states directly after the first contact between the two strands. As a consequence, the forward association process is highly favored and a very large K_{dim} can be realized. In a system that displays negative cooperativity, the duplex is likely to undergo complete dissociation when one pair of interactions is lost. In such cases, the backward rate constant is likely to dominate and a small K_{dim} is observed.

The folding and intertwining processes can be modulated by exposing the duplexes to different chemical environments, e.g., solvent,[7] temperature,[5a] and guest molecules.[8] These phenomena inspired chemists to modify the duplexes as a means to understand environmentally responsive cooperativity.

In this chapter, we review the studies aimed at the creation and control of duplexes and frame them using the concepts of cooperativity and synergy—two terms that mean "working together." We start by introducing basic structural features and general examples of duplexes, such as double helical biomolecules, metal-directed duplexes, and our main focus, self-assembled double helical foldamers. We then explore the synergistic effects and outcomes in the context of foldameric duplexes related to the stability of duplexes and also in the diverse contexts of chirality induction, foldaxanes (foldamer-based rotaxanes), reaction templates, anion-induced duplex formation, solvent-induced duplex formation, and other [2 + 2] duplex–guest complexes.

9.2 DUPLEXES: FROM DNA TO SELF-ASSEMBLED FOLDAMERS

9.2.1 BASIC STRUCTURES OF DUPLEXES

Duplexes usually consist of two intertwined strands with structural similarities and complementarities. Similar to the description of single helices, one can define the pitch distance and torsion angle between each turn of one strand within the duplex (Figure 9.2). When compared to a single helix, the condensation of interaction sites per unit volume in a duplex is achieved at the cost of increasing torsion angles (θ) and subsequent pitch distances (D) while reducing cavity size (d). This characteristic likely makes duplexes appropriate hosts for smaller guests or no guests[9] when compared to their corresponding single helices.

Multiple locally arrayed interactions are employed to guide and stabilize the formation of a duplex. Most frequently, interactions with high directionality such as metal–ligand interactions, hydrogen bonds, and salt bridges are designed in the horizontal directions of the single helices as a means to guide the recognition of two incoming strands. These interactions are most effective in non-polar solvents. Large aromatic surfaces are used frequently along the vertical directions to stabilize duplexes by providing π–π contacts that can gain even more importance when combined with the hydrophobic effect in polar protic media. These π contacts are

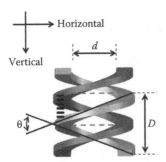

FIGURE 9.2 Generic structure of duplex. Symbols represent common physical parameters: θ = torsion angle; D = pitch distance; and d = cavity size. Both horizontal and vertical dashed lines represent interstrand interactions that contribute to duplex formation. (Source: Copyright (2011), Wiley-VCH, Weinheim. With permission.)

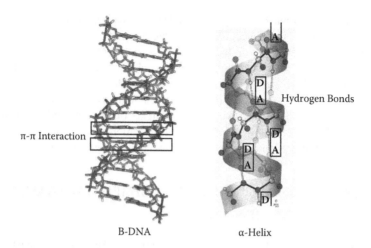

π-π Interaction

Hydrogen Bonds

B-DNA   α-Helix

FIGURE 9.3   Representations of (a) B-DNA and (b) α-helix.

less directional and are less frequently used to guide ordering. Exceptions in which donor–acceptor types of π-π interactions assist duplex formation in a selective manner have emerged.[6]

9.2.2   DOUBLE HELIX VERSUS SINGLE HELIX: NATURE'S CHOICE

Nature takes full advantage of helical structures for gene storage (DNA) and enzyme function (α-helices). Under physiological conditions, DNA forms a duplex structure and α-helices favor a single helix. In double helical DNA (Figure 9.3), hydrogen bonds between complementary base pairs (horizontal direction) selectively guide two appropriate strands into the duplex configuration.

In addition, π–π interactions between aromatic surfaces of base pairs (vertical direction) are believed to account for most of the thermal stability.[3a] In contrast, α-helices utilize three networks of hydrogen bonds between amide donors and acceptors (N–H•••O=C) as well as the constituent amide dipole moments to guide (vertical direction) and stabilize the single helical configuration.[10] Further stability of this structure in aqueous media is achieved by embedding α-helices inside a hydrophobic environment, e.g., protein interior or membrane,[11] which leads to stronger interactions when water is excluded.

9.2.3   METAL-DIRECTED DUPLEXES

During the creation of synthetic duplexes, transition metal ions were introduced as key organizational elements. One early example made use of Cu(II) and modified DNA bases to form artificial DNA (Figure 9.4a).[12] Another common approach takes advantage of the tendency of linear multi-dentate ligands to twist in a helical fashion.[13]

In the resulting duplexes, metal binding sites are appropriately arranged in two identical multi-dentate, bipyridyl-based ligands (**1**) to satisfy the metal–ligand

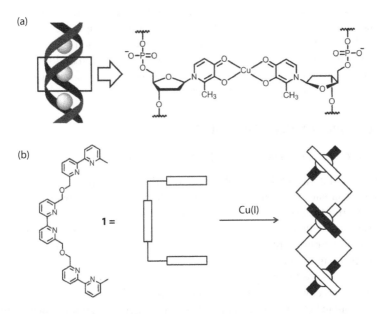

FIGURE 9.4 (a) Example of metal-directed DNA using Cu(II) and modified DNA bases. (b) Simplified example of ligand **1** designed for duplex formation with Cu(I). (*Source:* Copyright (1997), American Chemical Society. With permission.)

coordination geometry for the metal ions (Figure 9.4b).[1b] Multiple coordination centers within the complex are twisted relative to one other in the same direction to minimize both steric hindrance and strain. As a consequence, helical duplexes are formed instead of their linear isomers. These types of complexes now constitute an important family of duplexes and the reader is directed to selected reviews for more detail.[1b,1c,14]

9.2.4 SELF-ASSEMBLED DOUBLE HELICAL FOLDAMERS

In contrast to the many cases in which strong metal–ligand interactions are employed to drive the formation and organization of duplexes, nature has recourse to secondary interactions. The biomimetic challenge set by DNA is whether chemists can design duplexes based on purely organic components.

One common family that belongs to this class involves self-assembled hydrogen bonded arrays that are optimized to achieve extremely high homodimerization strengths in non-polar solvents.[2] However, the interstrand hydrogen bonding affinities of these complementary arrays decrease dramatically in polar media.[2b] Consequently, multiple secondary interactions are often introduced to stabilize duplexes across a broader variety of solvents. Just like DNA, hydrogen bonds and π–π stacking can work together.

In non-polar media, hydrogen bonds can stabilize a duplex while in polar media, solvophobic interactions that were often recognized by the appearance of π–π contacts can hold a duplex together. In some cases, ionic interactions were introduced to achieve robust performance. Similar strategies can be found in biology, where ionic interactions, e.g., interactions of alkali metals and negatively charged oxygen atoms, were found to be critical to enzyme functions.[15]

9.3 COOPERATIVITY OF DUPLEXES

9.3.1 FACTORS CONTRIBUTING TO COOPERATIVITY OF DUPLEXES

The thermal stability (ΔG) of duplexes inherently reflects the cooperativity among the interaction sites that run up and down each strand. A deeper understanding of (1) the specific factors that contribute to the thermal stability of duplexes, and (2) the existence of enhanced stability of the first few interactions that are formed will greatly improve our ability to tune and utilize cooperativity for the rational design of duplexes. Insights about overall stability are common, but characterizations of the first contacts have yet to be observed.

Yashima and coworkers reported that oligo-resorcinols (2, Figure 9.5) can dimerize into a double helix in water[7] and demonstrated their potential for switching[16] and sensing.[17] Interestingly, both solvent and chain length were found to play important roles in the stability of these duplexes. The duplex was shown to be driven by hydrophobic interactions; adding MeOH (25% v/v) to water reduced the stability by four orders of magnitude.

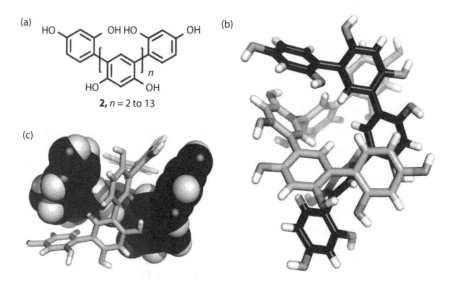

FIGURE 9.5 (a) Scheme of oligo-resorcinols **2**. (b) and (c) X-ray crystal structures of **2** (n = 3). Double helical configuration shows that π surfaces of each strand are buried.

TABLE 9.1

K_{dim} **Values for Oligoresorcinols with Various Chain Lengths (***n***)**

n	4	5	6	8	9	11	12	15
$\log K_{dim}$	2.5	3	3.4	5.2	6.1	7.8	7.3	7.7

In a subsequent paper,[18] the author demonstrated that mixtures of MeOH in water were needed to enable measurement of the dimerization constants (K_{dim} values) of oligo-resorcinols that had extremely large stabilities when using oligomers longer than the 6-mer. The chain length helps build overall positive structural cooperativity between hydrophobically driven π–π contacts during dimerization up to the 11-mer (Table 9.1). At this point, the authors believe that the entropy cost of merging two molecules overweighs the enthalpic benefit of burying π surfaces. As a result, the K_{dim} levels of and the stabilities of the 12-mer and 15-mer were shown to be slightly lower than those of the 11-mer.

Huc and coworkers used hydrogen bonding to preorganize pyridyl bisamides locally[5a] as a means to fix the structure of a foldamer into a conformation suitable for a double helix. However, none of the as-designed π–π interactions and hydrogen bonds were strong enough to hold the dimer together in competitive solvents such as dimethylsulfoxide (DMSO). To overcome this issue, the authors introduced more and multiple types of π–π contacts by attaching benzyl (Bn) substituents at the periphery of the foldamer (**3**, Figure 9.6). Consequently, the dimer was associated even in DMSO ($K_{dim} \sim 10^3$ M^{-1}) while all the other related double helices studied to that point had negligible K_{dim} values.[19]

In addition to chemical modification at the periphery, the backbone or the substituents facing the interior of the duplex can be engineered to stabilize it. Huc suggested[5] that π–π interactions drive duplex formation at the cost of extending the torsion angles of the foldamer to the degree required for dimerization. One strategy to overcome this penalty is to utilize tailored π surfaces such as diaza-anthracene that can enhance the driving force on account of its larger size and reduced torsion angle while still achieving the same pitch distance (Figure 9.7).[20]

This strategy turned out to be very successful. With the assistance of a diaza-anthracene backbone in addition to the benzyl substituents around the periphery, the K_{dim} of **4** was found to exceed 10^7 M^{-1} even in DMSO, which is four orders of magnitude stronger than the case where only benzyl substituents were used (see above).

If building blocks with intermediate π surfaces are used, the cost of expanding the torsion angle and the gain from dimerization almost cancel. This subtle energy balance can be manipulated to achieve new hybridized structures. For instance, more than two foldamers may intertwine to form multiplexes. In this process, larger torsion angles are always balanced by forming more π–π contacts; multiple strands can come together as long as they have enough space between the pitches for another π system. Consequently, steric factors can now play a role in determining the composition of the resulting complex where bulky groups inside the cavity would be expected to lower the maximum number of constituent strands within a multiplex.

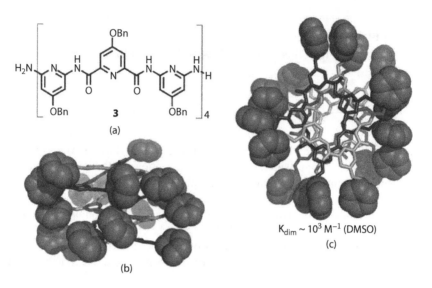

FIGURE 9.6 (a) Scheme of oligo-pyridyldiamides **3**. Benzyl side chains are emphasized to show interaromatic interactions in (b) side view and (c) top view of the x-ray crystal structure.

To demonstrate these hypotheses, 8-fluoroquinoline, a subunit with an intermediate surface size, was incorporated into a bisamide-based foldamer, and shown to form quadruplexes.[21] Foldamers constituted by 8-chloroquinoline bearing the larger chlorine atom formed cross-hybridized duplexes with the 8-fluoroquinoline-based foldamer.[22]

9.3.2 COOPERATIVITY IN CHIRAL DUPLEXES

Beyond demonstrations of how the solvent environment and the chemical structures of a foldamer can be used to tune the cooperativity (K_{dim}) of duplex formation, these same principles have been examined in the context of chirality. Helices are inherently chiral. Without any bias, however, helices with left and right handedness exist in equal populations and equilibrate via an intermediate random coil structure.

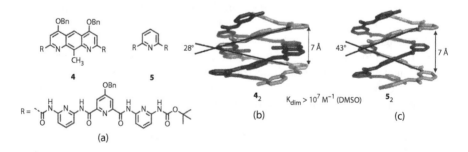

FIGURE 9.7 (a) Scheme of oligo-amides **4** and **5**. (b) Torsion angle of **4** in the solid state is smaller than that for (c) **5** (benzyl substituents are omitted).

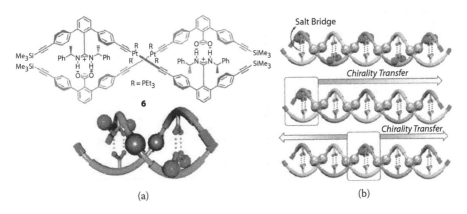

(a)   (b)

FIGURE 9.8 (a) Chemical and cartoon representations of double helix **6** formed between complementary carboxylates and amidiniums. (b) Chiral amplification can be achieved from both the termini and centers of duplexes. (*Source:* Copyright (2011), American Chemical Society. With permission.)

To obtain optically enriched double helices, two strategies have been investigated. Enantiopure chiral groups can be incorporated into a structure as a means to favor one-handedness over another. Alternatively, the equilibrium between M and P helices can be slowed when the interactions utilized to pair the two constituent strands into double helices are sufficiently strong. This situation allows for optical resolution. In an extreme case, a single enantiomer exists with one helical handedness coming from synergy between stability and chiral induction.

Among the many examples of duplexes reported by Yashima, the use of carboxylates and amidiniums is unique for their ability to form duplexes from complementary strands by *heterodimerization*.[23] Studies of chiral induction took advantage of chiral centers introduced readily into the amidinium moiety.[24] Because of the strong salt bridges between the two complementary strands and the geometry of the duplex, only one conformation of the amidinium was adopted. This resulting structural rigidity was expressed as the efficient amplification of the chiral sense from just one of the chiral amidinium subunits by its placement at either a central location or at the non-frayed end of the duplex sequence (Figure 9.8).

While the strong, positive cooperativity between the heterodimer of the strands bearing carboxylates and amidiniums can be applied to effect chiral induction, an even stronger interaction is needed to lock the handedness of the helicity as a means to access unidirectional rotatory switching of the pitch distance.[25] First, the reaction between boron hydride and hexaphenol yielded the racemic mixture of a double helix that was confirmed by inspection of its crystal structure.[26] Optical resolution was achieved using a chiral ammonium salt, (–)-N-dodecyl-N-methyl-ephedrinium bromide, to crystallize the anionic duplex in one-handed form out of solution with 100% ee, while the filtrate had a purity of 66% ee in favor of the other hand. This is the first example showing resolution of a mixture of duplexes into pure enantiomers that were shown to be configurationally stable in solution even after heating for prolonged periods of time.

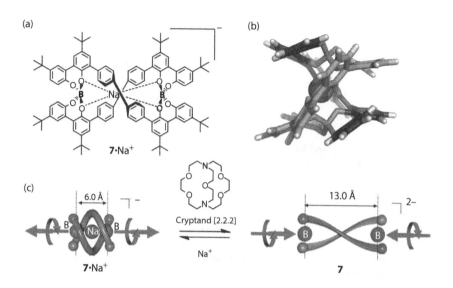

FIGURE 9.9 (a) Scheme of double-stranded oligo-phenol **7**•Na$^+$ bridged by spiroborates that sandwich a sodium cation. (b) X-ray crystal structure of **7**•Na$^+$ in its "stressed" configuration. (c) Duplex can undergo unidirectional rotatory motion by removing and adding sodium cation. (*Source:* Copyright (2010), Nature Publishing Group. With permission.)

Second, a sodium ion was discovered to be trapped as a guest inside the oligo-phenol duplex. It was subsequently found that the pitch distance of the tetraphenol duplex responded to the sodium ion (Figure 9.9). The length of the duplex doubled when the sodium ion was removed by cryptand [2.2.2] and it shrunk back to its original size by the reverse process. This molecule is optically pure and can rotate only by following its helical sense. Rotation in the opposite direction would have yielded its enantiomer and led to racemization. As a consequence, the tetraphenol duplex performs reversible unidirectional rotations using sodium cations as triggers.

9.3.3 Cooperativity in Time: Foldaxanes

While the factors influencing the stability of duplexes have been thoroughly studied, the dynamic features are rarely reported. Such dynamics can be examined in switching and self-sorting processes that respond to external stimuli. For example, duplexes can dissociate in response to introduced guest molecules. Fast dissociation of less stable duplex–guest complexes would allow evolution to the most stable complex. Slow dissociation of this desired complex then allows for further functions such as, switching. The duplex–guest interaction should cooperate reversibly with the host–host interaction within the duplex to accomplish folding and unfolding processes that take place at different rates.

Foldaxanes were first defined by Huc and Jiang as rotaxanes in which interlocked macrocycles are replaced by foldamers to fulfill the role of shuttling along dumbbell-like guest molecules.[27] Later, foldaxanes were synthesized with duplexes (Figure 9.10a).[8] In this case, a foldamer based on 8-fluoroquinoline was used to form a duplex with linear

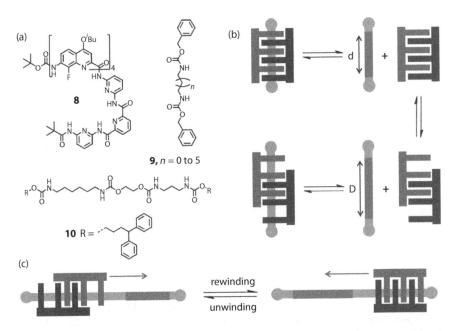

FIGURE 9.10 (a) Scheme of oligo-aromatic amide **8** that can form duplexes around dumb-bells (**9** and **10**) of variable lengths. Duplex **8₂** can (b) equilibriate between dumbbells bearing shorter or longer stations by following an association–dissociation pathway or (c) by shuttling along a dumbbell bearing two identical stations.

biscarbamates trapped inside the duplex's central cavity. As a foldaxane, the duplex is chemomechanically wound around the dumbbells such that the only way for the duplexes to become separated from the rods is to completely unwind.

When the duplex was combined with dumbbells of different lengths, the duplex **8₂** was found to exchange dynamically (Figure 9.10b) between different carbamate complexes. Temporary dissociation of **8₂** was believed to account for the slow exchange rates between these complexes. However, when using a dumbbell containing two different stations, one short and another long (**10**), the only movement observed was the translocation of the duplex along the length of the dumbbell (Figure 9.10c). In this case, the complete dissociation from the dumbbell was not observed. The net outcome of the shuttling is a partial unwinding and rewinding of the duplex that is concomitant with its association with the long or short stations, respectively.

When the stability of the complex formed between the duplex and the dumbbell is sufficiently strong, the interactions holding the two strands together in the duplex can be sacrificed.[27] Yet, what happens when the duplex–carbamate interactions are much weaker than the strand–strand interactions? Huc identified a foldaxane (**11•12**, Figure 9.11)[28] as a kinetic product while the [2 + 2] host–guest complex of the duplex (**12•11₂•12**) was found to be the thermodynamically favored product.

In this case, the host–host interaction in duplex **11₂** was so strong that the energy provided by complexation with the guest molecule **12** was unable to break the duplex apart thermodynamically. Consequently, the kinetic foldaxane **11•12** product

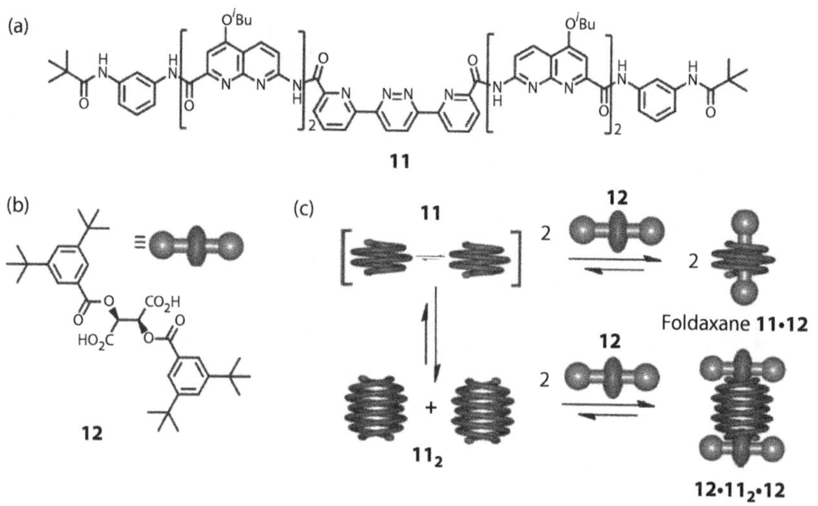

FIGURE 9.11 (a) Oligo-aromatic amide **11** is programmed to bind (b) guest thread **12** as a foldaxane. (c) Thermodynamic product is a [2 + 2] complex **12•11₂•12**. (*Source:* Copyright (2012), American Chemical Society. With permission.)

dissociates and reassembles into its duplex form 11_2 with the guest molecules becoming bound subsequently at the top and bottom regions of the duplex.

9.3.4 REACTIONS ON DUPLEXES

Biological duplexes like DNA not only have fascinating structures, but they also function. One challenge in the design of duplexes is to template or catalyze reactions. Studies to achieve this goal will enhance our understanding of duplex chemistry and open up the possibility of creating functional materials.

The unique pyridyl bisamide duplex reported by Huc (13_2, Figure 9.12a) was shown to undergo "self-promoted" N-oxidations.[29] The reaction rate was far faster in the duplex (Figure 9.12b) than in any related derivatives. The exact origin of this acceleration was not clear, although the author proposed that cation–π and dipolar interactions may play a role.

Another elegant design reported by Yashima became an important step for mimicking the function of DNA by using carboxylate–amidinium duplexes. Early studies have shown the ability to form this type of duplex through dynamic imine bond formations with a template.[30] Recently, template-directed reactions with diastereoselectivity have been demonstrated (Figure 9.13). An amidinium template is used to bind to two carboxylates, each of which bears one aldehyde functional group. The chiral bisamine was then introduced to form imine bonds in a step-wise manner with the two aldehydes to eventually form a duplex.

The chiral environment is determined by the first imine bond formation involving the chiral bisamine. Consequently, the relative arrangement of the unreacted aldehyde and amine in the ternary intermediate complex is programmed to favor

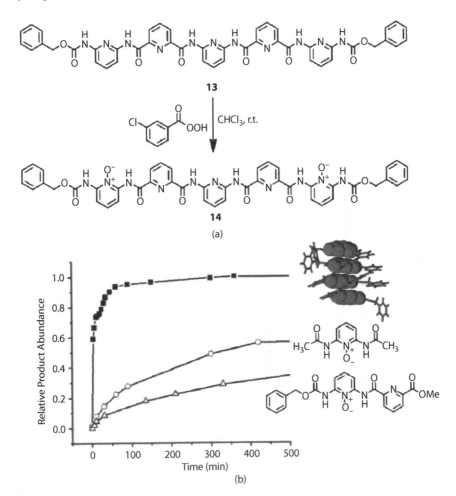

FIGURE 9.12 (a) Oligo-aromatic amide **13** can be oxidized into **14**. (b) Rate of oxidation is facilitated by formation of duplexes (black squares) when compared to monomeric controls (open circles, open triangles). (*Source:* Copyright (2005), American Chemical Society. With permission.)

one reaction pathway kinetically over the other. The diastereoselectivity is acquired during the second imine bond formation.

This templated reaction can be controlled by light utilizing an azobenzene moiety introduced into the linker.[32] The reaction can be turned on (*trans*) and off (*cis*) with a fivefold difference in rate upon photoisomerization of the azobenzene. This report further illustrates the utility of this class of salt-bridge duplex heterodimers in the realm of templated reactions.

9.3.5 ANION-INDUCED DUPLEX FORMATION: COOPERATIVITY OF BINDING

Anions are ubiquitous in biological systems.[33] To achieve further understanding of how anions are regulated and transported in biological systems, studies of anion

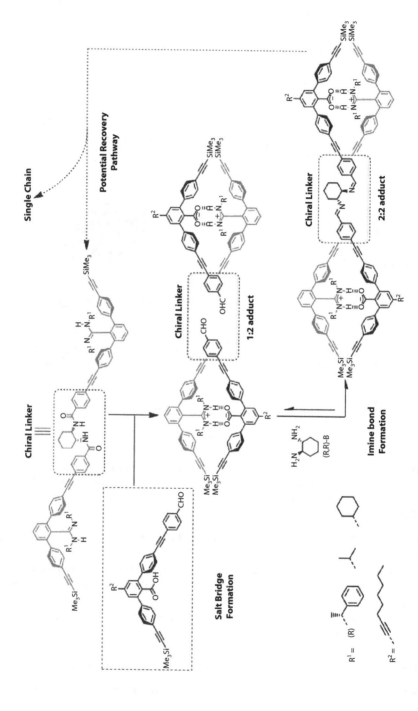

FIGURE 9.13 Scheme of diastereoselective imine bond formation promoted by chiral template capable of forming duplex with product.

receptors with advanced functionality are essential. The number of helicates, both duplexes and others, found to bind anions is still small even though many other examples have been shown to bind cations (metal-directed duplexes)[14] and neutral molecules (foldaxanes).[8,27,28]

The biggest challenge associated with duplex formation and anion binding is the compromise between interstrand interactions and providing enough room for the anions. Additionally, the possibility of forming cavities with elliptical shapes inside the double helix may not match with a spherical or pseudo-spherical anion. Not surprisingly, multiple anions are often favored during the formation of optimal host–guest contacts. The anions also assist the folding and intertwining processes required for formation of anion-duplex complexes. However, to accommodate more than one anion in a small internal volume requires the host–guest interactions to be sufficiently strong; otherwise, the anions would repel each other.

Interestingly, the first example of a duplex–anion complex is also the earliest report of a synthetic duplex.[34] Reported by de Mendoza, two bicyclic guanidiniums (Figure 9.14a) were shown to wrap around two sulfate anions to form complexes in solution. This behavior was not reproduced by using the corresponding chloride salt,

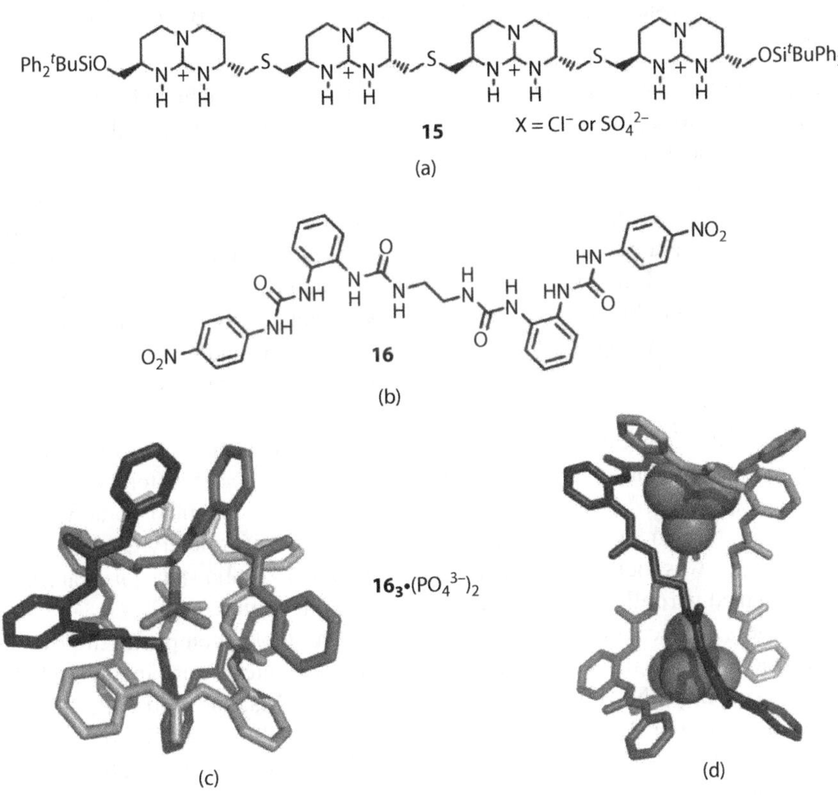

15 X = Cl⁻ or SO₄²⁻

(a)

16

(b)

16₃•(PO₄³⁻)₂

(c)

(d)

FIGURE 9.14 (a) Scheme of oligo-guanidinium **15**. (b) Scheme of oligo-aromatic urea **16**. (c) Top view of x-ray crystal structure of **16**. (d) Side view.

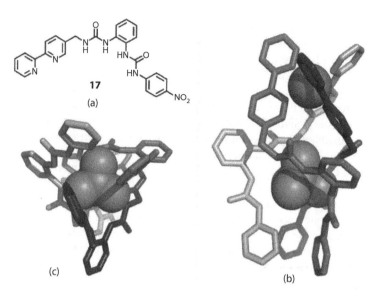

FIGURE 9.15 (a) Modified oligo-aromatic urea **17** and its x-ray crystal structure as triple helix with phosphate anion and nickel(II) cation. (b) Side view. (c) Bottom view.

suggesting that the stereochemical features of the sulfate[35] defined by its tetrahedral shape and its lone pairs of electrons may cooperate with the structural constraints of **15** to form the double helix.

Consistent with the challenge of analyzing such anion complexes, it was more than 10 years later that two phosphate anions were observed to be bound by three neutral oligourea-based receptors in both the solid state and solution phases (Figure 9.14b and c). This is the only anion-templated helicate reported in both phases. The stability of this complex was attributed to the tight coordination between strong hydrogen bond donors (oligourea N–H) and highly charged anions (PO_4^{3-}). In all other anion helicates,[36–38] the complexes were observed only in solution or in the solid state.

The relatively weak interaction between anions and receptors can be circumvented by using transition metal ions to help preorganize the receptor. Following this idea, a triplex of **17**[39] (Figure 9.15) was shown to be formed by incorporating bispyridyl, a metal binding moiety, with bisurea, an anion binding moiety. The resulted ion pair helicate was found to be stable in both solution and solid states; its stability can be attributed partially to the use of metal–ligand interactions.

For less charged anions such as halides, anion–duplex complexes are expected to be more challenging to create because host–guest interactions involving halides are weaker than those with sulfate and phosphate. Martin[36] and Gale[38] reported chloride and fluoride complexes, albeit only in the solid state (Figure 9.16); solution phase studies were not performed. Crystal structures of **18** with bromide were also reported.[40]

Maeda reported a [2 + 2] oligopyrrole-chloride duplex in the solid state[37] although subsequent solution phase studies excluded the existence of this complex in the

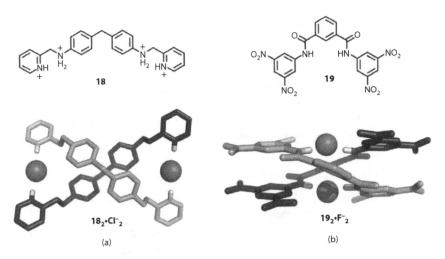

FIGURE 9.16 (a) Molecular and crystal structures of diammonium-bis-pyridinium ligand **18** forming [2 + 2] complex with Cl⁻. (b) Molecular and crystal structures of isophthalamide ligand **19** forming [2 + 2] complex with F⁻.

dichloromethane solutions. The authors believed that there was a lower driving force for assembling two [1 + 1] complexes into the duplex in solution even though the [1 + 1] complex was still formed. This observation suggests that crystal packing forces help favor duplex structures in the solid state when compared to single helices.

A ditopic oligopyrrole receptor (Figure 9.17) was later shown by Maeda[41] to form a solution phase [2 + 2] duplex with chloride at –50°C as a stable intermediate during the titration from a [1 + 1] complex to a [1 + 2] (receptor:chloride) complex. The

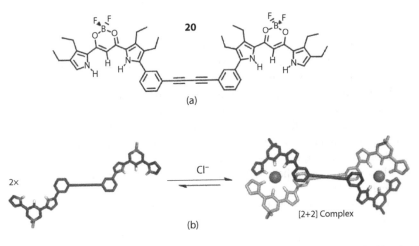

FIGURE 9.17 (a) Ditopic oligopyrrole **20** can bind chloride (b) in a [2 + 2] fashion in the solid state.

intertwining nature of the complex resembles Yashima's carboxylate–amidinium duplexes. The authors also examined the effects of the linker on the stability of the [2 + 2] duplex. They found that the duplex is favored when using less sterically bulky linkers and when using appropriate torsion angles to accommodate the two terminal binding sites. The first requirement is consistent with Yashima's report[6] that steric groups prevent receptors from coming together. The second argument echoes the design feature that receptors assemble into duplexes when the interactions that order the two strands are employed along the horizontal orientation (see above).

9.3.6   Hydrophobic Collapse in Host–Guest Complexation

Beyond the hydrogen bonds and electrostatic contacts that can be formed between anions and hosts, another strategy to facilitate complexation with anions is to take advantage of the hydrophobic character of the π-based units. This approach may be more important for singly charged anions in which the energies gained from hydrogen bonding are weaker than those formed with phosphate and sulfate. The tactic of using contacts between hydrophobic units, both aromatic and aliphatic, has been seen in a few host–guest complexes[42] and also in the formation of duplexes in aqueous solutions.[7]

The presence of hydrophobes provides a basis to employ solvophobic driving forces. In water, for example, hydrophobic collapse[43] is used to stabilize folded proteins and the structure of DNA. Moreover, when hydrogen bonds are buried inside a relative hydrophobic microenvironment with a low dielectric constant, the strength of the interactions is intrinsically stronger. The formation of complexes driven by hydrophobic collapse benefits from the liberation of water molecules between the apolar surfaces.

A report from Flood[44] employed the concept of hydrophobic collapse to enable the binding of chloride to an aryl-triazole foldamer, **21** (Figure 9.18), with high stabilities ($B_2 \approx 10^{12} M^{-2}$) in 50:50 water:acetonitrile. The authors found that a [2 + 1] (host:chloride) double helical complex was formed in acetonitrile and that its stability increased dramatically when the water content increased from 0 to 50%.

To illustrate the increasing positive cooperativity of duplex formation, the stepwise [1 + 1] and [2 + 1] binding constants (K_1 and K_2) were measured. In a noncooperative statistical formation of duplex, the ratio (K_2:K_1) was calculated to be 0.25. Values greater or less than 0.25 indicate positive or negative cooperativity. For duplex $21_2 \cdot Cl^-$, the ratio of K_2:K_1 changed from the negative cooperative value of 1:240 in acetonitrile, to the positive cooperative value of 165 in 50% water solution. The increasing cooperativity was shown to result from the increasing importance of the hydrophobic effect, which was turned on by the addition of water helping to bury more of the π surfaces of the double helix (Figure 9.18c) when compared to the single helix (Figure 9.18b).

Fortuitously, the hydrophobic collapse counteracted the decrease of K_1 attributed to the penalty of Cl^- desolvation. Consequently, an overall chloride binding affinity of $10^{12} M^{-2}$ in 50:50 water:acetonitrile solutions was achieved. This value exceeds all others reported for Cl^- in semi-aqueous solutions by more than six orders of magnitude. The high fidelity of this duplex (K_2:K_1 = 165) suggests that hydrophobic

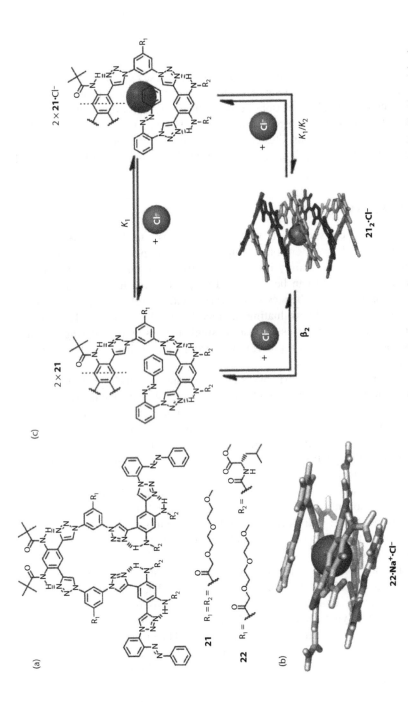

FIGURE 9.18 (a) Molecular structures of foldamers **21** and **22**. (b) Crystal structure of helical complex **22** formed with Cl⁻ (shown) in the center and Na⁺ in a glycol extracavity pocket (not shown); side chains are omitted for clarity. (c) Equilibria associated with binding tetrabutylammonium chloride in 50% water–acetonitrile. The [2 + 1] duplex (molecular model) has enhanced stability over the [1 + 1] complex.

collapse is a very successful strategy of forming anion helicates in polar solvents such as water, where the other host-guest interactions may begin to weaken.

9.3.7 OTHER 2 + 2 COMPLEXES

Other 2 + 2 complexes have been reported.[45] Most of these examples also employed ditopic receptors, but the following characteristics excluded them from our discussion: (1) direct interstrand interactions between the two receptors could not be identified; (2) no structural intertwining occurred; and (3) the [2 + 2] complexes resulted from the need for two binding sites constituted from the two receptors to optimize host–guest interactions with large anions,[45b] instead of being driven by interstrand contacts.

9.4 PERSPECTIVES

Relative to random coils, cooperativity is often intrinsic to the stability of single helix foldamer[46] and, by extension, to duplexes. The cooperativity we considered in this chapter, however, mainly involved stability of the duplexes relative to the single helices. The effect can be quantified by considering how the equilibrium between single and double helices (K_{dim}) varies with incremental changes in the design features and also by evaluating the relative stability in molecular recognition of double helical complexes against the single helices by considering the ratio $K_2:K_1$. We now comment on representative examples that display different types of cooperativities.

Yashima's oligoresorcinols[7,18] (Figure 9.5) take advantage of hydrophobically driven π–π interactions that represent straightforward examples showing the degree to which π surfaces cooperate with each other. The K_{dim} is seen to increase when chain length increases from 4 to 11 aromatic rings; the stability, log K_{dim}, increases linearly with the number of π units. This observation suggests that additional π units neither change the nature of the underlying attraction nor do they bias the strength of neighboring units.

The non-orientational character of π–π interactions reduces the possibility of the neighboring π surfaces influencing each other's stacking preferences. As a result, the incremental cooperativity gained from increasing chain length is similar across the series. Compared to the statistical scenario, linear growth still shows positive cooperativity between the π surfaces, presumably arising from the prefolded structures of oligoresorcinols in water.

Yashima's story has an unexpected end; when the length reaches beyond the 11-mer, a drastically different and complicated trend in log K_{dim} was observed, suggesting a change in some of the factors that control this behavior. This example illustrates the underlying complexity of these systems and the difficulty of identifying a single reason or parameter that controls them. The origin of the cooperativity observed with these and other duplexes still needs to be understood.

In a host–guest system, the cooperativity of forming double helices can be coupled to the cooperativity of step-wise guest complexation processes such as in Flood's chloride helicates (Figure 9.18). The cooperativity of double helix formation is quantified by the ratio $K_2:K_1$ and is shown to be enhanced dramatically with increasing water content. This evidence suggests that hydrophobic effects promote positive cooperativity associated with folding and intertwining two foldamers into double helices and forming a stable structure when binding the central chloride anion. This strategy, in which cooperativity of the double helices is turned on by solvent composition even in a small molecule, demonstrates that the structural properties of larger biological systems can be applied effectively to design functional molecules on the nanoscale.

Some studies have characterized the kinetics associated with duplexes. In a study of double helical foldaxane, Huc was able to distinguish two mechanistic pathways, i.e., translational shuttling and dissociation–association by which a double helix moves along the dumbbell.[27] The rates of the two pathways are so different that two-dimensional exchange spectroscopy (2D EXSY) shows only that the double helix followed the translational shuttling pathway. This qualitative experiment evaluates the positive cooperativity of forming double helices in comparison to that of forming a host–guest complex between a double helix and a dumbbell.

Overall, quantitative attempts to reveal cooperativity in double helices are still rare. We believe that more studies showing how these chemical components affect the cooperativity of double helix formation are needed to help advance our understanding.

9.5 CONCLUSIONS

Cooperativity is crucial in designing molecular constituents for the functional supramolecular self-assemblies. Especially when increasing numbers of components are involved, high degrees of positive cooperativity are usually demanded for the self-assembly of a desired architecture with high fidelity and specific performance. The double helix is one such functional assembly.

Although most of the conformational considerations can be found in the designs of single helices, knowledge of supramolecular chemistry and the intelligent recruitment of secondary interactions between single helices are needed to direct two oligomers into more complex duplexes. The complexity introduced by a second oligomer presents an extra challenge for molecular design. Because double helices are increasingly attractive targets of the supramolecular community, a plethora of chemistries have been investigated: chiral duplexes, foldaxanes, biomimetic duplexes, and anion-binding duplexes. The versatility of molecular moieties available for duplex chemistry has begun to connect their structural features to potent application-directed demonstrations of functionalities, such as chiral switches, self-templated reactions, and mechanically interlocked molecules.

Although the field of double helical foldamers is yet to be fully explored, we believe that the creation of novel structures with unique and advanced functions such as circularly polarized luminescence and artificial transmembrane ion channels may stem from the double helical topology. As a result, the combinations of multiple chemical components present in the intertwined helices that form when two oligomers are instructed to self-assemble together provides a coherent platform on which supramolecular chemists will stand and lead an orchestra of secondary interactions.

ACKNOWLEDGMENTS

We gratefully acknowledge the support by the Chemical Sciences, Geosciences, and Biosciences Division, Office of Basic Energy Sciences, Office of Science, U.S. Department of Energy.

REFERENCES AND NOTES

1. (a) Lehn, J. M.· *Supramolecular Chemistry*, VCH, Weinheim, 1995. (b) Piguet, C., Bernardinelli, G., and Hopfgartner, G. *Chem. Rev.* 1997, *97*, 2005. (c) Albrecht, M. *Chem. Rev.* 2001, *101*, 3457.
2. Selected papers about hydrogen bonded arrays: (a) Sijbesma, R. P., Beijer, F. H., Brunsveld, L. et al. *Science* 1997, *278*, 1601–1604. (b) Beijer, F. H., Sijbesma, R. P., Kooijman, H. et al. *J. Am. Chem. Soc.* 1998, *120*, 6761–6769. (c) Schmuck, C. and Wienand, W. *Angew. Chem. Intl. Ed.* 2001, *40*, 4363–4369. (d) Wang, X. Z., Li, X. Q., Shao, X. B. et al. *Chem. Eur. J.* 2003, *9*, 2904–2913. (e) Prabhakaran, P., Puranik, V. G., and Sanjayan, G. J. *J. Org. Chem.* 2005, *70*, 10067–10072. (f) Zhao, Y. P., Zhao, C. C., Wu, L.-Z. et al. *J. Org. Chem.* 2006, *71*, 2143–2146. (g) Spencer, E. C., Howard, J. A. K., Baruah, P. K. et al. *Cryst. Eng. Commun.* 2006, *8*, 468–472. (h) Lafitte, V. G. H., Aliev, A. E., Horton, P. N. et al. *J. Am. Chem. Soc.* 2006, *128*, 6544–6545. (i) Li, J., Wisner, J. A., and Jennings, M. C. *Org. Lett.* 2007, *9*, 3267–3269. (j) Todd, E. M. and Zimmerman, S. C. *Tetrahedron* 2008, *64*, 8558–8570. (k) Tang, Q., Liang, Z., Liu, J. et al. *Chem. Commun.* 2010, *46*, 2977–2979. (l) Zhang, P., Chu, H., Li, X. et al. *Org. Lett.* 2011, *13*, 54–57. (m) Groger, G., Meyer-Zaika, W., Bottcher, C. et al. *J. Am. Chem. Soc.* 2011, *133*, 8961–8971. (n) Mudraboyina, B. P. and Wisner, J. A. *Chem. Eur. J.* 2012, *18*, 14157–14164. (o) Wang, H. B., Mudraboyina, B. P., and Wisner, J. A. *Chem. Eur. J.* 2012, *18*, 1322–1327.
3. (a) Maurizot, V., Léger, J. M., Guionneau, P. et al. *Russ. Chem. B* 2004, *53*, 1572–1576. (b) Saito, N., Terakawa, R., Shigeno, M. et al. *J. Org. Chem.* 2011, *76*, 4841–4858. (c) Sugiura, H., Amemiya, R., and Yamaguchi, M. *Chem. Asian J.* 2008, *3*, 244–260.
4. Selected reviews: (a) Furusho, Y. and Yashima, E. *Chem. Rec.* 2007, *7*, 1–11. (b) Furusho, Y. and Yashima, E. *J. Polym. Sci. Pol. Chem.* 2009, *47*, 5195–5207. (c) Furusho, Y. and Yashima, E. *Macromol. Rapid Comm.* 2011, *32*, 136–146.
5. (a) Berl, V., Huc, I., Khoury, R. G. et al. *Nature* 2000, *407*, 720–723. (b) Acocella, A., Venturini, A., and Zerbetto, F. *J. Am. Chem. Soc.* 2004, 126, 2362–2367.
6. Yamada, H., Wu, Z. Q., Furusho, Y. et al. *J. Am. Chem. Soc.* 2012, *134*, 9506–9520.
7. Goto, H., Katagiri, H., Furusho, Y. et al. *J. Am. Chem. Soc.* 2006, *128*, 7176–7178.
8. Ferrand, Y., Gan, Q., Kauffmann, B. et al. *Angew. Chem. Intl. Ed.* 2011, *50*, 7572–7575.
9. Berni, E., Garric, J., Lamit, C. et al. *Chem. Commun.* 2008, 1968–1970.
10. Pauling, L., Corey, R. B., and Branson, H. R·*Proc. Natl. Acad. Sci. USA* 1951, *37*, 205–211.
11. Popot, J. L. and Engelman, D. M. *Annu. Rev. Biochem.* 2000, *69*, 881–922.
12. (a) Tanaka, K., Tengeiji, A., Kato, T. et al. *Science* 2003, *299*, 1212–1213. (b) Guido H. C., Corinna K., and Thomas C. *Angew. Chem. Intl. Ed.* 2007, *46*, 6226–6236.
13. Young, N. J. and Hay, B. P. *Chem. Commun.* 2013, *49*, 1354–1379.
14. (a) Lehn, J. M., Rigault, A., Siegel, J. et al. *Proc. Natl Acad. Sci. USA* 1987, *84*, 2565–2569. (b) Pfeil, A. and Lehn, J. M. *J. Chem. Soc., Chem. Commun.* 1992, *11*, 838–840. (c) Kramer, R., Lehn, J. M., and Marquis-Rigault, A. *Proc. Natl. Acad. Sci. USA* 1993, *90*, 5394–5398. (d) Swiegers, G. F. and Malefetse, T. J. *Chem. Rev.* 2000, *100*, 3483–3537.
15. (a) Williams, L. R. and Leggett, R. W. *Phys. Med. Biol.* 1987, *32*, 173–90. (b) Irnius, A., Speiciene, D., Tautkus, S. et al. *Mendeleev Commun.* 2007, *17*, 216–217.

16. Goto, H., Furusho, Y., and Yashima, E. *J. Am. Chem. Soc.* 2007, *129*, 109–112.
17. Goto, H., Furusho, Y., and Yashima, E. *J. Am. Chem. Soc.* 2007, *129*, 9168–9174.
18. Goto, H., Furusho, Y., Miwa, K. et al. *J. Am. Chem. Soc.* 2009, *131*, 4710–4719.
19. Haldar, D., Jiang, H., Leger, J. M. et al. *Angew. Chem. Intl. Ed.* 2006, *45*, 5483–5486.
20. (a) Berni, E., Kauffmann, B., Bao, C. et al. *Chem. Eur. J.* 2007, *13*, 8463–8469. (b) Berni, E., Dolain, C., Kauffmann, B. et al. *J. Org. Chem.* 2008, *73*, 2687–2694.
21. Gan, Q., Bao, C., Kauffmann, B. et al. *Angew. Chem. Intl. Ed.* 2008, *47*, 1715–1718.
22. Gan, Q., Li, F., Li, G. et al. *Chem. Commun.* 2010, *46*, 297–299.
23. (a) Tanaka, Y., Katagiri, H., Furusho, Y. et al. *Angew. Chem. Intl. Ed.* 2005, *44*, 3867–3870. (b) Furusho, Y., Tanaka, Y., and Yashima, E. *Org. Lett.* 2006, *8*, 2583–2856. (c) Ito, H., Furusho, Y., Hasegawa, T. et al. *J. Am. Chem. Soc.* 2008, *130*, 14008–14015. (d) Iida, H., Shimoyama, M., Furusho, Y. et al. *J. Org. Chem.* 2010, *75*, 417–423. (e) Wu, Z. Q., Furusho, Y., Yamada, H. et al. *Chem. Commun.* 2010, *46*, 8962–8964.
24. Ito, H., Ikeda, M., Hasegawa, T. et al. *J. Am. Chem. Soc.* 2011, *133*, 3419–3432.
25. Miwa, K., Furusho, Y., and Yashima, E. *Nat. Chem.* 2010, *2*, 444–449.
26. (a) Katagiri, H., Miyagawa, T., Furusho, Y. et al. *Angew. Chem. Intl. Ed.* 2006, *45*, 1741–1744. (b) Furusho, Y., Miwa, K., Asai, R. et al. *Chem. Eur. J.* 2011, *17*, 13954–13957.
27. Gan, Q., Ferrand, Y., Bao, C. et al. *Science* 2011, *331*, 1172–1175.
28. Gan, Q., Ferrand, Y., Chandramouli, N. et al. *J. Am. Chem. Soc.* 2012, *134*, 15656–15659.
29. Dolain, C., Zhan, C., Leger, J. M. et al. *J. Am. Chem. Soc.* 2005, *127*, 2400–2401.
30. Yamada, H., Furusho, Y., Ito, H. et al. *Chem. Commun.* 2010, *46*, 3487–3489.
31. (a) Yamada, H., Furusho, Y., Ito, H. et al. *Chem. Commun.* 2010, *46*, 3487–3489. (b) Yamada, H., Furusho, Y., and Yashima, E. *J. Am. Chem. Soc.* 2012, *134*, 7250–7253.
32. Tanabe, J., Taura, D., Yamada, H. et al. *Chem. Sci.* 2013, *4*, 2960–2966.
33. Alvarez-Leefmans, F. J. and Delpire, E., Eds. *Physiology and Pathology of Chloride Transporters and Channels in the Nervous System.* Academic Press, San Diego, 2009.
34. Sánchez-Quesada, J., Seel, C., Prados, P. et al. *J. Am. Chem. Soc.* 1996, *118*, 277–278.
35. Li, S., Jia, C., Wu, B. et al. *Angew. Chem. Intl. Ed.* 2011, *50*, 5721–5724.
36. Keegan, J., Kruger, P. E., Nieuwenhuyzen, M. et al. *Chem. Commun.* 2001, 2192–2193.
37. Haketa, Y. and Maeda, H. *Chem. Eur. J.* 2011, *17*, 1485–1492.
38. Coles, S. J., Frey, J. G., Gale, P. A. et al. *Chem. Commun.* 2003, *39*, 568–569.
39. Custelcean, R., Bonnesen, P. V., Roach, B. D. et al. *Chem. Commun.* 2012, *48*, 7438–7440.
40. Selvakumar, P. M., Jebaraj, P. Y., Sahoo, J. et al. *RSC Advances* 2012, *2*, 7689–7692.
41. Maeda, H., Kitaguchi, K., and Haketa, Y. *Chem. Commun.* 2011, *47*, 9342–9344.
42. (a) Diederich, F. *Angew. Chem. Intl. Ed.* 1988, *27*, 362–386. (b) Kubik, S., Goddard, R., Kirchner, R. et al. *Angew. Chem. Intl. Ed.*, 2001, *40*, 2648–2651. (c) Kubik, S. and Goddard, R. *Proc. Natl. Acad. Sci. USA*, 2002, *99*, 5127–5132. (d) Kubik, S., Kirchner, R., Nolting, D. et al. *J. Am. Chem. Soc.*, 2002, *124*, 12752–12760. (e) Rodriguez-Docampo, Z., Pascu, S. I., Kubik, S. et al. *J. Am. Chem. Soc.*, 2006, *128*, 11206–11210. (f) Suk, J. M. and Jeong, K. S. *J. Am. Chem. Soc.* 2008, *130*, 11868–11869. (g) Kubik, S. *Chem. Soc. Rev.* 2010, *39*, 3648–3663. (h) Juwarker, H. and Jeong, K. S. *Chem. Soc. Rev.* 2010, *39*, 3664–3674.
43. Nicholls, A., Sharp, K. A., and Honig, B. *Proteins Struct. Func. Gen.* 1991, *11*, 281–296.
44. Hua, Y., Liu, Y., Chen, C. H. et al. *J. Am. Chem. Soc.* 2013, *135*, 14401–14412.
45. (a) Lee, H. N., Xu, Z., Kim, S. K. et al. J. Am. Chem. Soc. 2007, 129, 3828–3829. (b) Chen, C. Y., Lin, T. P., Chen, C. K. et al. *J. Org. Chem.* 2008, *73*, 900–911.
46. Hill, D. J., Mio, M. J., Prince, R. B. et al. *Chem. Rev.* 2001, *101*, 3893–4011.

10 Ion Pair Templation Strategies for Synthesis of Interlocked Molecules

Graeme T. Spence and Bradley D. Smith

CONTENTS

10.1 MECHANICALLY INTERLOCKED MOLECULES

The most common classes of interlocked molecules are catenanes and rotaxanes[1] derived from the Latin *catena*, *rota*, and *axis* (meaning *chain*, *wheel*, and *axle*, respectively). As depicted in Figure 10.1, a catenane is composed of interlocking rings (a). A rotaxane contains a macrocyclic component looped around an axle, held in place by two bulky stoppers (b). A related supramolecular structure is the pseudo-rotaxane (c)—an inclusion complex in which an axle without stoppers resides inside a macrocycle. This interpenetrated, but not interlocked, assembly is often a precursor to catenanes and rotaxanes.

It should be noted that in addition to interlocked structures consisting of two components, higher-order [n]-catenanes and [n]-rotaxanes and even polymeric interlocked structures may be constructed. However, this chapter will focus on the synergistic concepts behind the preparation and behavior of [2]-catenanes and [2]-rotaxanes that may be applied to larger systems.

The possibility of constructing mechanically interlocked systems has been recognized since the 1950s. Initial research was driven by academic curiosity and the technical challenges of making these exotic molecules. The first successful preparation

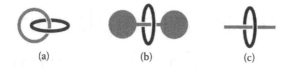

FIGURE 10.1   (a) Catenane. (b) Rotaxane. (c) Pseudorotaxane.

of an interlocked molecule was reported by Wasserman in 1960,[2] and was based on a statistically unlikely process, that is, a macrocyclization reaction occurring while the precursor chain was threaded through a preformed ring (Scheme 10.1). As expected for such a low probability event, catenane **1** was formed in a yield far below 1%. The inherent low yields of this statistical method could not be overcome, and although some alternative strategies were investigated, the area of interlocked molecules lay largely dormant for two decades.

10.1.1   TEMPLATED SYNTHESIS

The break-through that re-ignited interest in interlocked molecules was the develop-ment of new templated reactions led by the pioneering work of Sauvage in the 1980s. A chemical template, as defined by Busch,[4] "organises an assembly of atoms, with respect to one or more geometric [positions] in order to achieve a particular linking of atoms." By replacing *atoms* with *molecular components*, we can see how this definition may apply to interlocked structures.

Accordingly, Sauvage utilized the orthogonal arrangement of bidentate phen-anthroline ligands around a copper(I) center to template the synthesis of catenane **2** (Scheme 10.2).[5] With a hydroxy-functionalized phenanthroline ligand threaded through a similar macrocycle using copper(I) templation, a cyclization reaction afforded the interlocked product in a yield of 42%—far greater than results from any statistical approach.

The advantages of metal templation are the strength of metal–ligand bonds and the well-defined geometries of the resulting complexes. As a result, many metals and geometries have been employed in the templated synthesis of a plethora of distinct interlocked structures.[6] More recently, the wide-ranging catalytic abilities of transi-tion metals have been exploited in a synthetic process called *active metal* templation, in which the metal mediates both the association of the components and catalyzes the reaction to form the interlocked product.[7]

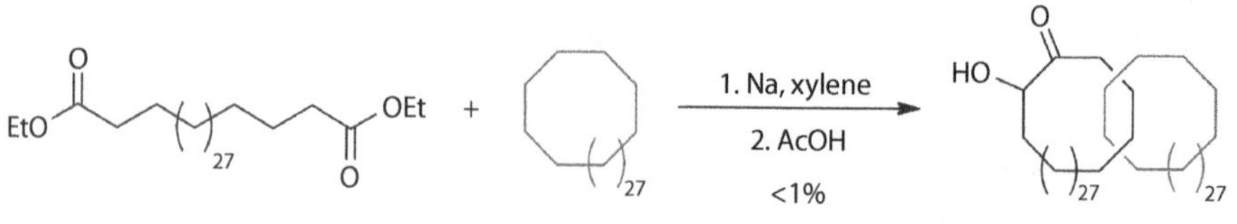

SCHEME 10.1   Synthesis of the first reported catenane **1** which relied on a statistical approach.[3]

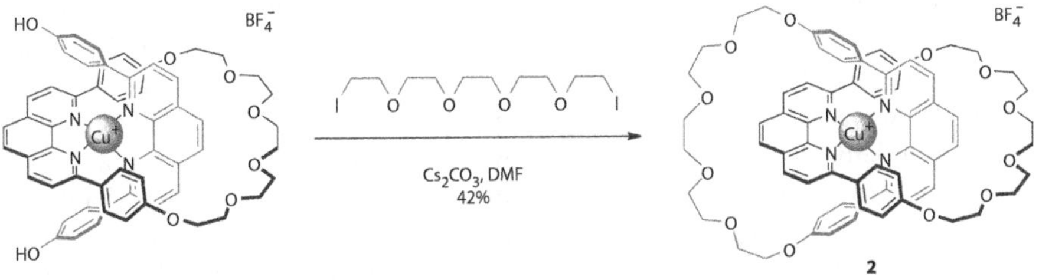

SCHEME 10.2 Synthesis of Sauvage's copper(I) templated catenane **2**.

Beyond metal templation, the favorable non-covalent interactions of organic cations have also been exploited to template the formation of interlocked structures with the group of Stoddart and co-workers providing major contributions. For example, the attractive charge transfer donor–acceptor interactions between an electron-accepting positively charged bis(4,4'-bipyridinium) macrocycle and an electron-donating dioxynaphthalene thread were used to form a strongly complexed pseudorotaxane (Scheme 10.3).[8] A copper-catalyzed alkyne-azide cycloaddition (CuAAC) reaction was subsequently employed to form the stoppered dioxynaphthalene axle and yield the target rotaxane **3**.

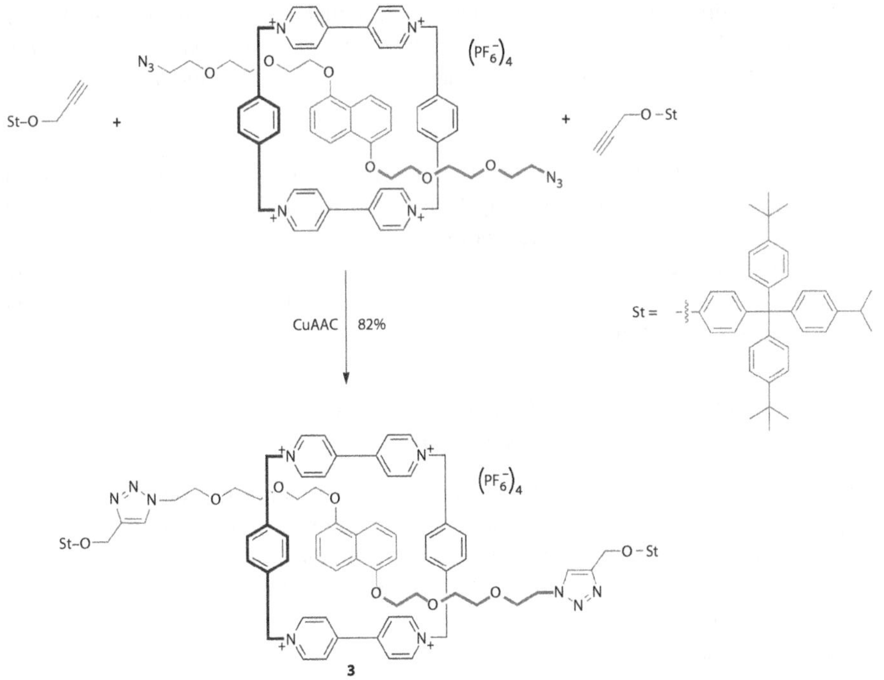

SCHEME 10.3 Synthesis of Stoddart's rotaxane **3** templated using charge transfer donor–acceptor interactions and stoppered dioxynaphthalene using copper-catalyzed alkyne-azide cycloaddition (CuAAC).

The topic of templated reactions has now expanded to include successful methods for templating the interlocked syntheses of uncharged components using hydrogen bonding, neutral donor–acceptor interactions, and the hydrophobic effect.[9–11] This chapter, however, focuses on the construction of mechanically interlocked molecules in which a definite ion pair interaction plays a key role in templated synthesis. Ion pairing is both a generalizable strategy to drive component association and a versatile method of producing interlocked structures with functional properties.

Recent advances highlight the vastly unexplored potential of interlocked structures for a broad range of nanotechnological applications—from selective sensing devices to molecular-scale switches. Crucial to these applications is the concept of synergy, that is, the functional performance of an interlocked molecule is greater than the sum of the individual free components. The central hypothesis of this chapter is that ion pair templation is an effective strategy for producing molecules that exhibit synergistic behaviors.

By definition, ion pairs are low energy assemblies, and removal of one of the ion pair partners after templated synthesis places an interlocked structure in a high energy state. Restoration of the ion pair interaction is highly favorable and produces a spring-like driving force that can be exploited for dynamic molecular functions.

The goal of this chapter is not to present an exhaustive collection of all of the research on this topic, but to provide a clear tutorial-style overview of the fundamental concepts aided by illustrative examples and informative schematics. Three types of ion pair templation strategies are described, namely, the use of integrated ionic components, the use of discrete ion templates, and traceless ion pair templation. These approaches can produce systems capable of two different functional behaviors by virtue of their synergistic properties.

First, the removal of a discrete ion pair partner places a structure in a high energy state, and the resulting assembly can retain strong affinity for that ion, most often due to a highly selective binding cavity. Consequently, such structures are well suited for small molecule recognition technologies. Secondly, as the ion pair interaction dictates the relative positions of the components within the interlocked structure, the removal and subsequent restoration of this interaction can result in controlled dynamic molecular motion.

10.1.2 CLASSIFICATION OF INTERLOCKED TEMPLATION MODES AND SYNTHETIC STRATEGIES

Templates are widely used in supramolecular chemistry and classified several ways. In general, they can be classed by whether a reaction is under thermodynamic or kinetic control[12] or by the precise topology of the templation.[13]

For interlocked structures specifically, it is possible for a template to be integrated within the components and thus be incorporated inside the resulting interlocked framework (Figure 10.2a).

Conversely, the template responsible for molecular association can be discrete, separate from the interlocked structure, and thus removable (Figure 10.2b). The templating charge-transfer donor–acceptor in rotaxane 3 (Scheme 10.3) is an example of integrated templation. The copper(I) metal in catenane 2 (Scheme 10.2) is strictly

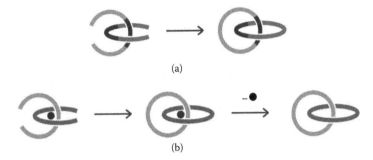

(a)

(b)

FIGURE 10.2 Different modes of interlocked templation. (a) Integrated within resulting framework. (b) Discrete and removable.

a discrete template as it can be removed from the molecule, leaving only the inter-locked phenanthroline macrocycles.[14]

Synthetically, there are a number of ways of approaching the task of inter-locked structure formation, with the most widely employed strategies illustrated in Figure 10.3. Catenanes are commonly synthesized using (a) *cyclization* (for example, **1** in Scheme 10.1 and **2** in Scheme 10.2); (b) *double cyclization* in which one or both macrocycles are formed in the fabrication of the interlocked product. Rotaxane formation is usually achieved by either (a) *clipping*, the cyclization of a macrocycle around an axle; (b) *stoppering* (for example, **3** in Scheme 10.3), the addition of suit-ably large stopper groups to a pseudorotaxane; or (c), *snapping* the covalent joining of two half axles within a macrocycle.

Applying these general strategies to the preparation of interlocked molecules requires the incorporation of appropriate functional groups to enable covalent

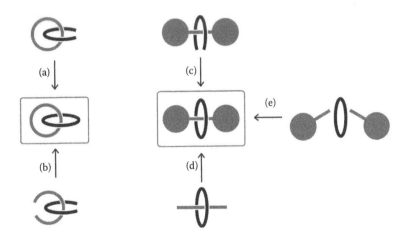

FIGURE 10.3 Common methods of interlocked structure formation. Cyclization (a) and double cyclization (b) of catenanes. Clipping (c), stoppering (d), and snapping (e) approaches for rotaxane synthesis.

capture of the target structure. Versatile, quantitative, and efficient couplings are ideal for this application, with CuAAC and Grubbs-catalyzed ring-closing metathesis (RCM) the two most common reactions exploited in this manner.[15,16]

10.2 ION PAIR TEMPLATION OF INTERLOCKED MOLECULES

10.2.1 INTEGRATED ION PAIR TEMPLATION USING IONIC COMPONENTS

The most conceptually simple example of an ion pair templated synthesis involves an electrostatic association of oppositely charged components (Figure 10.4). After they are associated in the correct alignment, a suitable reaction can then be used to convert the ion pair assembly into a permanently interlocked product.

This approach, in which both the cation and anion are integrated within opposing components of the target interlocked structure, has been applied successfully in several systems and is illustrated by Furusho and Yashima's "salt bridge" catenane **6** from 2010 (Scheme 10.4).[17]

A salt bridge is a recognition motif composed of both a strong electrostatic attraction and directional hydrogen bonding; it can be formed between amidine and carboxylic acid groups upon proton transfer. Both starting components **4** and **5** contain crescent-shaped *m*-terphenyl structural frameworks functionalized with the respective amidine and carboxylic acid motifs and terminal vinyl groups for RCM reactions. Thus, when salt bridge formation mediates the association of **4** and **5**, the vinyl groups are suitably aligned for catenane formation *via* double cyclization.

To promote association, the reaction was undertaken in a toluene non-polar solvent and the interlocked product was isolated in 68% yield. The salt bridge was initially retained in the rigid catenane framework due to the integrated nature of this templating interaction. Protonation of this motif using trifluoroacetic acid removed the ion pair interaction and unlocked the macrocycles to allow relatively free rotation with respect to each other.

Subsequent deprotonation using *N,N*-diisopropylethylamine restored the salt bridge, re-locking the macrocycles. As noted previously, the ability to remove the templating ion pair interaction within an interlocked structure can lead to controlled dynamic molecular motion. In this case, the result is acid–base switching between locked and unlocked states.

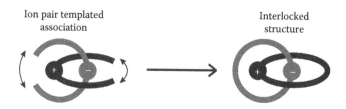

Ion pair templated association   Interlocked structure

FIGURE 10.4 Fundamental approach to interlocked structure formation using integrated ionic templation.

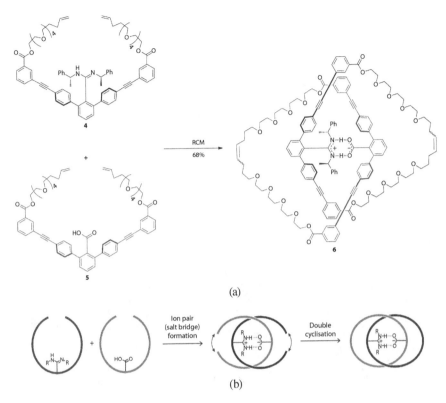

SCHEME 10.4   Amidinium–carboxylate salt bridge templated catenane **6**.

While the above example unavoidably contains a significant hydrogen bonding contribution to the templation reaction, other systems rely on purely non-directional electrostatic ion pair interactions. A rather impressive demonstration is Leigh and Winpenny's hybrid organic–inorganic rotaxane **7** also from 2010 (Figure 10.5).[18]

The axle component of **7** is composed of a dialkylammonium group with bulky 3,5-di-*tert*-butyl benzene stopper groups. The macrocycle is a heterometallic ring containing one cobalt(II) and seven chromium(III) metals, bridged by multiple fluoride and alkyl carboxylate anions. The synthesis of **7** involved the 33-component dynamic assembly of the monoanionic inorganic macrocycle around the cationic axle to afford product in a yield of 23%. Remarkably, this hybrid system exhibited good solubility in non-polar organic solvents. The solubilization of the anionic octametallic core was achieved by both charge balance with the cationic axle and the presence of a highly lipophilic periphery arising from extensive *tert*-butyl functionalization on both interlocked components.

10.2.2   ION PAIR TEMPLATION USING DISCRETE IONS

The use of discrete ion templates facilitates the exchange or removal of one of the templating ion pair partners after interlocked structure formation, placing

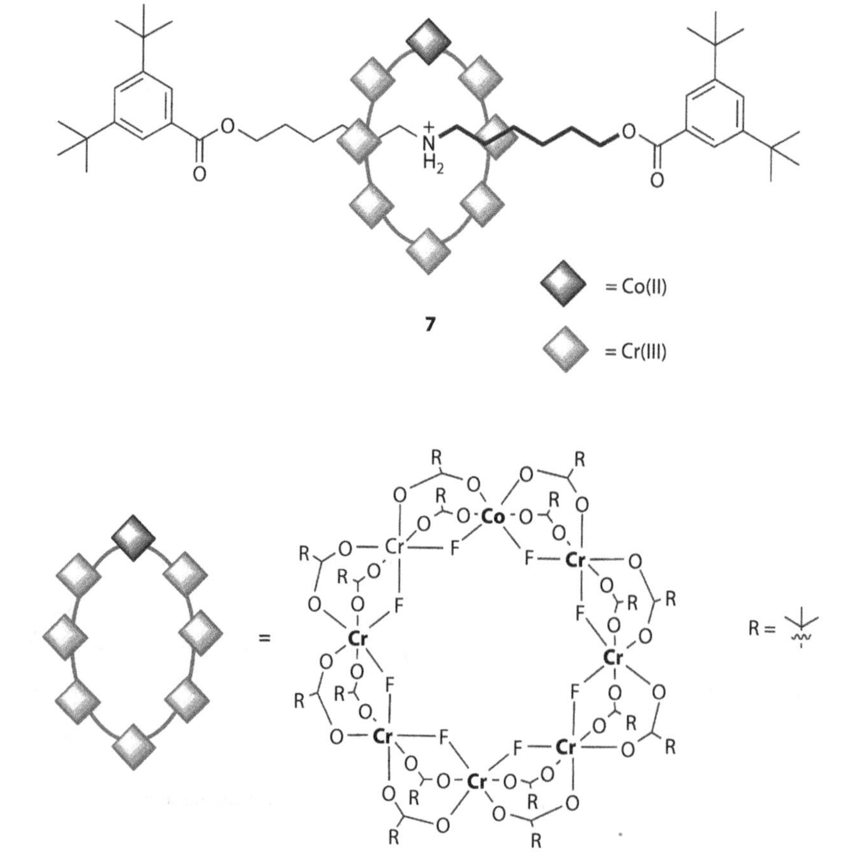

◇ = Co(II)

◇ = Cr(III)

7

R = ⋎

FIGURE 10.5 Hybrid organic–inorganic rotaxane **7** containing a cationic ammonium axle and an anionic heterometallic marcocycle.

the system in a higher energy state primed for spring-like functional behavior. To illustrate this point, consider rotaxane **10**, which was synthesized using the chloride salt of a positively charge pyridinium axle (**8**) and a *bis*-vinyl-appended macrocycle precursor (**9**) containing an anion binding isophthalamide cleft (Scheme 10.5).[19]

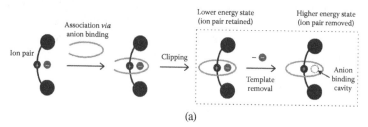

Lower energy state (ion pair retained)   Higher energy state (ion pair removed)

Association *via* anion binding

Ion pair

Clipping

Template removal

Anion binding cavity

(a)

SCHEME 10.5 Beer's anion binding rotaxane **10** synthesized using discrete anion templation. *(continued)*

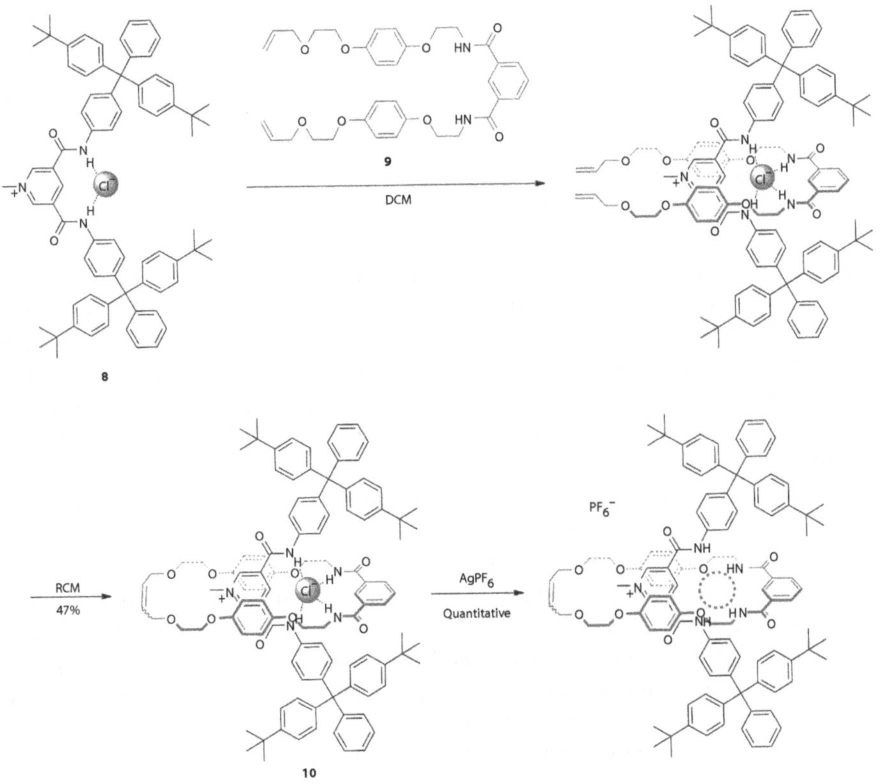

SCHEME 10.5 *(continued)* Beer's anion binding rotaxane **10** synthesized using discrete anion templation.

Association of these two components is mediated by the chloride anion *via* N-H···Cl⁻ and C-H···Cl⁻ hydrogen bonding and further stabilized by donor–acceptor interactions between the electron-rich hydroquinone groups of **9** and electron-deficient pyridinium axle **8**. Conversion of this orthogonal association complex into a permanently interlocked rotaxane was achieved by RCM clipping to afford **10** in 47% yield.

In this form, rotaxane **10** retains the tight pyridinium–chloride ion pair and additional convergent hydrogen bonding interactions. Silver salt precipitation, however, provides a facile method to remove the choride template from the interlocked cavity and replace it with a non-coordinating hexafluorophosphate anion. Although an ion pair interaction is formed between the rotaxane cation and new counter anion, it is not as strong as the previous chloride-containing ion pair, and this larger, more diffuse anion is not able to reside in the constrained anion binding cavity.

A return to the previous lower energy state is highly favorable, and thus chloride binding to the rotaxane host is strong. Furthermore, the inability of other discrete anions to fit within the interlocked binding cavity, which would maximize the desirable ion pair and hydrogen bonding interactions, provides selectivity to this

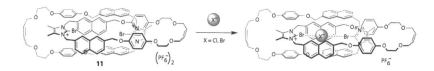

SCHEME 10.6 Bromo-imidazolium homocatenane **11** exhibiting halogen bond-mediated sensing of chloride and bromide anions.

recognition event. Namely, selective chloride binding is observed over the more basic oxonanions, dihydrogen phosphate, and acetate; 1:1 association constants of 1130, 300, and 100 M^{-1}, respectively, were obtained by ^1H NMR titration experiments in 1:1 $CDCl_3:CD_3OD$.[19] Importantly, this is a complete reversal in the selectivity exhibited by the free pyridinium axle **8**. The enhanced chloride recognition function is achieved by the synergistic binding behaviors of the two interlocked components and control over the ion pair interaction.

This strategy for constructing effective anion receptors has been employed extensively by Beer since 2002,[20,21] and more recently by a small number of other groups.[22,23] In addition, the incorporation of an optical or electrochemical reporter group within the interlocked structure enables these systems to be further extended into the area of anion sensing.[22–24,26] A recent example of this from 2012 is Beer's bromide-templated catenane **11** (Scheme 10.6) in which each macrocyclic component contains a bromo-imidazolium group for anion binding and a naphthalene reporter group for fluorescent sensing.[24] Anion binding occurs *via* halogen bonding—the highly directional interaction between covalently bound electron-deficient halogen atoms and Lewis basic anions, which is a new and promising addition to the supramolecular chemist's tool kit of solution-phase non-covalent interactions.[27,28]

In MeCN, host structure **11** exhibited significant emission spectral changes upon chloride and bromide complexation, while a large range of other relevant anions did not induce such responses. This impressive discrimination demonstrates the potential of ion pair templated interlocked hosts for application in molecular sensing devices.

Ion pair templation of interlocked molecules using discrete ions can also produce systems capable of controlled dynamic intercomponent motion. One such example is another ion pair templated rotaxane from Leigh and co-workers[29] which exhibited reversible shuttling behavior induced by a removable chloride template. The templating ion pair is created by displacing the neutral MeCN ligand of palladium(II)-based macrocycle **13** with the chloride counter anion of pyridinium thread **12** (Scheme 10.7). This reaction generates a negative charge associated with the macrocycle component, inducing electrostatically driven threading onto the cationic thread. The resulting pseudorotaxane is subjected to a CuAAC reaction with a suitable azide stopper to capture the target interlocked rotaxane **14** in 64% yield.

While this system could be classified alternatively as integrated templation because the negative charge exists within the macrocycle component, the ion pair interaction is entirely dependent on the chloride ligand. The shuttling behavior of **14** arises from the ability to strip this anion from the palladium(II) metal by silver

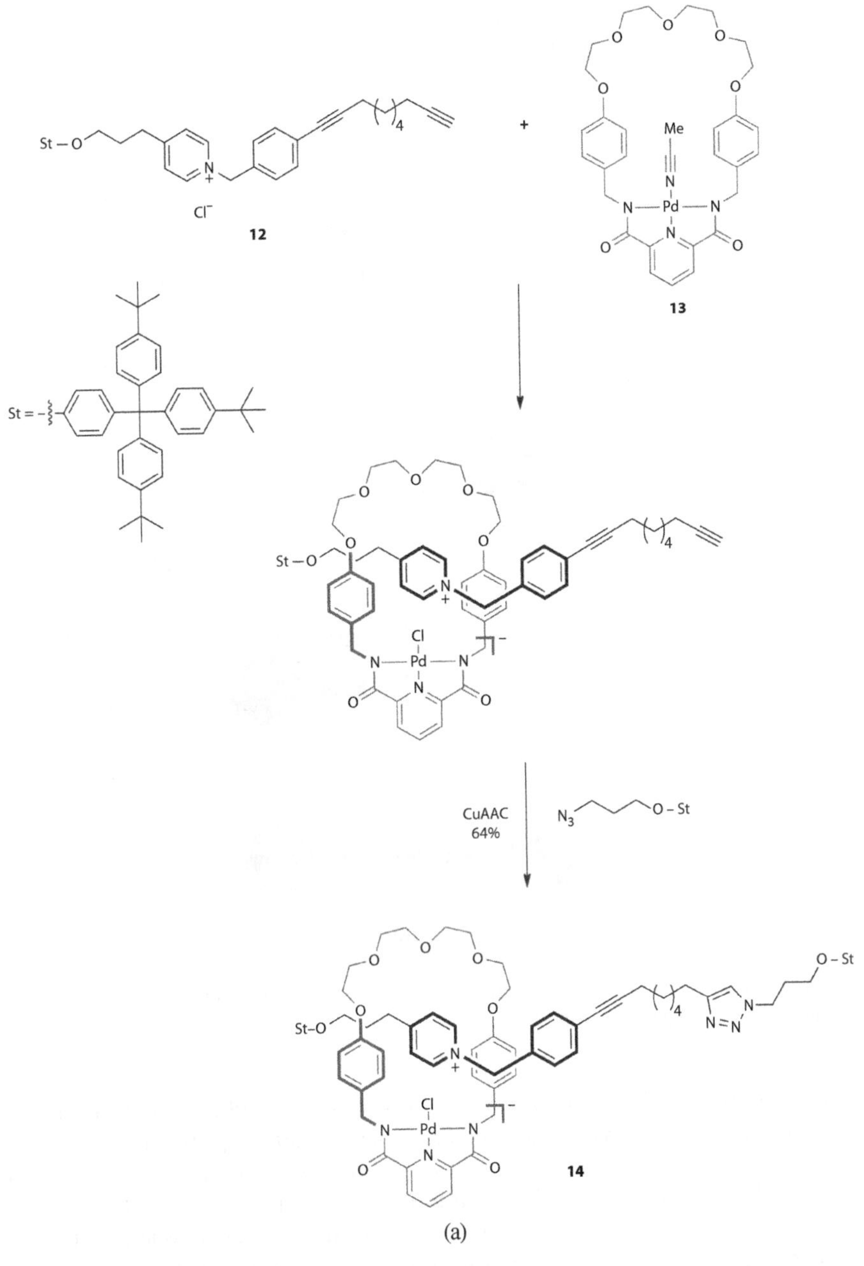

SCHEME 10.7 Threading-followed-by-stoppering synthesis of Leigh's palladium(II)-based ion pair rotaxane **14**. *(continued)*

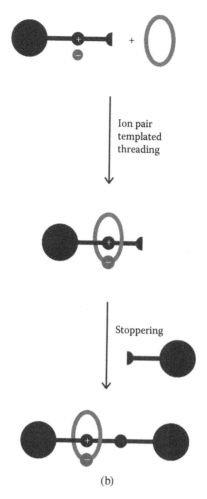

(b)

SCHEME 10.7 *(continued)* Threading-followed-by-stoppering synthesis of Leigh's palladium(II)-based ion pair rotaxane **14**.

salt precipitation (Scheme 10.8). This removes the negative charge of the macrocycle component and, as a consequence, the favorable ion pair templating interaction.

The axle, however, contains an alternative "station" for the macrocycle. Specifically, the triazole group formed in the stoppering reaction is able to chelate to the newly vacant coordination site on the metal upon macrocycle translocation. Despite this stabilization, this secondary rotaxane co-conformation is still higher in energy than its original ion paired state, and hence the molecular motion is completely reversed upon chloride addition.

In addition to systems constructed by utilizing halide counter anions as discrete templates, other anions can play similarly vital roles in controlling the formation of interlocked structures. For example, the formation of Sekiya and Kuroda's quadruply interlocked assembly **16** from palladium(II)-based coordination cage **15** is highly

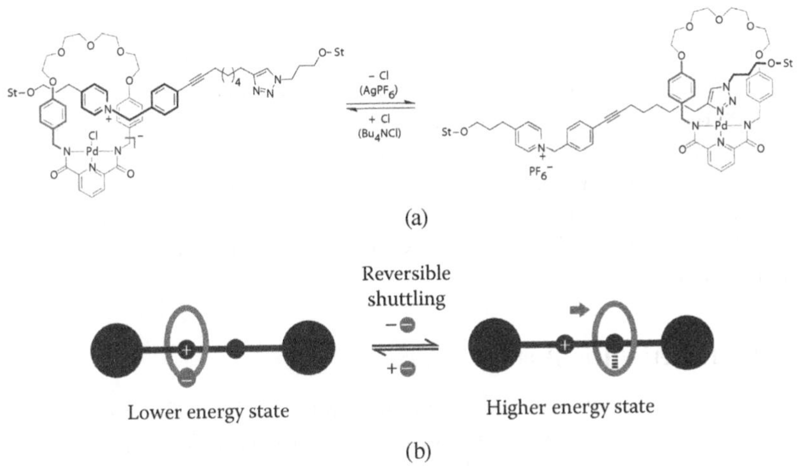

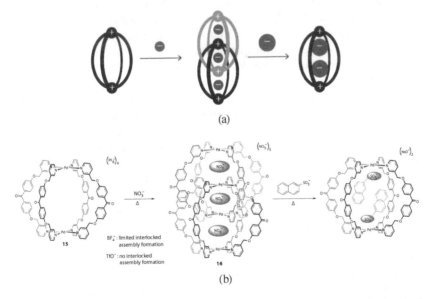

(a)

(b)

SCHEME 10.8 Reversible shuttling behavior of rotaxane **14**.

dependent on the exact nature of the anion template (Scheme 10.9).[30] In the presence of nitrate, the interlocked dimer is formed quantitatively upon heat-induced dynamic metal-ligand behavior. The new counter anions can be encapsulated completely within the three internal cavities and provide sufficient electrostatic stabilization of this tetra-metallic assembly.

A similar structure can be formed with the slightly larger tetrafluoroborate counter anions, however, in much reduced yields due to lower encapsulation affinity and

(a)

(b)

SCHEME 10.9 Anion-controlled formation and disassembly of quadruply interlocked assembly **16**.

less anion–cavity complementarity. For the even larger triflate, no interlocked product is possible. Moreover, when the nitrate anions within the interlocked assembly are exchanged for 2-naphthalenesulfonate, which exhibits higher affinity for the palladium(II) centers but cannot fit within the interlocked cavities, the dimer spontaneously disassembles into monomer **15** with two encapsulated sulfonate anions. This highly selective control of interlocked assembly is based on the powerfully simple concept of size complementarity between the cationic framework and its exchangeable counter anions, and is a feature of such coordination cages and their interpenetrated dimers.[31]

10.2.3 Traceless Ion Pair Templation

Receptors that bind both cations and anions can be employed for interlocked structure formation using a strategy best described as *traceless* ion pair templation. In this approach, such a receptor binds a reactive ion pair to produce a neutral product that retains neither of the templating ion pair partners. The product—interlocked if appropriately designed components are used—does, however, retain its ion pair binding ability.

Traceless ion pair templation of an interlocked molecule was first used by Smith in 2000 and focused on macrobicycle **17** that contains both cation and anion binding motifs in the forms of crown ether and isophthalamide groups (Figure 10.6).[32,33] As a result, this receptor binds salts, including potassium chloride, as solvent separated ion pairs. Such binding is synergistic, for example, the presence of potassium in the crown ether section was shown to enhance the binding of chloride to the isophthalamide cleft by a factor of seven (1:1 association constants of 50 and 350 M^{-1} for the binding of chloride in 3:1 d_6-DMSO:CD$_3$CN in the absence and presence of potassium, respectively).[33] The synthesis and investigation of ion pair receptors is of special interest due to a variety of important applications as salt extraction, salt solubilization, and membrane transport agents.

To exploit this ion pair receptor for rotaxane formation, Smith utilized a discrete potassium template bound within the crown ether group of **17** to aid the association of an anionic phenolate half-axle **19** in the adjacent isophthalamide cleft (Scheme 10.10). Successive esterification reactions between this complex, isophthaloyl dichloride (**18**), and a further equivalent of the half-axle **19** are necessary to form a fully stoppered axle component. The templated arrangement promotes formation of this axle within the macrocycle, yielding uncharged rotaxane **20**.

The cooperative behaviors of both the anionic and cationic templates are vital for efficient interlocked molecule formation. The anion templation component of this approach, in which the phenolate anion is held within an isophthalamide macrocycle and consumed in the reaction, was pioneered by Vögtle in 1999.[34] However, the macrobicycle design and presence of the bound alkali metal enabled rotaxane **20** to be synthesized in the polar solvent medium of 5:1 THF:DMF in a yield of 20%. Furthermore, the direct templating action of the potassium was demonstrated by the fact that no equivalent reaction occurred in its absence or with the larger cesium cation that was not able to bind efficiently in the crown ether.

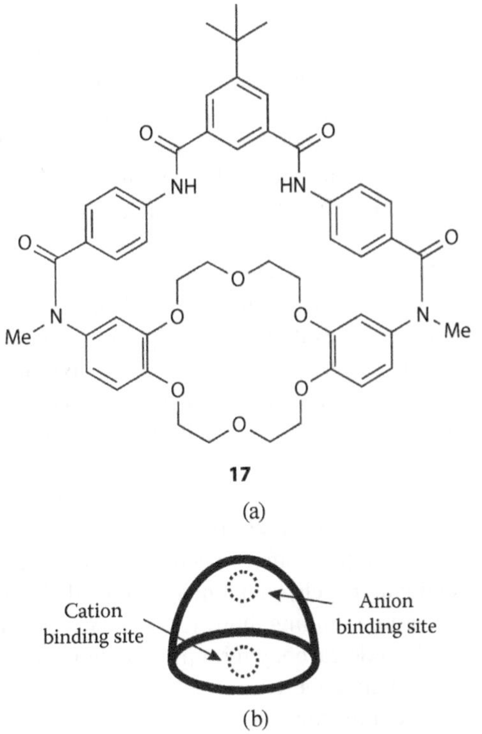

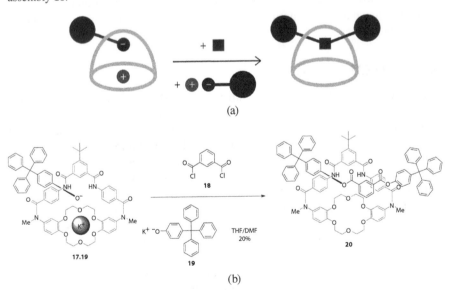

17

(a)

Cation binding site — Anion binding site

(b)

FIGURE 10.6 Anion-controlled formation and disassembly of quadruply interlocked assembly **16**.

(a)

17.19   **19**   THF/DMF 20%   **20**

18

(b)

SCHEME 10.10 Synthesis of salt binding rotaxane **20** using macrocycle **17** and traceless ion pair templation.

Unsurprisingly, the presence of the bulky interlocked axle within rotaxane **20** was found to have a substantial effect on the salt binding properties of the component macrocycle. First, the rotaxane displayed considerably stronger chloride binding compared to macrocycle **17** (300 and 50 M^{-1}, respectively, in the same d_6-DMSO:CD_3CN media).[33] This difference was attributed to DMSO complexation inside free macrocycle **17** that inhibited chloride binding. This DMSO competition cannot occur to the same extent in rotaxane **20** due to the presence of the interpenetrating axle.

Second, the same steric shielding effect in **20** is also responsible for a lack of observed cooperativity in ion pair binding. No enhancement in chloride affinity occurred upon potassium complexation to the adjacent crown ether ($K_a = 300$ M^{-1} with or without potassium cations present). Hence, the axle induced an insulating effect on the through-space electrostatic interaction between the bound potassium and chloride ions, resulting in deviation away from the synergistic behavior of macrocycle **17**.

Additional studies showed that ion pair binding has its own influence on the intercomponent dynamics of rotaxane **20**. In polar solvents such as d_6-acetone and d_6-DMSO, the broad ^{1}H NMR spectrum revealed dynamic co-conformational behavior between the axle and macrocycle components, most likely pirouetting and shuttling motions. Upon potassium binding, however, only one co-conformation existed in solution, with one of the axle carbonyl groups thought to coordinate to the bound alkali metal, halting the dynamic motion.

In contrast, chloride complexation to the macrocycle isophthalamide cleft had no apparent effect on the system's dynamics, presumably due to the absence of any potential axle–anion interactions to lock the rotaxane co-conformation. Overall, the work shows that effective traceless ion pair templation can produce uncharged mechanically interlocked molecules with ion-dependent switchable dynamic properties.

10.3 CONCLUSIONS

Since research into the templated synthesis of interlocked structures began over three decades ago, many strategies have been employed to produce, in remarkably high yields, a multitude of diverse architectures with increasing complexity and supramolecular functionality. In this chapter, we described three successful ion pair templation strategies, specifically the use of integrated cationic and anionic components, discrete ion templates, and reactive ion pairs bound by macrocyclic receptors, all to mediate interlocked structure formation.

As the supramolecular chemistry of interlocked structures progresses and researchers begin applying these elegant systems to a broad range of technologies, the focus on synergistic functional behavior will increase. Crucially, ion pair templation looks promising for achieving this goal. Mainly through the ability to exchange discrete ion templates, the templating ion pair interaction can be removed, and as a result the overall energy state of a system is perturbed.

This behavior can be exploited for the fabrication of spring-like devices in which a return to the ion paired state is highly favorable, for example, in the synthesis of interlocked molecules that exhibit strong and selective ion recognition properties through cavity binding and those capable of controllable ion-induced molecular motion.

The continuing development of such systems exhibits great potential in the areas of sensing, molecular switches, and nanoscale machines capable of doing work. Realization of these ambitious goals will likely require new synthetic methods to construct more complex hierarchical ion pair networks containing multiple aligned mechanically interlocked assemblies for amplified functionality and macroscopic device incorporation.

ACKNOWLEDGMENTS

We thank the English Speaking Union (UK) for a Postdoctoral Lindemann Fellowship (GTS) and the National Science Foundation (USA) for financial support.

REFERENCES AND NOTES

1. Sauvage, J. P. and Dietrich-Buchecker, C., *Molecular Catenanes, Rotaxanes and Knots: A Journey through the World of Molecular Topology.* Wiley-VCH: Weinheim, 1999.
2. Wasserman, E., *J. Am. Chem. Soc.* 1960, *82*, 4433–4434.
3. The preformed cycloalkane ring used by Wasserman was partially deuterated for IR characterization using C-D stretching frequencies.
4. Busch, D. H., *J. Inclusion Phenom.* 1992, *12*, 389–395.
5. Dietrich-Buchecker, C. O., Sauvage, J. P., Kintzinger, J. P., *Tetrahedron Lett.* 1983, *24*, 5095–5098.
6. Beves, J. E., Blight, B. A., Campbell, C. J. et al. *Angew. Chem., Intl. Ed.* 2011, *50*, 9260–9327.
7. Crowley, J. D., Goldup, S. M., Lee, A. L. et al. *Chem. Soc. Rev.* 2009, *38*, 1530–1541.
8. Dichtel, W. R., Miljanić, O. Š., Spruell, J. M. et al. *J. Am. Chem. Soc.* 2006, *128*, 10388–10390.
9. Hunter, C. A., *J. Am. Chem. Soc.* 1992, *114*, 5303–5311.
10. G. Hamilton, D., K. M. Sanders, J., E. Davies, J. et al., *Chem. Commun.* 1997, 897–898.
11. Wenz, G., Han, B. H., and Müller, A., *Chem. Rev.* 2006, *106*, 782–817.
12. Thompson, M. C. and Busch, D. H., *J. Am. Chem. Soc.* 1964, *86*, 213–217.
13. Anderson, S., Anderson, H. L., and Sanders, J. K. M., *Acc. Chem. Res.* 1993, *26*, 469–475.
14. Dietrich-Buchecker, C. O., Sauvage, J. P., and Kern, J. M., *J. Am. Chem. Soc.* 1984, *106*, 3043–3045.
15. Mohr, B., Weck, M., Sauvage, J. P. et al. *Angew. Chem., Intl. Ed.* 1997, *36*, 1308–1310.
16. Hanni, K. D. and Leigh, D. A., *Chem. Soc. Rev.* 2010, *39*, 1240–1251.
17. Nakatani, Y., Furusho, Y., and Yashima, E. *Angew. Chem., Intl. Ed.* 2010, *49*, 5463–5467.
18. Lee, C. F., Leigh, D. A., Pritchard, R. G. et al. *Nature* 2009, *458*, 314–318.
19. Wisner, J. A., Beer, P. D., Drew, M. G. B. et al., *J. Am. Chem. Soc.* 2002, *124*, 12469–12476.
20. Lankshear, M. D. and Beer, P. D., *Acc. Chem. Res.* 2007, *40*, 657–668.
21. Spence, G. T. and Beer, P. D., *Acc. Chem. Res.* 2013, *46*, 571–586.
22. Chae, M. K., Suk, J. M., and Jeong, K. S., *Tetrahedron Lett.* 2010, *51*, 4240–4242.
23. Zhao, Y., Li, Y., Li, Y. et al., *Chem. Commun.* 2010, *46*, 5698–5700.
24. Caballero, A., Zapata, F., White, N. G. et al., *Angew. Chem., Intl. Ed.* 2012, *51*, 1876–1880.

25. Hancock, L. M., Marchi, E., Ceroni, P. et al., *Chem. Eur. J.* 2012, *18*, 11277–11283.
26. Evans, N. H., Rahman, H., Leontiev, A. V. et al., *Chem. Sci.* 2012, *3*, 1080–1089.
27. Cavallo, G., Metrangolo, P., Pilati, T. et al., *Chem. Soc. Rev.* 2010, *39*, 3772–3783.
28. Chudzinski, M. G., McClary, C. A., and Taylor, M. S., *J. Am. Chem. Soc.* 2011, *133*, 10559–10567.
29. Barrell, M. J., Leigh, D. A., Lusby, P. J. et al., *Angew. Chem., Intl. Ed.* 2008, *47*, 8036–8039.
30. Sekiya, R., Fukuda, M., and Kuroda, R., *J. Am. Chem. Soc.* 2012, *134*, 10987–10997.
31. Freye, S., Michel, R., Stalke, D. et al., *J. Am. Chem. Soc.* 2013, *135*, 8476–8479.
32. Shukla, R., Deetz, M. J., and Smith, B. D., *Chem. Commun.* 2000, 2397–2398.
33. Deetz, M. J., Shukla, R., and Smith, B. D., *Tetrahedron* 2002, *58*, 799–805.
34. Hübner, G. M., Gläser, J., Seel, C. et al., *Angew. Chem., Intl. Ed.* 1999, *38*, 383–386.

11 Synergy of Reactivity and Stability in Nanoscale Molecular Architectures

Kei Goto

CONTENTS

11.1 INTRODUCTION

Development of reaction environments that furnish both reactivity and stability is one of the most important topics of organic chemistry. Resolution of this difficult conundrum is attained routinely in nature. In the active sites of enzymes, reactive species that are very unstable in artificial systems are sometimes stabilized and play important roles in chemical transformations with extraordinarily high activity.

Such remarkable features of enzymatic reactions largely rely on the synergy of reactivity and stability in geometrically isolated environments. The reactive functionality in the active site of an enzyme usually is incorporated in a cavity or cleft and geometrically isolated from other reactive groups in the same and/or different subunits by topology and higher-order structures of proteins.

Because of such isolation from other functionalities that could cause decomposition of the species, the reactive species can have long lifetimes without affecting their reactivities toward substrate molecules. In host–guest chemistry, stabilization of highly reactive species by isolation from the bulk phase has been attained by utilizing a variety of molecular capsules.[1–3]

Guest molecules encapsulated in the inner phases of capsules are protected effectively from collisions and reactions with other molecules. Basically, this mode of isolation is intended to constrict any possible reaction pathways of the encapsulated guest species, which is in contrast with the reactive species in the active site of enzymes exhibiting high reactivity toward substrates.

For modeling the isolated environments of enzymes, molecular systems with open cavities containing covalently embedded functionalities should be desirable. This overview describes the design and development of nanoscale molecular cavities for regulation of the reactivities of endohedral functional groups, especially for kinetic stabilization of highly reactive species with their intrinsic reactivities unaffected.

11.2 MODELING OF GEOMETRICALLY ISOLATED ENVIRONMENTS BY BOWL-SHAPED MOLECULAR CAVITIES

Cysteine and selenocysteine residues in proteins are critical components for regulation of a wide range of biological functions. Isolated cysteine and selenocysteine residues located in cavities or clefts within proteins are restrained sterically from forming interchain and/or intrachain disulfide or diselenide bridges.

The highly reactive intermediates generated by their reactions with reactive oxygen and nitrogen species play crucial roles in redox regulation and signaling.[4–8] Sulfenic acids (RSOH) and selenenic acids (RSeOH) are among the most important species of this class. Sulfenic acids are the reversibly oxidized forms of cysteine thiols (Scheme 11.1) that play critical roles as catalytic centers in enzymes and as sensors of oxidative and nitrosative stress in enzymes and transcriptional regulators.[9,10]

Selenenic acids have been well recognized as intermediates in the catalytic cycle of glutathione peroxidase (GPx) and are generated by oxidation of selenocysteine residues with peroxides (Scheme 11.1).[5,6,11,12] However, much of our knowledge of the chemistry of these species has been obtained in an indirect and speculative manner because the species are notoriously unstable in artificial systems due to very rapid self-condensation leading to corresponding thiosulfinates or selenoseleninates (Scheme 11.2).

In particular, selenenic acids are so unstable that no example of isolation of a selenenic acid in pure form appeared until our reports were published. For elucidation of the roles of these acids in biological functions, stable reference compounds with which basic information on their properties and reactivities can be obtained experimentally are required.

Steric protection using bulky substituents is one of the most effective methods for stabilization of reactive species such as dimerization and self-condensation that undergo facile bimolecular decomposition. Typical examples of highly reactive

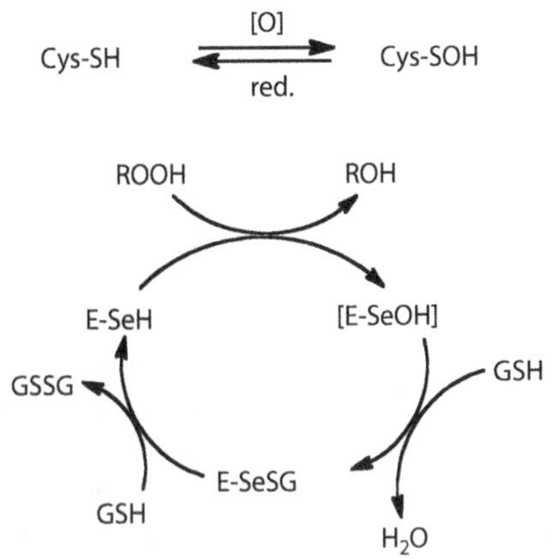

SCHEME 11.1   Redox reactions involving a sulfenic acid and a selenenic acid.

$$2R\text{-}YOH \longrightarrow R\text{-}Y(O)\text{-}Y\text{-}R$$
$$(Y = S, Se)$$

SCHEME 11.2   Self-condensation of sulfenic acids and selenenic acids.

species stabilized by steric protection include disilene **1**,[13] diphosphene **2**,[14] and germanethione **3**[15] shown in Figure 11.1. It was reported, however, that the bimolecular decomposition of a selenenic acid cannot be prevented even with a 2,4,6-tri-*tert*-butyl group (denoted as Mes* hereafter), one of the bulkiest steric protection groups; selenenic acid **4** bearing a Mes* group disproportionated to the corresponding diselenide and seleninic acid via a self-condensation reaction (Scheme 11.3).[16]

Steric protection of highly reactive sulfur or selenium species such as sulfenic acids and seleninic acids is generally difficult, mainly because only one bulky substituent can be introduced to such chalcogen-containing species unlike compounds carrying multiple bulky groups shown in Figure 11.1.

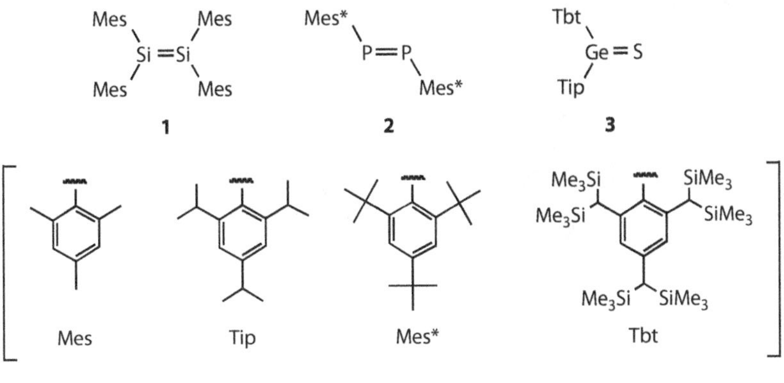

FIGURE 11.1   Examples of reactive species stabilized by steric protection.

$$3\ Mes^*SeOH \xrightarrow[\text{4\% } D_2O/CD_3CN]{25°C,\ 2h} Mes^*SeSeMes^* + Mes^*SeO_2H$$

4

SCHEME 11.3　Decomposition of selenenic acid 4.

It should be possible to enhance steric protection effects by further increasing the steric bulkiness of the *ortho*-subsituents of the Mes* group (Figure 11.2). However, it would severely inhibit the reactivity of the protected species toward other molecules in exchange for the improved stability. Furthermore, severe steric shielding of the reactive functionality often causes undesired intramolecular side reactions with the protecting group in the vicinity, as shown in the intramolecular cyclization of ger-manethione **5** (Scheme 11.4).[17]

For modeling the essential features of the geometrically isolated environ-ments in the active sites of enzymes, we designed the molecular bowls shown in Figure 11.3.[18,19] The functional groups X and Y are isolated from each other by the steric repulsion between the peripheral moieties of the bowls. Their reactivity with other molecules is not inhibited so much because of the relatively large space around the central functionality.

Such a mode of steric protection that furnishes both stability and reactivity can be referred to as *peripheral steric protection* in contrast with the conventional *proxi-mal steric protection* shown in Figure 11.2. For this purpose, we sought to develop nanoscale bowl-shaped molecular cavities large enough to serve as a reaction envi-ronment for the functionality of and secure enough to suppress the bimolecular decomposition processes of the endohedral reactive species.

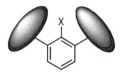

FIGURE 11.2　Proximal steric protection.

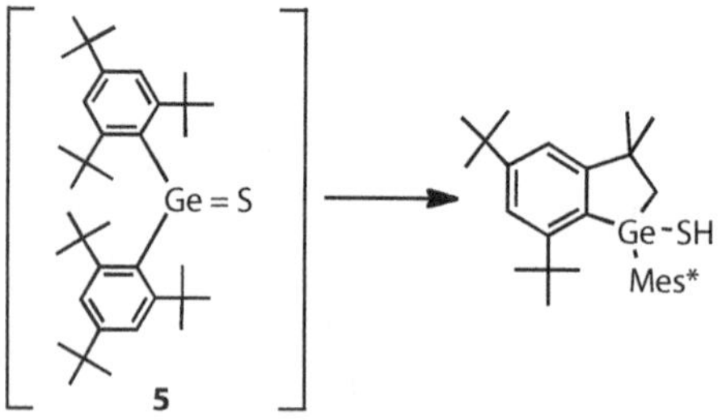

SCHEME 11.4　Intramolecular reaction of germanethione 5.

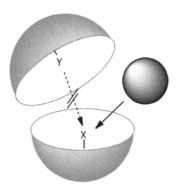

FIGURE 11.3   Peripheral steric protection by bowl-shaped molecules.

11.3   BOWL-SHAPED MOLECULAR CAVITIES BASED ON BIMACROCYCLIC FRAMEWORKS

11.3.1   BIMACROCYCLIC CYCLOPHANES

Hart et al. reported the syntheses of a series of cavity-shaped molecules based on *m*-terphenyl frameworks, including cuppedophanes.[20,21] Lüning et al. proposed the concept of concave reagents bearing endohedral functional groups in the concave positions of bicyclic cyclophanes.[22,23] These reagents are designed so that the concave shielding of the central functionality would increase the selectivity of its reactions.

We synthesized the bowl-shaped bicyclic cyclophane, Ar¹X (Figure 11.4) and applied it to the stabilization of a sulfenic acid.[24] X-ray analysis of bromide **6** indicated that this cyclophane has a shallow bowl-shaped structure with a diameter of ~1.4 nm (Figure 11.5). The arenesulfenic acid **8** was synthesized by solid-state pyrolysis of butyl sulfoxide **7** and isolated as stable crystals (Scheme 11.5). The effect of peripheral steric protection due to the bowl-shaped framework was illustrated by the

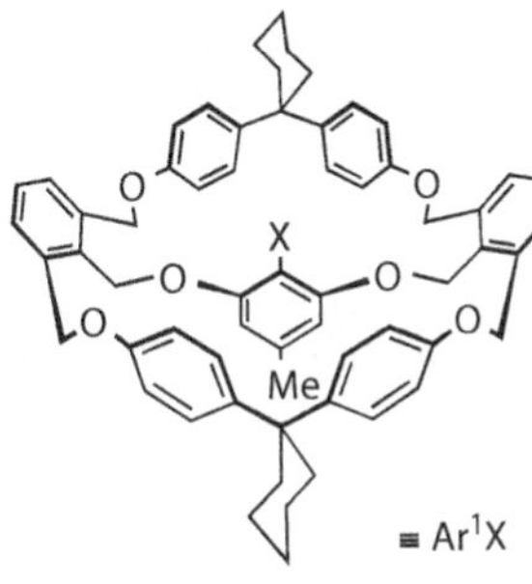

$\equiv$ Ar¹X

6: X = Br

FIGURE 11.4   Bowl-shaped cyclophane.

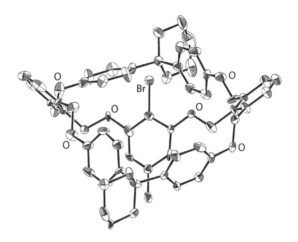

FIGURE 11.5 ORTEP drawing of **6** (30% probability).

high thermal stability of sulfenic acid **8**; only slight decomposition was observed even after heating at 90°C for 12 hr in toluene-d_8.

While the bimolecular decomposition of **8** is effectively prevented by the framework, **8** readily reacted with small molecules such as methyl propiolate and 1-butanethiol to yield sulfoxide **9** and disulfide **10**, respectively. These results validate the concept of molecular bowls depicted in Figure 11.3.

11.3.2 BRIDGED CALIX[6]ARENES

Calixarenes are cyclic compounds consisting of aromatic rings connected by methylene groups in meta positions. They have been widely utilized as versatile platforms in molecular recognition and supramolecular chemistry.[25–27] Among the family of calixarenes, calix[6]arenes have been recognized as important host molecules that possess cavities large enough to accommodate organic molecules.

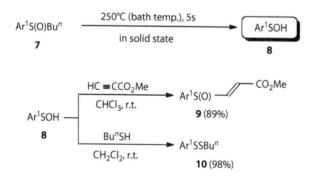

SCHEME 11.5 Synthesis and reactions of sulfenic acid 8.

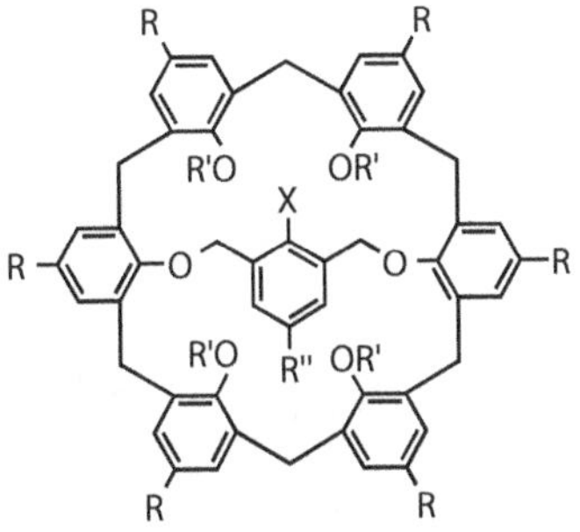

11

FIGURE 11.6 Bowl-shaped bridged calix[6]arene.

One of the major drawbacks of calix[6]arenes in view of their application as molecular cavities is the large conformational flexibility of their frameworks. Much attention has been paid to designing functionalized molecular systems based on rigid and well-defined calix[6]arene frameworks. To develop a bowl-shaped molecular cavity utilizing a calix[6]arene macrocycle, we designed a bridged calix[6]arene **11** bearing an endohedral functionality on the bridging *m*-xylenyl unit (Figure 11.6).[28–32]

Lüning et al. also independently reported functionalized bridged calix[6] arenes bearing a similar framework.[23,33] The bridging of the calix[6]arene macrocycle with a *m*-xylenyl unit constructs a bowl-shaped cavity with an endohedral functional group and considerably reduces the conformational flexibility of the framework. Bridged calix[6]arenes of type **11** with various functional groups were prepared by building bridges with appropriate *m*-xylenyl units over the distal positions of a parent calix[6]arene and subsequent alkyl capping of the remaining hydroxy groups.

The usefulness of the bridged calix[6]arene framework for the stabilization of reactive species was demonstrated by the synthesis of a stable sulfenic acid.[30] Thermolysis of *tert*-butyl sulfoxide **12** in toluene produced the corresponding sulfenic acid **13** almost quantitatively; the acid was isolated as stable crystals (Scheme 11.6).

The crystallographic analysis of **13** indicated that the calixarene macrocycle of **13** adopts the 1,2,3-alternate conformation and surrounds the benzenesulfenic acid

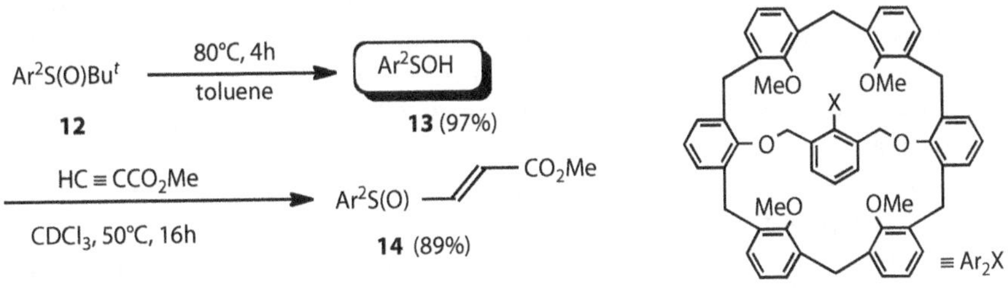

SCHEME 11.6 Synthesis and reaction of sulfenic acid 13.

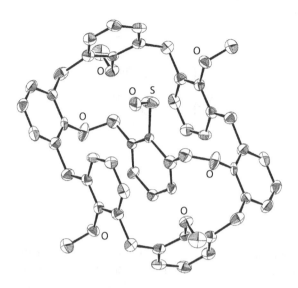

FIGURE 11.7   ORTEP drawing of **13** (30% probability).

moiety (Figure 11.7). Thus, the SOH functionality resides in an environment preventing it from self-condensation. The reactivity of **13** toward other reagents was found to be lower than that of **8** bearing a bimacrocyclic cyclophane framework[24] as indicated by its much slower reaction with methyl propiolate to produce **14**. The central functionality of the bridged calix[6]arene **13** is considered to be more strongly shielded by the surrounding macrocycle.

The conformational mobility of the bridged calix[6]arenes of type **11** is largely dependent on the capping groups of the lower rim. In solution, compound **13** bearing four methyl caps in the lower rim was found to undergo slow conformational interconversion on the NMR timescale at room temperature, although it adopted the 1,2,3-alternate conformation in the crystalline state.[30]

At elevated temperatures, the ArOMe units of **13** undergo rapid flipping motion with the OMe groups passing through the annulus. In contrast, it was found that several conformationally frozen isomers can be obtained when arylmethyl groups were introduced as the capping units in the lower rim of **11** (Figure 11.8).[29,34] In most cases, the arylmethylation of the four hydroxy groups in the lower rim afforded the cone isomer (a) and the 1,2,3-alternate isomer (b) as the major isomers.

For some compounds, the partial cone isomer (c) is obtained as a minor product. When the central functionality X is a large group such as a butylseleno or phenylethynyl group, the formation of the inverted cone isomer (a′) is preferred kinetically to the normal cone isomer (a). These conformational isomers were separable by silica gel chromatography. The results indicate that the bridged calix[6]arene of type **11** presents a good solution for the problem associated with the conformational flexibility of calix[6]arenes.

The first isolable selenenic acid was also synthesized by taking advantage of the bridged calix[6]arene framework bearing four benzyl caps in the lower rim.[31] Selenenic acid **16a** fixed in the cone conformation was synthesized by oxidation of

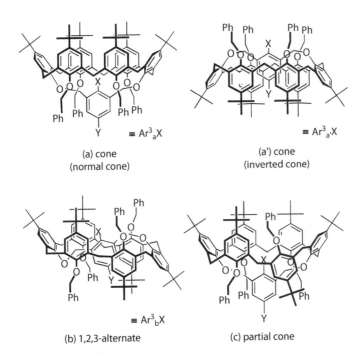

(a) cone
(normal cone)

(a') cone
(inverted cone)

(b) 1,2,3-alternate

(c) partial cone

FIGURE 11.8 Conformational isomers of bridged calix[6]arene.

butyl selenide **15a′** bearing the inverted cone conformation with mCPBA and subsequent thermolysis of the resulting selenoxide (Scheme 11.7). It is probable that the conformational change took place from the inverted cone (a′) to the normal cone (a) with the flipping of the central bridging unit when the butyl selenoxide derived from **15a′** was converted to **16a** by elimination of 1-butene.

Selenenic acid **16a** was isolated in the form of colorless crystals, and its structure was established by crystallographic analysis (Figure 11.9). This is the first example of x-ray analysis of a selenenic acid. It was revealed that the SeOH functionality is deeply embedded in the cavity of the p-*tert*-butylcalix[6]arene macrocycle with a cone conformation and the self-condensation of the selenenic acid moiety seems to be extremely difficult. The thermal stability of **16a** was remarkable both in the crystalline state and in solution. Heating **16a** in $CDCl_2CDCl_2$ at 120°C for 5 hr caused no decomposition. Such stability is extraordinary, considering the report that even selenenic acid **4** bearing a Mes* group undergoes decomposition completely within 2 ht in 4% D_2O/CD_3CN at 25°C (Scheme 11.3).[16]

Selenenic acid **16b** fixed in the 1,2,3-alternate conformation was also synthesized by a similar procedure (Scheme 11.7).[32] Intriguingly, it was found that the difference in the conformation of the calix[6]arene macrocycle affects the properties and reactivity of the SeOH functionality. While the 1,2,3-alternate isomer **16b** showed comparable stability to the cone isomer **16a**, they showed different spectral properties (Table 11.1).

For example, the OH absorption bands of **16a** and **16b** in the IR spectra suggest intramolecular interactions in **16b** while no such significant interactions exist in **16a**.

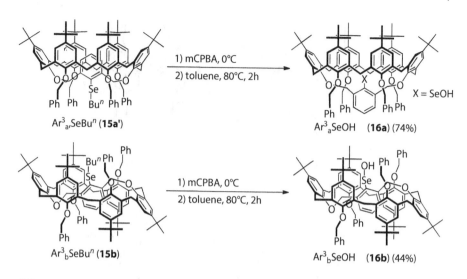

SCHEME 11.7   Syntheses of selenenic acids 16a and 16b.

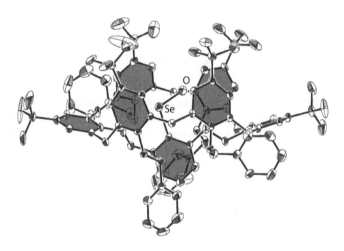

FIGURE 11.9   ORTEP drawing of **16a** (20% probability).

TABLE 11.1

Spectral Data of Selenenic Acids 16a and 16b

	$\delta_{Se}{}^a$	$\delta_{OH}{}^a$	$\nu_{OH}/cm^{-1}{}^b$
16a	1133.8	−0.05	3523
16b	1097.5	4.34	3392

[a] In $CDCl_3$.
[b] In CH_2Cl_2.

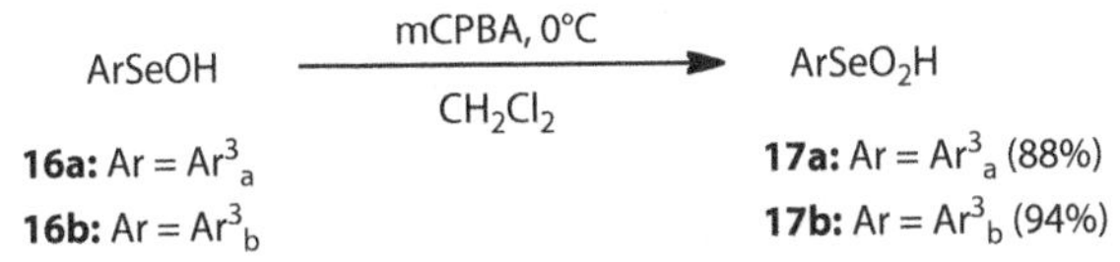

$$\text{ArSeOH} \xrightarrow[\text{CH}_2\text{Cl}_2]{\text{mCPBA, 0°C}} \text{ArSeO}_2\text{H}$$

16a: Ar = Ar^3_a **17a:** Ar = Ar^3_a (88%)

16b: Ar = Ar^3_b **17b:** Ar = Ar^3_b (94%)

SCHEME 11.8 Oxidation of selenenic acids 16a and 16b.

$$Ar^3_a\text{SeOH} \xrightarrow[\text{CDCl}_3,\ 25°C,\ 4h]{\text{10 eq Bu}^n\text{SH}} Ar^3_a\text{SeSBu}^n$$

16a   **18a** (78%)

$$Ar^3_b\text{SeOH} \xdashrightarrow[\text{CDCl}_3,\ 50°C,\ 12h]{\text{10 eq Bu}^n\text{SH}}\!\!\!/\!\!\!\to Ar^3_b\text{SeSBu}^n$$

16b

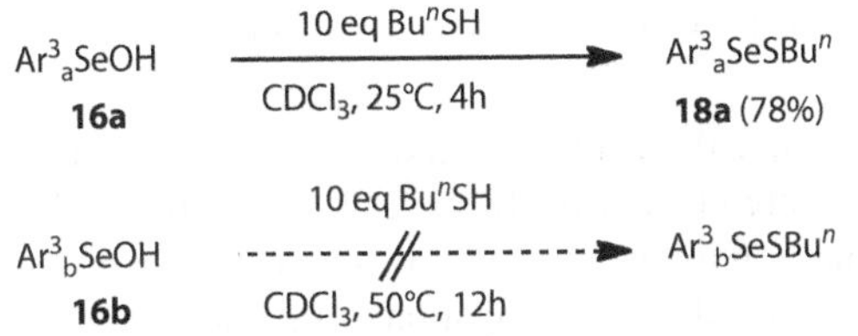

SCHEME 11.9 Reactions of selenenic acids 16a and 16b with a thiol.

In the ^{1}H NMR spectrum, the signal of the OH proton of **16a** was observed at δ −0.05, indicating that the OH group is highly shielded by the calix[6]arene macrocycle in the cone conformation. The OH signal of the 1,2,3-alternate isomer **16b** appeared at δ 4.34, which is consistent with the existence of the intramolecular hydrogen bonding in this conformational isomer.

The influence of the conformation of the calix[6]arene macrocycle was also found in the reactivities of both isomers.[32] While both **16a** and **16b** reacted with mCPBA to produce the corresponding seleninic acids **17a** and **17b**, respectively (Scheme 11.8), they showed different reactivities toward a thiol. Although the cone isomer **16a** reacted with an excess of 1-butanethiol at room temperature to give selenenyl sulfide **18a**, the 1,2,3-alternate isomer **16b** underwent no reaction even at 50°C (Scheme 11.9).

The low reactivity of **16b** toward the nucleophile can be explained by assuming that the intramolecular hydrogen bonding of the OH group reduces the electrophilicity of the selenium center of the SeOH functionality, although the steric congestion around the SeOH functionality may also be responsible. These results clearly demonstrate that the calix[6]arene macrocycle of each conformation provides a considerably different environment for the endohedral functionality, regulating its properties and reactivity.

11.4 BOWL-SHAPED MOLECULAR CAVITY BASED ON ALL-CARBON AND ACYCLIC FRAMEWORK

11.4.1 Molecular Design

The bimacrocyclic frameworks described above are versatile molecular platforms because various functionalizations and derivatizations can be performed to endow them with particular properties. For example, water-soluble derivatives of the

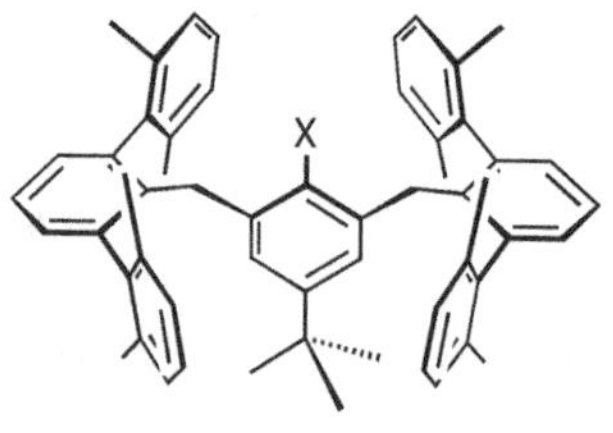

≡ BmtX

FIGURE 11.10 Bowl-shaped molecule with all-carbon and acyclic framework.

cyclophane Ar¹X[35] and the bridged calix[6]arene **11**[36] were synthesized and their unique properties were elucidated. In the case of the bridged calix[6]arene **11**, several conformationally fixed isomers are available to expand the variety of the scaffold. However, in terms of the investigation of the intrinsic properties of the reactive species, the presence of the oxygen atoms in their frameworks is unfavorable because they may cause electronic perturbation of the species.

The development of molecular cavities based on a more inert framework that does not contain heteroatoms is desired for the kinetic stabilization of reactive species without perturbation of their intrinsic properties. For prevention of bimolecular decomposition of reactive species such as dimerization and self-condensation, it is considered essential to surround the functional group from all sides and from a distance in any conformation of the molecule. A simpler acyclic molecule with larger conformational flexibility could work as a molecular cavity for the stabilization of reactive species if it meets these criteria.

As such a molecular cavity, a bowl-shaped molecule with an all-carbon and acyclic framework shown in Figure 11.10 (denoted as BmtX hereafter) was designed.[37] BmtX carries two rigid *m*-terphenyl units in the positions that correspond to the bridgeheads of bicyclic molecules so that the central functionality X is effectively surrounded in any conformation of the molecule. The steric shielding of the central functionality in BmtX is considered much less severe than that of the bridged calix[6]arene **11**. These features are expected to be suitable for investigation of the intrinsic reactivity of the endohedral reactive species toward other substrates without causing bimolecular decomposition of the species.

11.4.2 STABILIZATION OF SULFENIC ACID AND SELENENIC ACID

By utilizing a molecular cavity, the chemical transformations associated with a sulfenic acid were extensively investigated. Sulfenic acids are generally assumed to be transient intermediates in the oxidation of thiols, both to disulfides and to sulfinic acids (Scheme 11.10).[38] From a biological view, the redox processes between cysteine thiols and cysteine sulfenic acids play key roles in the redox regulation systems.[9,10]

However, the chemical evidence for these processes has been entirely circumstantial because of the instability of sulfenic acids. In particular, oxidation of a thiol to a sulfenic acid, the most fundamental process for the formation of a sulfenic acid, has

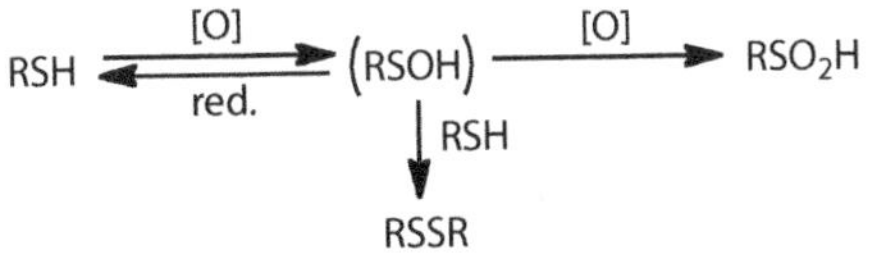

$$RSH \xrightleftharpoons[\text{red.}]{[O]} (RSOH) \xrightarrow{[O]} RSO_2H$$

$$\downarrow RSH$$

$$RSSR$$

SCHEME 11.10   Reaction processes involving sulfenic acids.

never been demonstrated experimentally. No observation of an intermediary sulfenic acid in the direct oxidation of a thiol has been reported.

When thiol **19** bearing a Bmt group was treated with iodosobenzene, a mild oxidant that usually converts thiols to disulfides, sulfenic acid **20** was formed and isolated as stable crystals (Scheme 11.11).[39] The peripheral steric effect of the Bmt group is considered to prevent the self-condensation of **20** and also the formation of the symmetrical disulfide by the reaction of the initially formed sulfenic acid **20** with thiol **19**.

On the other hand, sulfenic acid **20** readily reacted with other thiols such as 1-butanethiol and thiophenol to produce the corresponding disulfides **21a** and **21b**, respectively. Sulfenic acid **20** was oxidized to sulfinic acid **22** by oxaziridine **23**, and reduced to the parent thiol **19** by triphenylphosphine. The reduction of **20** to **19** was also effected by utilizing dithiothreitol (DTT) as a reducing agent via the intermediary disulfide **21c**. Thus, by taking advantage of a Bmt group, a stable sulfenic acid was synthesized by direct oxidation of a thiol for the first time and all the processes shown in Scheme 11.10 have been demonstrated conclusively. Furthermore, treatment of **20** with benzylamine resulted in the dehydrative condensation to sulfenamide **24**, indicating that a sulfenic acid exhibits electrophilic reactivity even in the presence of an amine.

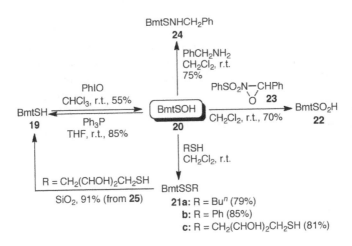

SCHEME 11.11   Synthesis and reactions of sulfenic acid 20.

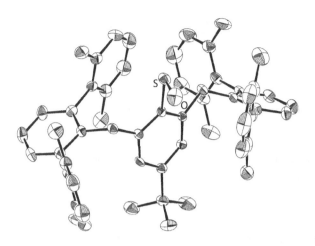

FIGURE 11.11 ORTEP drawing of **20** (30% probability).

X-ray crystallographic analysis revealed the structure of **20**, in which two con-cave *m*-terphenyl units surround the central SOH functionality like the brim of a bowl (Figure 11.11).[39] Two possible forms of structures of a sulfenic acid have been proposed: a sulfenyl form (Figure 11.12a) and a sulfoxide form (Figure 11.12b). The S–O bond length (1.67965 Å) of **20** is distinctly longer than those of sulfoxides (1.44 to 1.59 Å), suggesting that the sulfenyl form rather than the sulfoxide form is more probable for an arenesulfenic acid.

The synthesis of a stable selenenic acid was also achieved by utilizing a Bmt group and its reactivity was investigated in detail.[40] In the catalytic cycle of GPx and many of the synthetic mimics of GPx, it is thought that a selenenic acid intermediate is formed by oxidation of a selenol with peroxides.[5,41] However, no experimental evi-dence has been obtained for this important chemical transformation; even trapping of an intermediary selenenic acid has not been reported.

The reaction of selenol **25** bearing a Bmt group with an equimolar amount of hydro-gen peroxide afforded selenenic acid **26**, demonstrating this elementary chemical pro-cess for the first time (Scheme 11.12). The structure of **26** was established by X-ray crystallographic analysis. The SeOH functionality is incorporated in the cavity formed by *m*-terphenyl units, which is similar to the crystal structure of sulfenic acid **20**.

It is well accepted that the selenic acid form in the catalytic cycle of GPx is reduced to the selenol form by glutathione via the selenenyl sulfide form. Selenenic acid **26** readily reacted with various thiols to produce the corresponding selenenyl

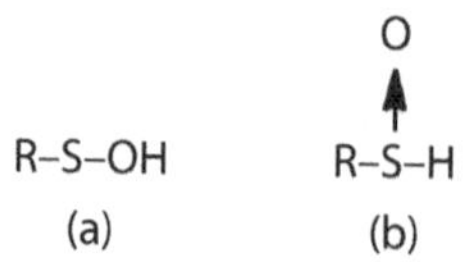

$$R\text{–}S\text{–}OH \qquad\qquad R\text{–}\overset{\displaystyle O}{\overset{\displaystyle \uparrow}{S}}\text{–}H$$

(a)   (b)

FIGURE 11.12 Two possible structures of sulfenic acid.

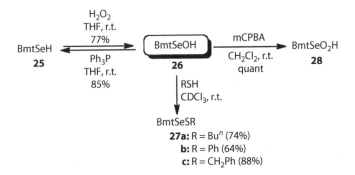

SCHEME 11.12   Synthesis and reactions of selenenic acid 26.

sulfides **27a** through **c**. However, treatment of these selenenyl sulfides with an excess of thiols in the presence of triethylamine did not convert them to selenol **25**; only the thiol exchange reaction at the selenium atom proceeded.

Reduction of selenenic acid **26** to selenol **25** via a selenenyl sulfide form was achieved by utilizing 1,4-dithiols (Scheme 11.13). The reaction of selenenic acid **26** with 1,4-butanedithiol or DTT afforded selenenyl sulfides **27d** and **27e**, respectively, that were converted to selenol **25** by treatment with triethylamine to release the corresponding cyclic disulfides. These results present the first experimental demonstration of all the three chemical transformations included in the catalytic cycle of GPx. Selenenic acid **26** was also reduced to selenol **25** by treatment with triphenylphosphine (Scheme 11.12). Oxidation of **26** with mCPBA afforded seleninic acid **28**.

Sulfenic acids and selenenic acids can be regarded as sulfur and selenium analogues of hydroperoxides (ROOH). Their properties as oxidizing agents are intriguing from the views of both basic chemistry and biochemistry. The mechanism of the oxidation of trivalent phosphorus compounds with hydroperoxides has been

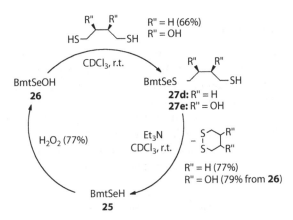

SCHEME 11.13   Modeling of chemical processes in the catalytic cycle of GPx.

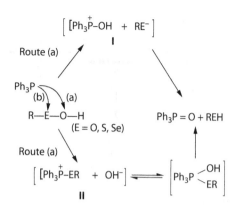

SCHEME 11.14 Possible mechanisms for reduction of sulfenic acids and selenenic acids by a phosphine.

elucidated in detail.[42] In contrast, no such mechanistic study has been performed on sulfenic acids and selenenic acids.

There are two possible mechanisms for oxidation of trivalent phosphorus compounds by these oxidizing agents (Routes (a) and (b) in Scheme 11.14). Route (a) involves the initial attack of the phosphorus on the hydroxylic oxygen atom to form the intermediate **I** while Route (b) involves the phosphonium hydroxide intermediate **II**. It was reported that reaction of hydroperoxides proceeds via Route (a) based on a tracer study using $H_2^{18}O$.[42] The mechanism for oxidation of triphenylphosphine with sulfenic acid **20** and selenenic acid **26** was investigated by similar methods. The reactions of **20** and **26** with triphenylphosphine were demonstrated to proceed via Route (b) in both cases.[43] These results indicate that the relatively bulky phosphine molecule attacks the sulfur and selenium atoms of **20** and **26** despite lower accessibility to these atoms than that of the hydroxylic oxygen atom, confirming that the sulfur and selenium atoms of sulfenic and selenenic acids possess highly electrophilic characters.

11.4.3 Stabilization of *S*-Nitrosothiol

In biological systems, *S*-nitrosothiols (RSNO) play key roles as reagents for the storage and transport of nitric oxide (NO).[8,44,45] However, they are usually unstable due to facile bimolecular decomposition to the corresponding disulfide and nitric oxide (Scheme 11.15).[46] Several examples of aliphathic *S*-nitrosothiols that have been isolated and structurally characterized to date.[47–50] In contrast, *S*-nitrosothiols bearing aromatic substituents are much less stable than aliphatic derivatives, and we know of no example of the isolation of an aromatic *S*-nitrosothiol. It was reported that the half-lives of ArSNOs (Ar = phenyl, *p*-methoxyphenyl, *p*-nitrophenyl,

$$2\ RSNO \longrightarrow RSSR + 2NO$$

SCHEME 11.15 Bimolecular decomposition of S-nitrosothiols.

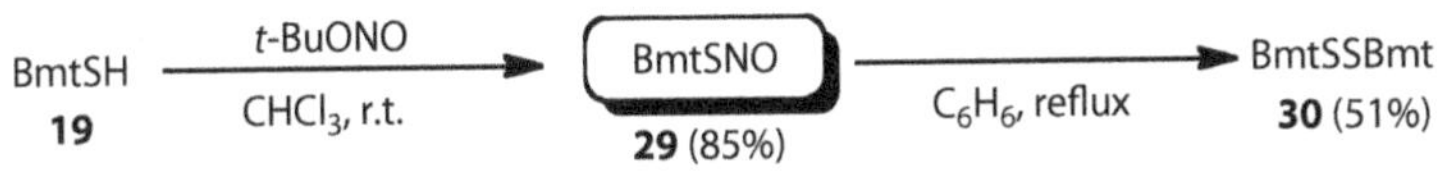

SCHEME 11.16 Synthesis and reaction of S-nitrosothiol **29**.

3,5-di-*tert*-butyl-4-hydroxyphenyl) were 7 to 14 min in dichloromethane at room temperature.[51]

Okazaki et al. reported the synthesis of a stable aromatic *S*-nitrosothiol by taking advantage of a Bmt group.[52] Treatment of thiol **19** with *tert*-butyl nitrite afforded *S*-nitrosothiol **29** that was isolated as deep green crystals (Scheme 11.16). X-ray crystallographic analysis revealed that the C–S–N–O linkage adopts the syn conformation in contrast with *S*-nitrosothiols bearing tertiary alkyl groups that have the anti-conformation.[47,48] *S*-Nitrosothiol **29** showed very high thermal stability; no reaction took place in benzene-d_6 at 50°C for 12 hr, whereas the decomposition was completed in refluxing benzene for 75 hr to produce the corresponding disulfide **30**.

11.5 DENDRIMER-TYPE MOLECULAR CAVITY BASED ON EXTENDED *M*-TERPHENYL FRAMEWORK

11.5.1 MOLECULAR DESIGN

One of the most difficult problems in the stabilization of reactive species is their decomposition via intramolecular processes. When intermolecular decomposition processes of reactive species such as dimerization and self-condensation are prevented by steric protection, they tend to undergo intramolecular side reactions with a substituent nearby.

It is well recognized that reactive species bearing Mes* groups are liable to cause side reactions involving the *ortho-tert*-butyl group as shown in Scheme 11.4.[17] The Bmt group shows relatively low propensity to undergo such intramolecular side reactions, considering its remarkable steric effects to prevent intermolecular decomposition processes. This is mainly because it protects the central functional group from peripheral positions. In some special cases, however, the Bmt group suffers from the involvement of the *ortho*-methylene group in reaction with the central functionality.

For example, *N*-thiosulfinylaniline **31** undergoes intramolecular cyclization to **32** upon heating for a prolonged time (Scheme 11.17).[53] A more inert molecular cavity

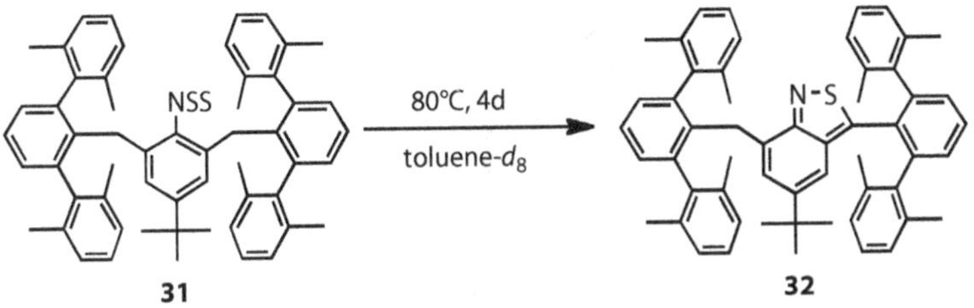

SCHEME 11.17 Intramolecular reaction of N-thiosulfinylaniline **31**.

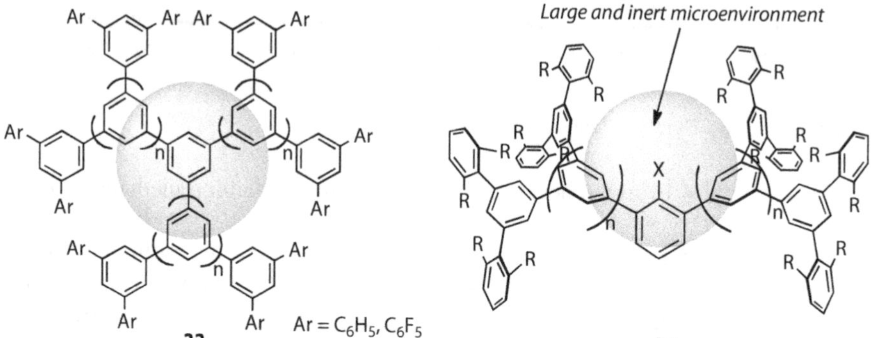

FIGURE 11.13 Dendrimers based on phenylene frameworks.

with no alkyl groups in the vicinity of the reactive functionality is desired because those groups tend to be involved in undesired side reactions. Recently, the inner spaces of dendrimers have attracted increasing attention as reaction environments for various purposes.

In the core position of the 1,3,5-phenylene-based dendrimer **33** reported by Miller et al., there is an inert and large space without any alkyl groups (Figure 11.13).[54] We applied this molecular motif to the development of a novel dendrimer-type molecular cavity **34** in which the extended *m*-terphenyl framework forms a large cleft to provide stabilizing environment for a reactive functionality in the central position.[55] The cavity size of **34** can be readily enlarged by increasing the generation of the dendrimer. The molecular cavity shown in Figure 11.14 (denoted as BpqX hereafter) was designed as the minimum generation model of dendrimer **34**, which was applied to stabilization of reactive species that have been difficult of access by conventional methods.

The inert nature of the Bpq group against the intramolecular side reaction was demonstrated by the remarkable thermal stability of *N*-thiosulfinylaniline **35** in comparison with the Bmt derivative **31**; upon heating **35** in benzene-d_6 at 80°C for 11 days and subsequent heating at 100°C for 7 days in a sealed tube, no decomposition of **35** was observed (Scheme 11.18).[55]

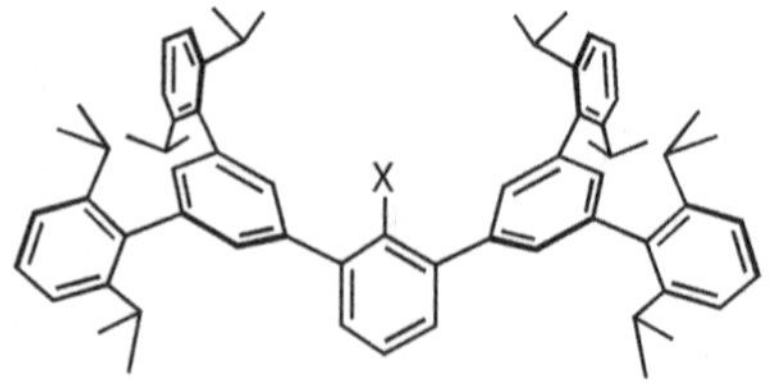

$\equiv$ BpqX

FIGURE 11.14 Dendrimer-type molecular cavity.

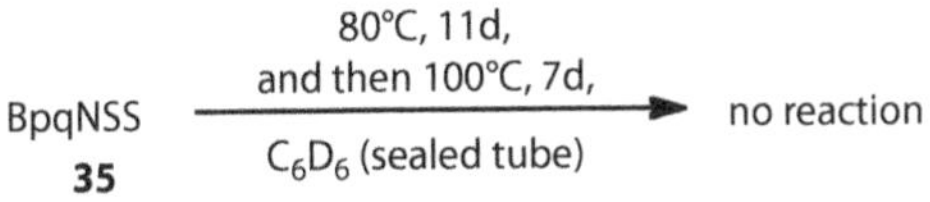

SCHEME 11.18 Thermal stability of N-thiosulfinylaniline 35.

11.5.2 STABILIZATION OF S-NITROSOTHIOL AND THIONITRATE

Independently of the synthesis of BmtSNO (**29**) by Okazaki et al.,[52] we reported the first synthesis of a stable aromatic S-nitrosothiol by taking advantage of a Bpq group.[56] Treatment of thiol **36** bearing a Bpq group with ethyl nitrite afforded the corresponding S-nitrosothiol **37**, which was isolated as brownish-green crystals (Scheme 11.19). X-ray crystallographic analysis established that the C–S–N–O linkage of **37** has the syn conformation, which is similar to **29** bearing a Bmt group.

Although **37** is an example of an aromatic S-nitrosothiol, and S-nitrosothiols are recognized to be less stable than aliphatic derivatives, **37** showed much higher thermal stability than observed early for S-nitrosothiols. Even after heating in benzene-d_6 at 80°C for 60 hr, 38% of **37** remained unchanged; the remainder of **37** was converted to the dibenzothiophene derivative **38** and thiol **36** (Scheme 11.20). In this reaction, no formation of the corresponding disulfide **39** was observed.

These results suggests that the Bpq group effectively suppresses the reaction of the initially formed thiyl radical **40** with a second molecule of S-nitrosothiol **37** to afford **39**, which enabled the slow reaction to afford **38** and **36** to take place. As described in Section 11.4.3, BmtSNO (**29**) was converted to the symmetrical disulfide **30** upon heating under similar conditions.[52] The framework of the Bpq group is apparently more effective for suppressing the bimolecular decomposition of reactive species than that of the Bmt group. In spite of such high thermal stability, S-nitrosothiol **37**

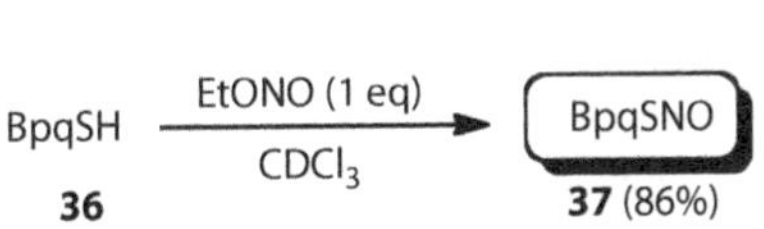

SCHEME 11.19 Synthesis of S-nitrosothiol 37.

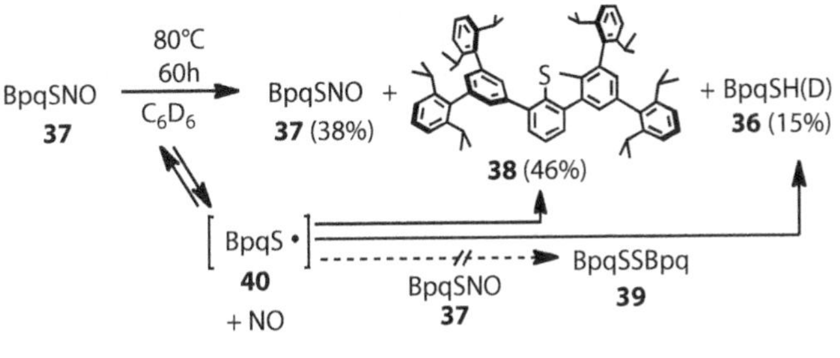

SCHEME 11.20 Thermal reaction of S-nitrosothiol 37.

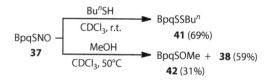

SCHEME 11.21   Reactions of S-nitrosothiol 37.

reacted with various reagents such as 1-butanethiol and methanol to produce **41** and **42**, respectively (Scheme 11.21).

Thionitrates ($RSNO_2$) have attracted increasing attention as key intermediates in the bioactivation of organic nitrates ($RONO_2$) such as nitroglycerin and isosorbide nitrates that are known to exert vasodilating action through the transformation to nitric oxide.[57–59] However, thionitrates are usually very labile because of facile bimolecular decomposition and many of the chemical processes proposed for the biotransformations involving thionitrate intermediates have remained hypothetical.

Although the fundamental chemistry of thionitrates was studied intensively by Oae and coworkers by utilizing *tert*-butyl thionitrate (*t*-BuSNO$_2$), which is relatively stable in solution,[46] there had been no example of isolation and crystallographic analysis of a stable thionitrate until we reported the synthesis of a bulky triarylmethyl thionitrate.[50] The synthesis and reactivity of an aromatic thionitrate bearing a Bpq group were then investigated.

Oxidation of *S*-nitrosothiol **37** with *tert*-butyl nitrite or N_2O_4 afforded the corresponding thionitrate **43**, which was isolated as stable crystals (Scheme 11.22).[56] The structure of **43** was established by crystallographic analysis, revealing that the SNO_2 functionality is incorporated in a cleft formed by two convex *m*-terphenyl units (Figure 11.15a).[60] Treatment of **43** with triphenylphosphine reduced **43** to *S*-nitrosothiol **37**, while the reaction with 1-butanethiol afforded the unsymmetrical disulfide **41**.[56]

Biological studies have proposed that hydrolysis of thionitrates produces sulfenic acids and nitrite (NO_2^-) by nucleophilic substitution at sulfur atom (Scheme 11.23).[61,62] However, no chemical evidence for this process has been reported because of the intrinsic instability of both $RSNO_2$ substrates and RSOH products.

This chemical process was investigated by utilizing the Bpq derivatives.[63] Although treatment of thionitrate **43** with water under weakly acidic or neutral conditions resulted in no reaction, hydrolysis of **43** proceeded under alkaline conditions to afford the corresponding sulfenic acid **44** (Scheme 11.24), providing experimental evidence for the chemical transformation shown in Scheme 11.23. These results

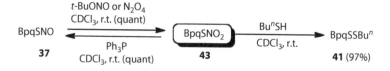

SCHEME 11.22   Synthesis and reactions of thionitrate 43.

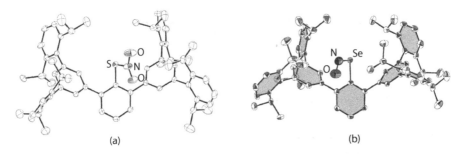

FIGURE 11.15 (a) ORTEP drawing of **43** (50% probability). (b) ORTEP drawing of **50** (50% probability).

corroborate the possible formation of cysteine sulfenic acids by similar transformation in the clefts of proteins.

11.5.3 STABILIZATION OF SE-NITROSOSELENOL

The S-nitrosation of cysteine residues to produce S-nitrosothiols is one of the most important NO-mediated modifications of proteins. In addition, it has been suggested that the interactions of NO (or NO-derived species) with selenocysteine residues are also involved in NO-mediated biofunctions.[64-67] For example, GPx is inactivated by treatment with an S-nitrosothiol as well as by endogenous NO, presumably through Se-nitrosation of the SeH groups of selenoproteins to produce Se-nitrososelenols (RSeNO).[64,66,67]

However, essentially no chemical information about this species has been available despite its potential importance from a biological view. For elucidation of the role of Se-nitrososelenols in NO-mediated modification of selenoproteins, the synthesis of a stable reference compound of this species is indispensable.

Du Mont et al. reported the Se-nitrosation of selenol **45** bearing an extremely bulky alkyl substituent (Scheme 11.25).[68] When selenol **45** was treated with *tert*-butyl nitrite at −78°C, the formation of the corresponding Se-nitrososelenol **46** was indicated by infrared (IR) spectroscopy. However, above −78°C, **46** decomposed, forming NO, diselenide **47**, and triselenide **48**.

$$RSNO_2 + H_2O \longrightarrow RSOH + NO_2^- + H^-$$

SCHEME 11.23 Hydrolysis of thionitrates to produce sulfenic acids.

$$BpqSNO_2 \xrightarrow[\text{2) aq NH}_4\text{Cl}]{\text{1) NaOH/THF-H}_2\text{O}} BpqSOH$$
43   **44** (66%)

SCHEME 11.24 Hydrolysis of thionitrate 43 to produce sulfenic acid 44.

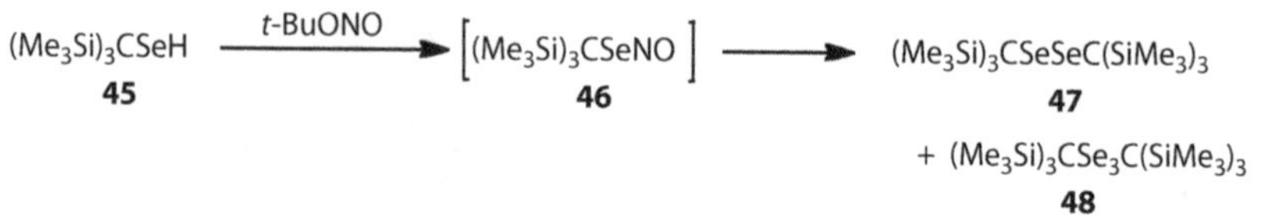

$$(Me_3Si)_3CSeH \xrightarrow{\text{t-BuONO}} \left[(Me_3Si)_3CSeNO\right] \longrightarrow (Me_3Si)_3CSeSeC(SiMe_3)_3$$

45   **46**   **47**

$$+ (Me_3Si)_3CSe_3C(SiMe_3)_3$$

48

SCHEME 11.25   Attempted synthesis of Se-nitrososelenol **46**.

We examined the reaction of selenol **49** bearing a Bpq group with nitrosating agents.[69] Treatment of **49** with ethyl nitrite led to the quantitative formation of *Se*-nitrososelenol **50**, isolated as reddish-purple crystals, presenting the first synthesis of a stable *Se*-nitrososelenol (Scheme 11.26). The structure of **50** was established definitely by X-ray crystallographic analysis (Figure 11.15b). The C–Se–N–O linkage was found to adopt the syn conformation, which is similar to *S*-nitrosothiol **37**.

Se-nitrososelenol **50** showed characteristic spectral features that are expected to be useful in identification of *Se*-nitrosated species in various situations. In particular, the ^{77}Se NMR (CDCl$_3$) signal of **50** was observed at an extraordinarily low field of δ 2229, suggesting the strong magnetic deshielding effect of the NO group. While *Se*-nitrososelenol **50** is stable at room temperature at least for 1 week in CDCl$_3$, thermolysis of **50** in benzene-d_6 at 80°C for 13 hr afforded symmetrical diselenide **51** quantitatively (Scheme 11.26).[70] Considering the thermal stability of S-nitrosothiol **37**,[56] the reactivity of *Se*-nitrososelenols toward bimolecular decomposition is obviously higher than that of S-nitrosothiols.

The model study of the inactivation of GPx by nitrosating agents was then examined. It was proposed that the inactivation of GPx by an *S*-nitrosothiol proceeds by a two-step mechanism (Scheme 11.27).[66] The first step is *Se*-nitrosation of selenocysteine to produce the *Se*-nitrososelenol intermediate that can be reduced to the active form by treatment with DTT. The second step involves the formation of an interbridge selenenyl sulfide and cannot be reversed by a reducing agent.

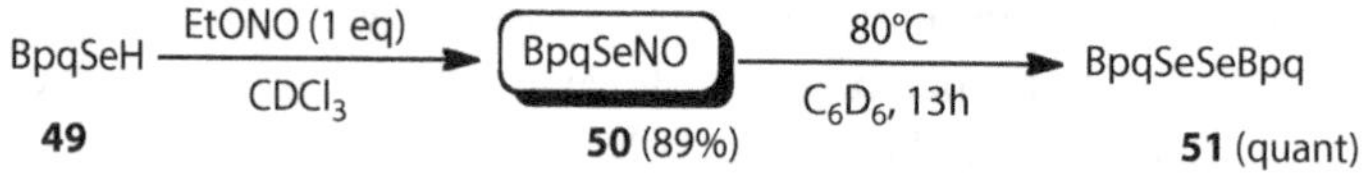

$$BpqSeH \xrightarrow[\text{CDCl}_3]{\text{EtONO (1 eq)}} \boxed{BpqSeNO} \xrightarrow[\text{C}_6\text{D}_6,\ 13h]{80°C} BpqSeSeBpq$$

49   **50** (89%)   **51** (quant)

SCHEME 11.26   Synthesis and reaction of Se-nitrososelenol **50**.

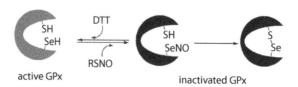

active GPx   inactivated GPx

SCHEME 11.27   Proposed mechanism for the inactivation of GPx by an S-nitrosothiol.

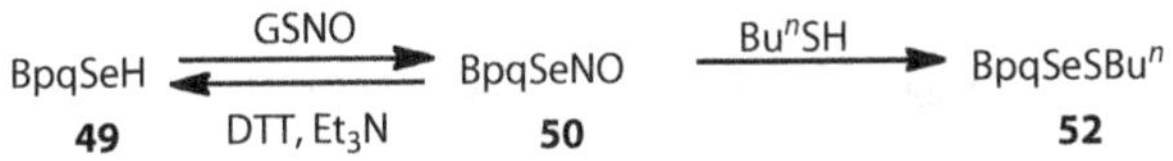

SCHEME 11.28   Modeling of the GPx inactivation processes.

Treatment of selenol **49** with *S*-nitrosoglutathione (GSNO) resulted in the NO transfer from sulfur to selenium to produce *Se*-nitrososelenol **50**, which reacted with 1-butanethiol to afford selenenyl sulfide **52** (Scheme 11.28).[69] While *Se*-nitrososelenol **50** was reduced to selenol **49** by DTT in the presence of base, selenenyl sulfide **52** showed no reaction under the same conditions. These results are supportive for the proposed mechanism for the GPx inactivation shown in Scheme 11.27, suggesting that *Se*-nitrosation of selenoproteins may play an important role in redox regulation systems.

11.5.4   MODEL STUDY OF IODOTHYRONINE DEIODINASE

Type-1 iodothyronine deiodinase (ID-1) is an enzyme containing a selenocysteine residue in its active site that catalyzes the conversion of a human thyroid prohormone (thyroxine **T4**) to a biologically active hormone (3,5,3′-triiodothyronine **T3**) through 5′-deiodination (Scheme 11.29).[71-73]

The proposed mechanism for the catalytic cycle of ID-1 involves a ping-pong bisubstrate reaction in which the selenol form of the enzyme (E-SeH) first reacts with **T4** to form a selenenyl iodide intermediate (E-SeI) with release of the deiodinated compound **T3**. The subsequent reaction between the selenenyl iodide intermediate and an unidentified thiol cofactor regenerates the selenol form.

The involvement of a selenenyl iodide as an intermediate in this deiodination process has been widely accepted, and many model studies on the catalytic cycle of ID-1 have assumed the formation of such intermediates.[74-76] However, no experimental evidence demonstrated the formation of a selenenyl iodide in the deiodination reaction of 2,6-diiodophenol derivatives by a selenol (Scheme 11.30a).

This is largely due to the instability of the intermediary selenenyl iodides in artificial systems; they usually undergo facile disproportionation to the corresponding diselenides and iodine (Scheme 11.30b).[77-79] For the demonstration of deiodination,

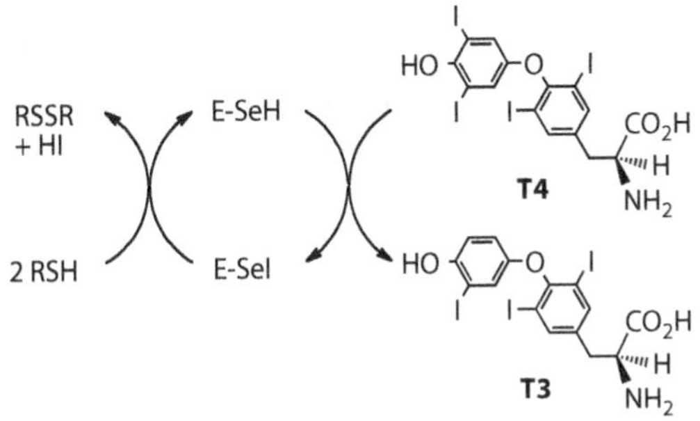

SCHEME 11.29   Proposed mechanism for the catalytic cycle of ID-1.

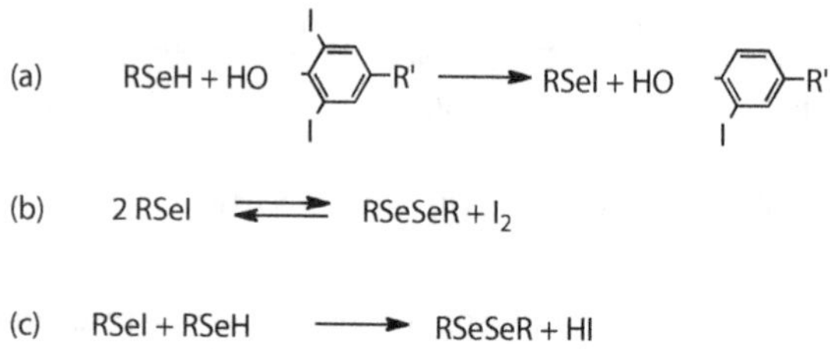

(a) $RSeH + HO$ ⟨benzene ring with I substituents and R'⟩ $\longrightarrow$ $RSeI + HO$ ⟨benzene ring with I substituent and R'⟩

(b) $2\ RSeI$ ⇌ $RSeSeR + I_2$

(c) $RSeI + RSeH$ $\longrightarrow$ $RSeSeR + HI$

SCHEME 11.30   Reactions involving selenenyl iodides.

it is essential for the selenenyl iodide intermediate *not* to undergo the disproportionation and *not* to react with its selenol form (Scheme 11.30c).

There have been several examples of the isolation of selenenyl iodides stabilized against disproportionation by introduction of an extremely bulky alkyl substituent (**53**)[80,81] or an intramolecular coordinating group (**54**)[82] (Figure 11.16). However, it was reported that even **54** readily reacts with the corresponding selenol to produce the symmetrical diselenide.

The synthesis of a stable selenenyl iodide and model studies on the deiodination process were examined by utilizing the Bpq group.[83] Treatment of selenol **49** with *N*-iodosuccinimide (NIS) led to the quantitative formation of selenenyl iodide **55** isolated as purple crystals (Scheme 11.31). The structure of **55** was established by X-ray crystallographic analysis. When selenenyl iodide **55** was heated at 100°C in toluene-d_8 for 1 week in a sealed tube, no decomposition was observed. Contrary to the well-accepted notion that selenenyl iodides are thermally unstable species, it was demonstrated that this species intrinsically has high thermal stability so far as the diselenide formation is suppressed.

Deiodination of **T4** model compounds, 2,6-diiodo-4-phenoxyphenol (**56a**) and *N*-butyrylthyronine methyl ester **56b** by selenol **49** was examined (Scheme 11.32). When diiodophenol **56a** was treated with an equimolar amount of selenol **49** in the presence triethylamine, deiodination proceeded slowly, and after 1 week 50% of **56a** was converted to monoiodophenol **57a**. Concomitantly, the formation of selenenyl iodide **55** (39%) was confirmed.

The reaction of thyronine derivative **56b** under the same conditions yielded the deiodinated product **57b** (65% from **56b**) and selenenyl iodide **55** (55% from **49**). These results provide the chemical evidence for the proposed deiodination of a 2,6-diiodophenol derivative by a selenol. It is notable that, in the reaction of **56b**, only deiodination at the outer ring took place—similar to the conversion of **T4** to **T3** in the ID-1 catalytic cycle.

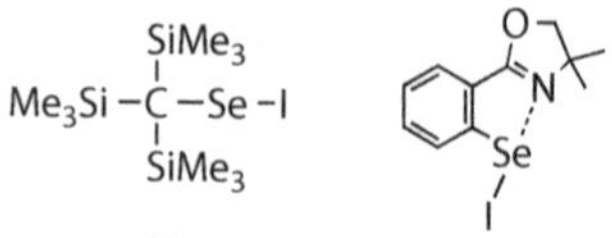

$$Me_3Si - \overset{\displaystyle SiMe_3}{\underset{\displaystyle SiMe_3}{\overset{|}{\underset{|}{C}}}} - Se - I$$

53

⟨structure of compound 54: benzene ring fused to a ring with C=O, N, and Se–I, with O–CH and isopropyl substituents⟩

54

FIGURE 11.16   Examples of isolable selenenyl iodides.

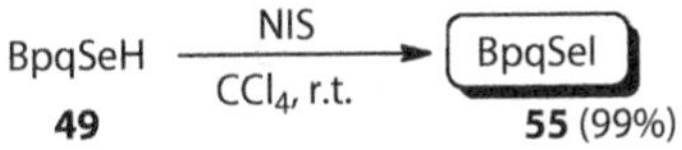

SCHEME 11.31 Synthesis of selenenyl iodide 55.

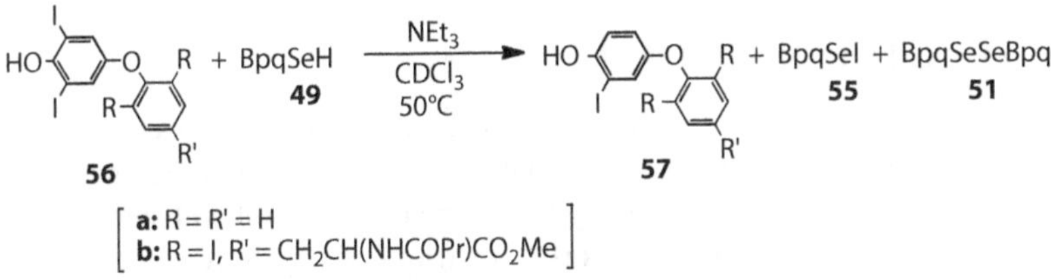

$$\left[\begin{array}{l} \textbf{a:}\ R = R' = H \\ \textbf{b:}\ R = I,\ R' = CH_2CH(NHCOPr)CO_2Me \end{array}\right]$$

SCHEME 11.32 Deiodination of diiodophenols by selenol 49.

Employment of BpqSH (**36**) instead of selenol **49** for the reaction with **56a** produced no change (Scheme 11.33), suggesting that a selenol functionality with higher nucleophilicity is essential for this deiodination reaction. Through control experiments utilizing several 2,6-diiodophenol derivatives, it was indicated that deiodination proceeds via nucleophilic attack of a selenol (or selenolate) with iodine on the keto forms of the diiodophenols (Scheme 11.34), and that a high affinity between selenium and iodine plays a key role in the formation of a selenenyl iodide.[84]

While the physiological cofactor of the second step of the ID-1 catalytic cycle has not been identified yet, it is customary to use DTT as the second substrate for in vitro experiments. Treatment of selenenyl iodide **55** with DTT in the presence of triethylamine afforded selenol **49** quantitatively (Scheme 11.35). Thus, all the chemical transformations included in the catalytic cycle shown in Scheme 11.29 were established experimentally, corroborating the involvement of a selenenyl iodide as an intermediate in the enzymatic reaction.

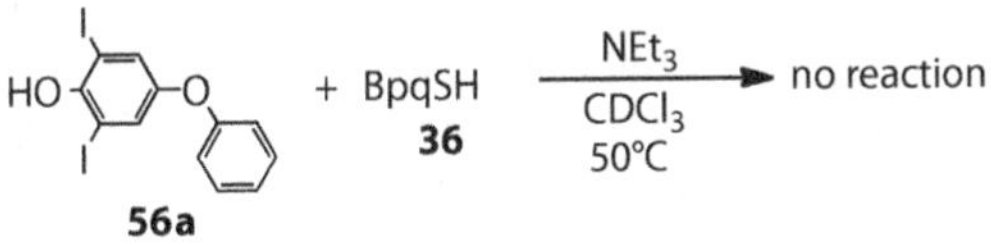

SCHEME 11.33 Reaction of a diiodophenol with thiol 36.

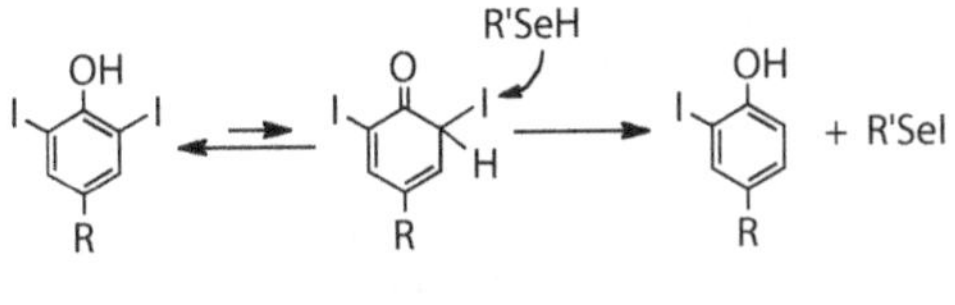

SCHEME 11.34 Proposed mechanism for the deiodination process via enol/keto tautomerization.

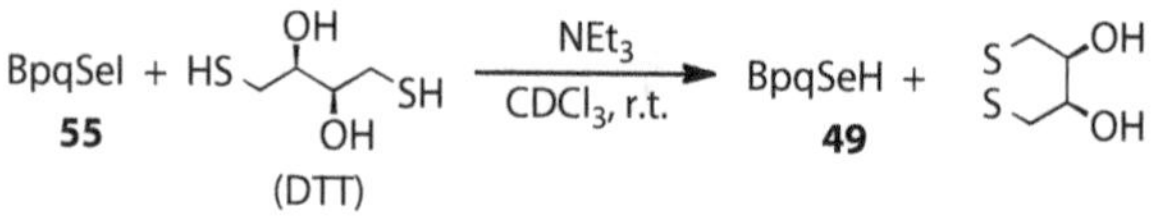

SCHEME 11.35 Reduction of selenenyl iodide 55 by DTT.

11.6 CONCLUDING REMARKS

Several types of bowl-shaped molecular cavities have been developed for modeling the essential features of geometrically isolated environments in proteins that furnish both stability and reactivity. By taking advantage of the peripheral steric protection endowed by these molecular frameworks, a variety of highly reactive species that were difficult of access by conventional methods have been synthesized as stable compounds. Their reactivities have been elucidated experimentally to demonstrate the chemical processes that have been proposed but remain hypothetical.

The synergy of high reactivity and stability in the well-defined isolated reaction environments of nanoscale molecular systems is expected to open undiscovered applications of reactive species to various purposes. Although the covalent frameworks of the molecular cavities described here are very secure and chemically stable, they are static in nature. Interest in stabilization of reactive species by using mechanical bonding systems such as rotaxanes has grown recently.[85–89]

Mechanically bonded complexes that undergo no dissociation have the same kinetic stability as covalent molecules, whereas their supramolecular structures generate a variety of dynamic and responsive functions. Use of such supramolecular motifs may prove a useful approach to construction of the cavity-shaped frameworks that are more relevant to the modeling of enzymes.

ACKNOWLEDGMENTS

The author is very grateful to all the collaborators, especially to Emer. Prof. Renji Okazaki and Emer. Prof. Takayuki Kawashima (both at the University of Tokyo), Emer. Prof. Gaku Yamamoto (Kitasato University), and Prof. Norihiro Tokitoh (Kyoto University). The contributions of the students and the Japan Society for the Promotion of Science (JSPS) Postdoctoral Fellows whose names are listed in the reference section are also acknowledged.

REFERENCES

1. Breiner, B. and Nitschke, J. R. In *Supramolecular Chemistry: From Molecules to Nanomaterials*, Vol. 4, Gale, P. A. S. and Jonathan, W., Eds. John Wiley & Sons, Chichester, 2012, pp. 1575–1588.
2. Warmuth, R. In *Molecular Encapsulation: Organic Reactions in Constrained Systems*, Brinker, U. H. and Jean-Luc, M., Eds. John Wiley & Sons, Chichester, 2010, pp. 227–268.
3. Yoshizawa, M., Klosterman, J. K., and Fujita, M., *Angew. Chem., Intl. Ed.* 2009, *48*, 3418–3438.

4. Forman, H. J., Fukuto, J. M., and Torres, M., *Am. J. Physiol.* 2004, *287*, C246–C256.
5. Brigelius-Flohe, R., Banning, A., and Schnurr, K., *Antioxid. Redox Signal.* 2003, *5*, 205–215.
6. Miyamoto, Y., Koh, Y. H., Park, Y. S. et al., *Biol. Chem.* 2003, *384*, 567–574.
7. Paulsen, C. E. and Carroll, K. S., *ACS Chem. Biol.* 2010, *5*, 47–62.
8. Heinrich, T. A., da Silva, R. S., Miranda, K. M. et al., *Br. J. Pharmacol.* 2013, *169*, 1417–1429.
9. Claiborne, A., Mallett, T. C., Yeh, J. I. et al., *Adv. Protein Chem.* 2001, *58*, 215–276.
10. Poole, L. B., Karplus, P. A., and Claiborne, A., *Annu. Rev. Pharmacol. Toxicol.* 2004, *44*, 325–347.
11. Flohe, L., Guenzler, W. A., and Schock, H. H., *FEBS Lett.* 1973, *32*, 132–134.
12. Rotruck, J. T., Pope, A. L., Ganther, H. E. et al., *Science* 1973, *179*, 588–590.
13. West, R., Fink, M. J., and Michl, J., *Science* 1981, *214*, 1343–1344.
14. Yoshifuji, M., Shima, I., Inamoto, N. et al., *J. Am. Chem. Soc.* 1981, *103*, 4587–4589.
15. Tokitoh, N., Matsumoto, T., Manmaru, K. et al., *J. Am. Chem. Soc.* 1993, *115*, 8855–8856.
16. Reich, H. J. and Jasperse, C. P., *J. Org. Chem.* 1988, *53*, 2389–2390.
17. Lange, L., Meyer, B., and du Mont, W. W., *J. Organomet. Chem.* 1987, *329*, C17–C20.
18. Goto, K. and Okazaki, R., *Liebigs Ann. Recl.* 1997, 2393–2407.
19. Goto, K. and Kawashima, T., *J. Synth. Org. Chem., Jpn.* 2005, *63*, 1157–1170.
20. Vinod, T. K. and Hart, H., *J. Am. Chem. Soc.* 1988, *110*, 6574–6575.
21. Vinod, T. K. and Hart, H., *J. Org. Chem.* 1990, *55*, 881–890.
22. Lüning, U., *Liebigs Ann. Chem.* 1987, 949–955.
23. Lüning, U. In *Molecular Encapsulation: Organic Reactions in Constrained Systems*, Brinker, U. H. M. et al., Eds. John Wiley & Sons, Chichester, 2010, pp. 175–199.
24. Goto, K., Tokitoh, N., and Okazaki, R., *Angew. Chem., Intl. Ed. Engl.* 1995, *34*, 1124–1126.
25. Gutsche, C. D., *Progr. Macrocyclic Chem.* 1987, *3*, 93–165.
26. Gutsche, C. D., *Calixarenes Revisited.* Royal Society of Chemistry: Cambridge, 1998.
27. Baldini, L., Sansone, F., Casnati, A. et al. In *Supramolecular Chemistry: From Molecules to Nanomaterials*, Vol. 3, Gale, P. A. S. and Jonathan, W. et al., Eds. John Wiley & Sons, Chichester, 2012, pp. 863–894.
28. Tokitoh, N., Saiki, T., and Okazaki, R., *J. Chem. Soc., Chem. Commun.* 1994, 1899–1900.
29. Saiki, T., Goto, K., and Okazaki, R. *Chem. Lett.* 1996, 993–994.
30. Saiki, T., Goto, K., Tokitoh, N. et al., *J. Org. Chem.* 1996, *61*, 2924–2925.
31. Saiki, T., Goto, K., and Okazaki, R., *Angew. Chem., Intl. Ed. Engl.* 1997, *36*, 2223–2224.
32. Goto, K., Saiki, T., Akine, S. et al., *Heteroatom. Chem.* 2001, *12*, 195–197.
33. Ross, H. and Lüening, U., *Angew. Chem., Intl. Ed. Engl.* 1995, *34*, 2555–2557.
34. Akine, S., Goto, K., and Kawashima, T., *Tetrahedron Lett.* 2003, *44*, 1171–1174.
35. Akine, S., Goto, K., and Okazaki, R., *Chem. Lett.* 1999, 681–682.
36. Akine, S., Goto, K., Kawashima, T. et al., *Bull. Chem. Soc. Jpn.* 1999, *72*, 2781–2783.
37. Goto, K., Holler, M., and Okazaki, R., *Tetrahedron Lett.* 1996, *37*, 3141–3144.
38. Hogg, D. R. In *The Chemistry of Sulphenic Acids and Their Derivatives*, Patai, S., Ed. John Wiley & Sons, Chichester, 1990, pp. 361–402.
39. Goto, K., Holler, M., and Okazaki, R., *J. Am. Chem. Soc.* 1997, *119*, 1460–1461.
40. Goto, K., Nagahama, M., Mizushima, T. et al., *Org. Lett.* 2001, *3*, 3569–3572.
41. Bhabak, K. P. and Mugesh, G. *Acc. Chem. Res.* 2010, *43*, 1408–1419.
42. Denney, D. B., Goodyear, W. F., and Goldstein, B. *J. Am. Chem. Soc.* 1960, *82*, 1393–1395.
43. Goto, K., Shimada, K., Nagahama, M. et al., *Chem. Lett.* 2003, *32*, 1080–1081.
44. Williams, D. L. H., *Acc. Chem. Res.* 1999, *32*, 869–876.
45. Hogg, N., *Annu. Rev. Pharmacol. Toxicol.* 2002, 42, 585–600.

46. Oae, S. and Shinhama, K., *Org. Prep. Proced. Intl.* 1983, *15*, 165–198.
47. Field, L., Dilts, R. V., Ravichandran, R. et al., *J. Chem. Soc., Chem. Commun.* 1978, 249–250.
48. Arulsamy, N., Bohle, D. S., Butt, J. A. et al., *J. Am. Chem. Soc.* 1999, *121*, 7115–7123.
49. Bartberger, M. D., Houk, K. N., Powell, S. C. et al., *J. Am. Chem. Soc.* 2000, *122*, 5889–5890.
50. Goto, K., Hino, Y., Kawashima, T. et al., *Tetrahedron Lett.* 2000, *41*, 8479–8483.
51. Petit, C., Hoffmann, P., Souchard, J. P. et al., *Phosphorus, Sulfur Silicon Relat. Elem.* 1997, *129*, 59–67.
52. Itoh, M., Takenaka, K., Okazaki, R. et al., *Chem. Lett.* 2001, 1206–1207.
53. Tan, B., Goto, K., Kobayashi, J. et al., *Chem. Lett.* 1998, 981–982.
54. Miller, T. M., Neenan, T. X., Zayas, R. et al., *J. Am. Chem. Soc.* 1992, *114*, 1018–1025.
55. Goto, K., Yamamoto, G., Tan, B. et al., *Tetrahedron Lett.* 2001, *42*, 4875–4877.
56. Goto, K., Hino, Y., Takahashi, Y. et al., *Chem. Lett.* 2001, 1204–1205.
57. Chen, Z. and Stamler, J. S., *Trends Cardiovasc. Med.* 2006, *16*, 259–265.
58. Mayer, B. and Beretta, M., *Br. J. Pharmacol.* 2008, *155*, 170–184.
59. Lang, B. S., Gorren, A. C. F., Oberdorfer, G. et al., *J. Biol. Chem.* 2012, *287*, 38124–38134.
60. Goto, K., Yoshikawa, S., Ideue, T. et al., *J. Sulfur Chem.* 2013, *34*, 705–710.
61. Chen, Z., Zhang, J., and Stamler, J. S., *Proc. Natl. Acad. Sci. USA* 2002, *99*, 8306–8311.
62. Carballal, S., Radi, R., Kirk, M. C. et al., *Biochemistry* 2003, *42*, 9906–9914.
63. Goto, K., Shimada, K., Furukawa, S. et al., *Chem. Lett.* 2006, *35*, 862–863.
64. Asahi, M., Fujii, J., Suzuki, K. et al., *J. Biol. Chem.* 1995, *270*, 21035–21039.
65. Freedman, J. E., Frei, B., Welch, G. N. et al., *J. Clin. Invest.* 1995, *96*, 394–400.
66. Asahi, M., Fujii, J., Takao, T. et al., *J. Biol. Chem.* 1997, *272*, 19152–19157.
67. Dobashi, K., Asayama, K., Nakane, T. et al., *Free Radic. Res.* 2001, *35*, 319–327.
68. Wismach, C., du Mont, W. W., Jones, P. G. et al., *Angew. Chem., Intl. Ed.* 2004, *43*, 3970–3974.
69. Shimada, K., Goto, K., Kawashima, T. et al., *J. Am. Chem. Soc.* 2004, *126*, 13238–13239.
70. Shimada, K., Goto, K., and Kawashima, T., *Chem. Lett.* 2005, *34*, 654–655.
71. Berry, M. J., Banu, L., and Larsen, P. R., *Nature (London)* 1991, *349*, 438–440.
72. Larsen, P. R. and Berry, M. J., *Annu. Rev. Nutr.* 1995, *15*, 323–352.
73. Kuiper, G. G. J. M., Kester, M. H. A., Peeters, R. P. et al., *Thyroid* 2005, *15*, 787–798.
74. Beck, C., Jensen, S. B., and Reglinski, J., *Bioorg. Med. Chem. Lett.* 1994, *4*, 1353–1356.
75. Vasil'ev, A. A. and Engman, L. *J. Org. Chem.* 1998, *63*, 3911–3917.
76. Mugesh, G., du Mont, W. W., Wismach, C. et al., *Chem BioChem* 2002, *3*, 440–447.
77. du Mont, W. W., Kubiniok, S., Peters, K. et al., *Angew. Chem., Intl. Ed. Engl.* 1987, *26*, 780–781.
78. du Mont, W. W., Martens, A., Pohl, S. et al., *Inorg. Chem.* 1990, *29*, 4847–4848.
79. du Mont, W. W., Martens-von Salzen, A., Ruthe, F. et al., *J. Organomet. Chem.* 2001, *623*, 14–28.
80. du Mont, W. W. and Wagner, I. *Chem. Ber.* 1988, *121*, 2109–2110.
81. Ostrowski, M., Wagner, I., du Mont, W. W. et al., *Z. Anorg. Allg. Chem.* 1993, *619*, 1693–1698.
82. Mugesh, G., Panda, A., Singh, H. B. et al., *Chem. Eur. J.* 1999, *5*, 1411–1421.
83. Goto, K., Sonoda, D., Shimada, K. et al., *Angew. Chem., Intl. Ed.* 2010, *49*, 545–547.
84. Sase, S., Ebisawa, K., Goto, K. *Chem. Lett.* 2012, *41*, 766-8.
85. Buston, J. E. H., Young, J. R., and Anderson, H. L., *Chem. Commun.* 2000, 905–906.
86. Oku, T., Furusho, Y., and Takata, T., *Org. Lett.* 2003, *5*, 4923–4925.
87. Arunkumar, E., Forbes, C. C., Noll, B. C. et al., *J. Am. Chem. Soc.* 2005, *127*, 3288–3289.
88. Barnes, J. C., Fahrenbach, A. C., Cao, D. et al., *Science* 2013, *339*, 429–433.
89. Winn, J., Pinczewska, A., and Goldup, S. M., *J. Am. Chem. Soc.* 2013, *135*, 13318–13321.

12 Supramolecular Catalyst with Substrate Binding Sites

Yoshinori Takashima and Akira Harada

CONTENTS

12.1 INTRODUCTION

Enzymes form products from substrates with high selectivity and efficiency at ambient temperatures. Enzymes feature well-defined binding pockets that bind substrates during reactions using non-covalent interactions. The behavior is analogous to a lock and key system in which the substrate (key) has a complementary size and shape to the binding site (lock). Enzymes selectively promote the formation of the transition state and intermediates of a particular reaction.

Recently, the three-dimensional structures of various enzymes have been determined using single crystal X-ray diffraction analysis. In concurrent studies, the determination of the positions of the catalytic groups and substrate on the enzymes helped clarify the route of enzymatic reaction. These structural analyses have provided information about the location of the substrate in relation to the catalytic groups and also the enzymatic mechanisms.

Inspired by biological enzymatic systems, host–guest chemistry has attempted to mimic enzyme catalysis. Many chemists developed innovative syntheses (supramolecular catalysts) using chemical processes that take advantages of biological systems.[1] Supramolecular catalysts employ non-covalent bonds in the substrate receptor recognition systems to achieve high selectivity and efficiency (higher reaction rates and turnover efficiency). An important feature of enzymatic systems is the presence of a binding site.

Most of these reactions are accelerated cooperatively by the formation of host–guest (host–substrate) complexes in which the energy of the transition state is lowered, thus accelerating rates of reaction. Inhibitor molecules that bind to active sites more strongly than substrates can significantly reduce reaction rates, often to a level similar to that of a reaction without a host catalyst. This chapter focuses on the development of novel supramolecular catalysts that utilize host–guest interactions that mimic certain aspects of enzymatic catalysts.

12.2 CYCLODEXTRIN DERIVATIVES AS ENZYME MODELS

Supramolecular catalysts using synthetic host molecules have been well researched. Early studies realized a hydrolysis reaction for ester derivatives using modified cyclodextrins (CDs).[2-5] CDs are suitable for the study of two-substrate supramolecular catalysis and for the synthesis of artificial enzymes. 2-Benzyimidazoleacetic acid–modified α-cyclodextrin hydrolyzes m-tert-butylphenyl acetate at an accelerated rate. The imidazole has a benzoate group in a position that imitates the function of the aspartate ion with the catalytic triad characteristics of serine proteases such as chymotrypsin (Figure 12.1).[6-8]

CDs were used as model catalysts for ribonuclease A (RNase A). RNase A is a pancreatic ribonuclease that cleaves single-stranded RNA. RNase A catalyzes the cleavage of the phosphodiester bond between the 5′-ribose of a nucleotide and the phosphate group attached to the 3′-ribose of an adjacent pyrimidine nucleotide. The hydrolysis using RNase A is facilitated by the imidazole moieties of two histidine units. One imidazole unit acts as a free base and another functions as an acid. To mimic RNase A, a βCD with two imidazole groups was prepared from βCD diiodide (Figure 12.2).[9] RNase A uses the two imidazole groups of histidines 12 and 119 as its principal catalytic groups in the hydrolysis of RNA.

Vitamin B plays important roles in cell metabolism. Vitamin B actually consists of eight chemically distinct biologically active agents that function as coenzymes. Pyridoxine, pyridoxal, and pyridoxamine or pyridoxine hydrochloride can all be called vitamin B6 as they are all converted to the active form. Pyridoxine is involved in the metabolism of amino acids and lipids. A vitamin B6-dependent enzyme model

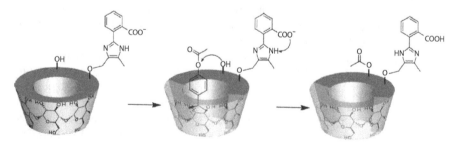

FIGURE 12.1 Cyclodextrin supramolecular catalyst designed to mimic catalytic triad in chymotrypsin.

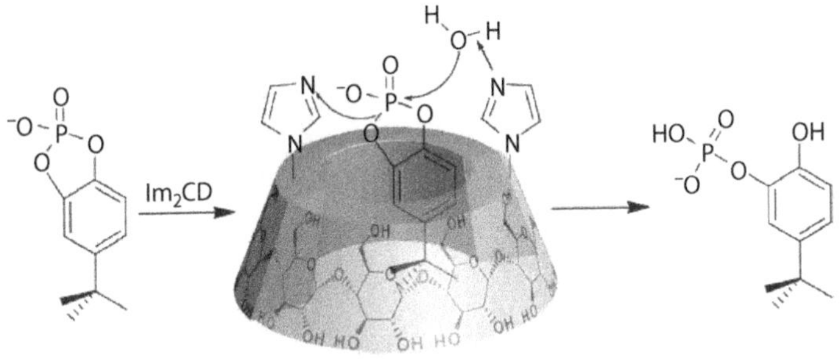

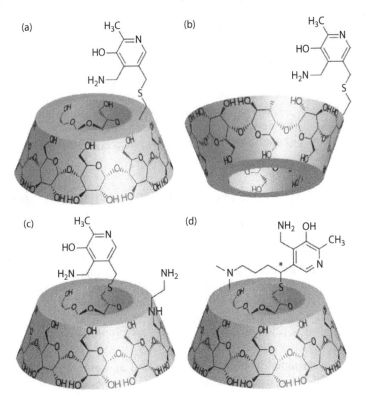

FIGURE 12.2   Cyclodextrin bis(imidazoles) catalyzing hydrolysis of phosphate substrate.

in which pridoxamine was linked to the primary face of a βCD (6-pridoxamine-βCD, Figure 12.3a) transforms α-keto acids into α-amino acids.

βCD with pridoxamine catalyzes the transamination of phenylpyruvic acid to pyruvic acid even though native pyridoxamine shows no selectivity. The molecular recognition property of βCD for the phenyl group provides the selectivity for substrates.[10]

FIGURE 12.3   Pridoxamine-βCDs as vitamin B6-dependent enzyme model.

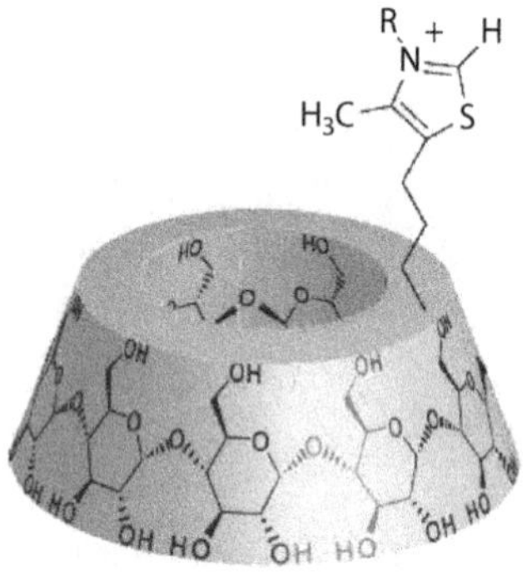

FIGURE 12.4   Thiazolium-βCDs as carboxylase enzyme models.

When the pyridoxamine analog was modified to the secondary hydroxyl group of βCD (2-pridoxamine-βCD, Figure 12.3b), 2-pridoxamine-βCD exhibited similar transamination catalysis activity to that of the 6-pridoxamine-βCD analogue.[11] A βCD carrying a pyridoxamine and an ethylenediamine on the primary hydroxyl group also afforded amino acids from keto acids with chiral selectivity (Figure 12.3c and d).[12,13]

Thiamine pyrophosphate functions as a coenzyme with enzymes such as carboxylase. A thiazolium ring with an appended CD moiety selectively recognizes substrates using the hydrophobic cavity of the CD. A thiazolium ring with appended γCD catalyzed the benzoin condensation of two benzaldehydes (Figure 12.4). The rate of condensation was 150-fold higher than that for a thiazolium salt without appended CD because the γCD-modified thiazolium ring effectively included benzaldehyde in the γCD cavity. The affinity of the benzoin product with γCD was smaller than that of benzylaldehyde, thus allowing the reaction to proceed without inhibition from the benzoin products.[14]

12.3   CYCLIC HOST MOLECULES AS METALLOENZYME MODELS

Chymotrypsin secreted by the pancreas is one of the enzymes that breaks up proteins and polypeptides. Zinc plays an important role in accelerating the catalytic reaction rate. These metalloenzymes operate through activation of a carbonyl group by coordinating with its oxygen and a water molecule that acts as a nucleophile. An αCD with an appended Ni[2+]-oxime complex at the secondary face of the αCD accelerated the reaction rate of hydrolysis of p-nitrophenyl acetate (Figure 12.5). The catalytic reaction rate is faster than that of the αCD without appended Ni[2+].[18] The reaction mechanism is reminiscent of the catalytic activity of the alkaline phosphatase enzyme.[19]

A βCD disubstituted with a Zn[2+] complex and an imidazole group effectively accelerated the hydrolysis of bound catechol cyclic phosphate. When the Zn[2+]

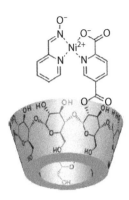

FIGURE 12.5   αCD with Ni²⁺-oxime complex as artificial enzyme.

complex was located at the opposite site of an imidazole group on the primary hydroxyl group of the βCD, the disubstituted βCD exhibited the fastest reaction rate. Although the hydrolysis reaction rates of catechol products A and B without catalysts were comparable, the disubstituted βCD selectively gave the catechol product A in preference to B due to the geometry of the inclusion complex (Figure 12.6).[20]

Cavitands can be functionalized with catalytic active sites such as Zn salen, porphyrin, phenanthrolin, and pyridone complexes.[21] These synthetic host molecules

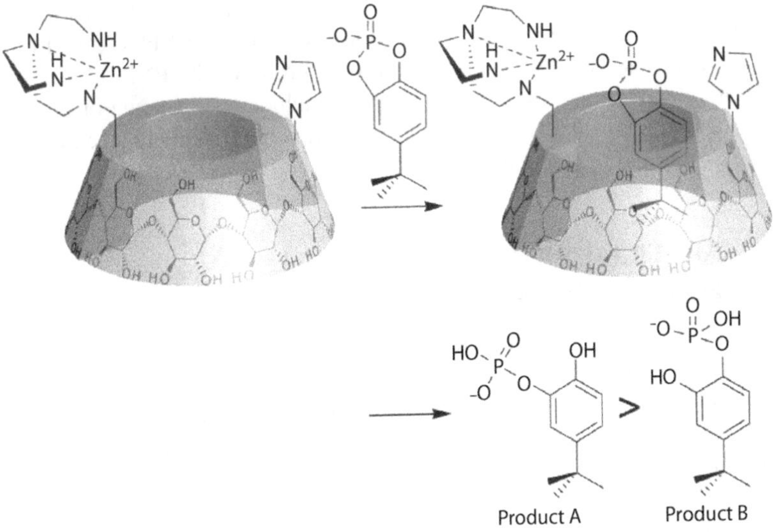

Product A   Product B

FIGURE 12.6   βCD with Zn²⁺ complex and imidazole group catalyzing hydrolysis of catechol phosphate substrate.

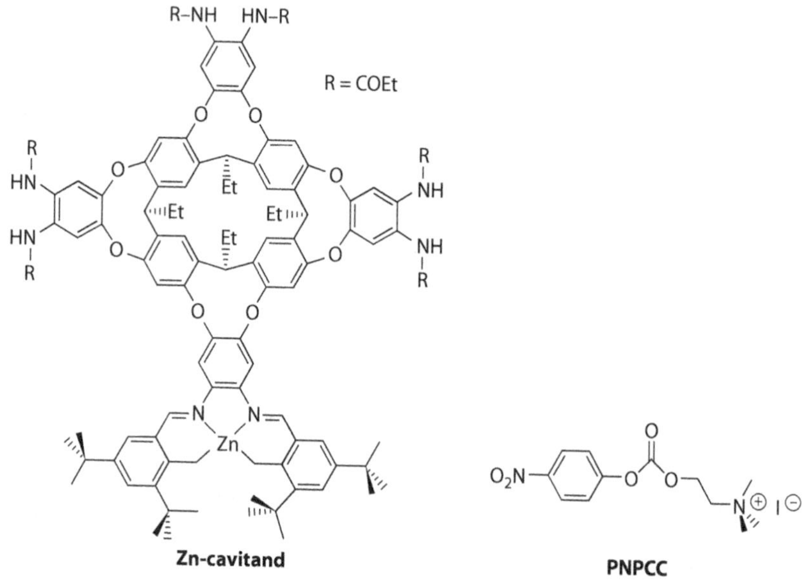

FIGURE 12.7 Chemical structures of Zn-cavitand with hydrolysis catalytic activity and *p*-nitrophenyl choline carbonate (PNPCC) as substrate.

have catalytically active functional groups, and the positions of catalytic groups determined by estimating the transition state are important for catalytic activities. Cavitands have affinities for alkylammonium ions. When a zinc(II) salen-functionalized cavitand hydrolyzes *p*-nitrophenyl choline carbonate (PNPCC), the Lewis acid zinc(II) activates the well-positioned carbonyl group of the PNPCC. The reaction rate is accelerated more than 50-fold with a stoichiometric amount of catalyst. The cation–π interactions and C = O···Zn coordination bond likely reduce the activation energy barrier in the reaction (Figure 12.7).[22,23]

12.4 ARTIFICIAL BIOMACROMOLECULES FOR ASYMMETRIC CATALYSTS

A metal complex with a host protein using a combination of avidin as a protein and biotin with a rhodium diphosphine complex was prepared as a supramolecular asymmetric catalyst. Avidin shows a high affinity for biotin ($K = \sim 10^{15}$ M^{-1}), indicating that the rhodium diphosphine complex quantitatively binds into the chiral space of avidin.

Hydrogenation of *N*-acetamidoacrylate with the avidin–biotin and rhodium–diphosphine complexes shows a moderate enantioselectivity, yielding (S)-*N*-acetamidoalanine at 41% *ee* (Figure 12.8).[24] Subsequently, an artificial metalloenzyme system using a combination of biotinylated diphosphine with mutated streptavidine was prepared. This catalytic system quantitatively yielded highly enantioselective products ((R) = 94%).[25]

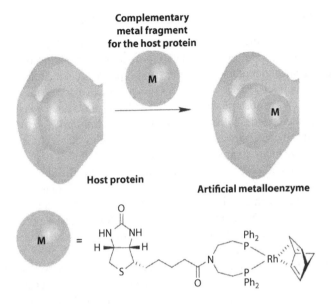

FIGURE 12.8 Strategy to incorporate catalytically active artificial metalloenzymes with avidin.

An artificial metalloenzyme for asymmetric hydrogenation has been created successfully. Monoclonal antibodies as host proteins exhibit extremely high affinities for their respective antigens. A monoclonal antibody for an achiral rhodium complex was prepared. Although the achiral rhodium complex alone did not show enantioselectivity for the hydrogenation of 2-acetamidoacrylic acid, the complex in the presence of the antibody showed catalytic hydrogenation to produce *N*-acetyl-L-alanine in a high (>98%) enantiomeric excess (Figure 12.9).[26]

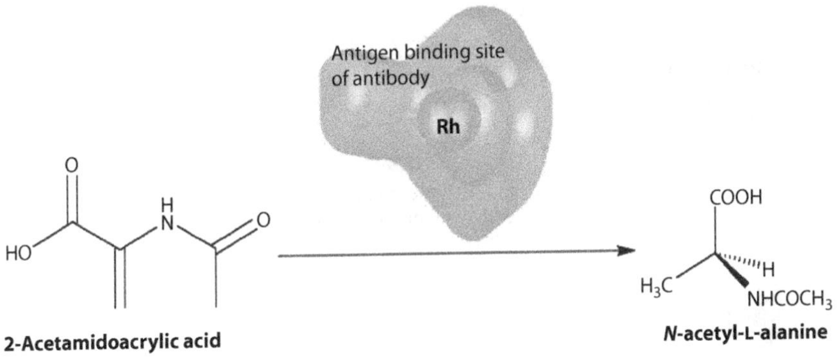

FIGURE 12.9 Representation of complex of antibody 1G8 and Rh catalyst to show substrate specificity.

12.5   SUPRAMOLECULAR PROCESSING CATALYSIS

Supramolecular structures such as rotaxanes occasionally play important roles in enzymatic activities. For example, DNA polymerases contain sliding clamps in which the ring-shaped protein assemblies form supramolecular complexes. The ring-shaped clamp peptides of DNA (or RNA) polymerases participate in binding two polynucleotide chains for replication.[27–30] The clamps have no active sites and replication of polynucleotides is difficult without clamps. Similarly, cyclodextrins (CDs) and other host molecules are ring-shaped host molecules that include various guests to form supramolecular complexes such as rotaxanes.[31–35]

Nolte, Rowan, and coworkers prepared cavity-functionalized porphyrin catalysts with diameter of about 9 Å. The catalytic macrocycles strongly bound viologen derivatives in their own cavities through electrostatic and π–π stacking interactions ($K_a = 10^5$ to 10^6 M^{-1}). The catalytic macrocycle with manganese porphyrin strongly binds a polymer substrate (polybutadiene) that is included in the cavity, and the macrocycle oxidizes polymer substrate complexes.

The bulky ligand (*tert*-butylpyridine) of the catalytic macrocycle is attached outside the porphyrin to prevent the macrocycle from oxidizing the substrate outside the cavity and effectively shielding the catalyst from forming the inactive μ-oxo dimeric species. Polybutadiene ($M_n = 3,000$, *cis* = 98%) was completely converted into polyepoxide (*trans:cis* = 80:20) by the addition of an oxygen donor (iodosylbenzene or hypochloride). This indicates that the catalytic macrocycle with manganese porphyrin slides on the polymer chain during the epoxidation reaction (Figure 12.10).[36]

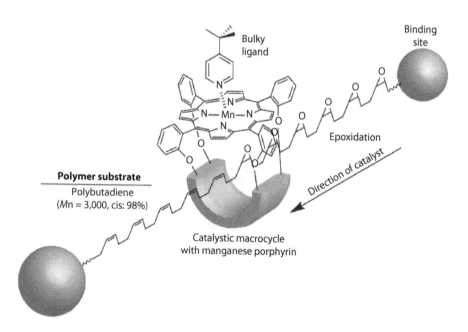

FIGURE 12.10   Toroidal Mn catalyst encircling polybutadiene substrate and epoxidation of polybutadiene.

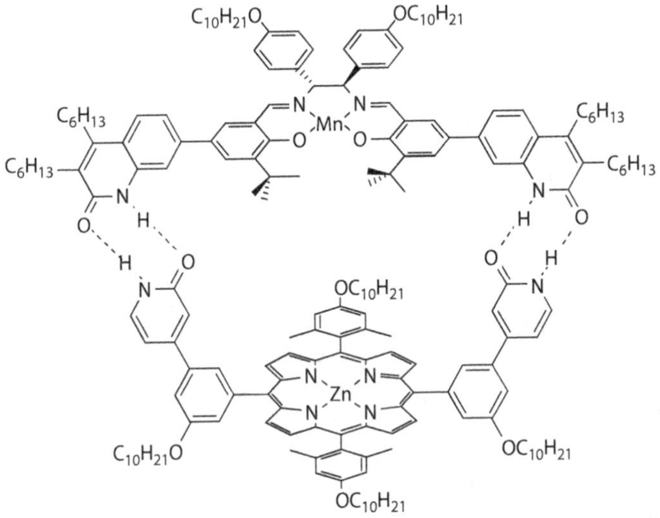

FIGURE 12.11 Chemical structure of supramolecular macrocyclic heterodimer Mn catalyst.

Supramolecular catalysts that mimic the ability of processive enzymes feature macrocycles with metal porphyrins as the catalytically active sites. Wärnmark and coworkers reported a dynamic supramolecular combination of a Mn (III) salen complex and a Zn(II) porphyrin through hydrogen bonding. The supramolecular complex has a binding cavity halfway between the Mn (III) salen and the Zn(II) porphyrin. The Mn (III) salen unit functions as the catalytic center for the epoxidation of olefin derivatives. However, reaction is not limited to substrates included within the cavity. Excluded substrates are also epoxidized by the supramolecular complex (Figure 12.11).[37,38]

Previous reports of supramolecular catalysts focused on hydrolysis and coupling reactions to yield low-molecular-weight molecules. However, supramolecular catalysts for the preparation of polymers reminiscent of DNA polymerases were not reported. Harada and coworkers reported that CDs selectively form inclusion complexes with some lactones that act as starting materials of the polyesters. The hydrolysis of ε-caprolactone (ε-CL) in the presence of αCD is suppressed in preference to ε-CL alone but is promoted in the presence of βCD. Conversely, the hydrolyses of δ-valerolactone (δ-VL) in the presence of αCD, βCD, and γCD, respectively, are suppressed, indicating that the hydrolysis of lactones depends on the sizes of the CDs and lactones (Figure 12.12).[39]

The secondary hydroxyl groups of CDs have reactivity to the carbonyl carbons of lactones due to the high basicity of those groups. This suggests that if lactones are heated with CDs in bulk without water, they may form polymers because hydrolysis does not take place.

When a mixture of βCD dried at 80°C in vacuum and δ-VL was heated at 100°C under an argon atmosphere, poly(δ-VL) was obtained in high yield (>98% after 48 hr) although δ-VL did not yield polymers under the same conditions without CDs (Figure 12.13a). γCD also generated a polymer but in a lower yield.

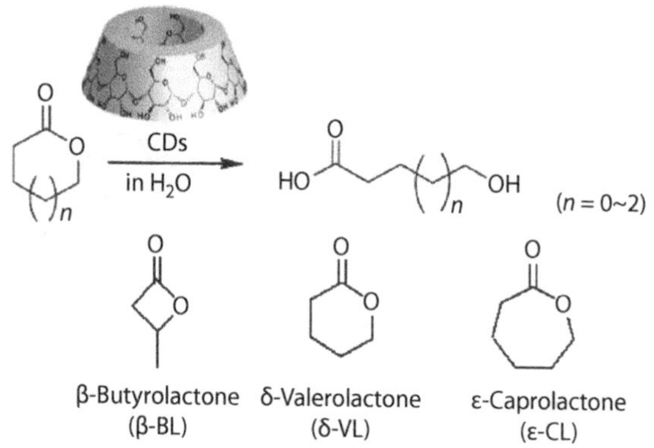

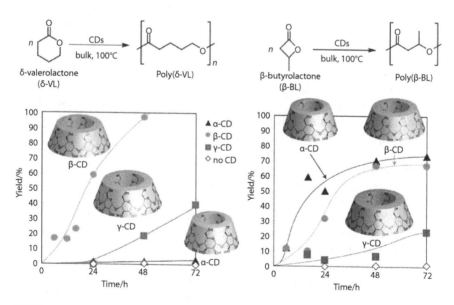

FIGURE 12.12 Hydrolysis of lactones in water.

In contrast, the activity of αCD is considerably lower under the same conditions. Furthermore, the order of the polymer yield of β-butyrolactone (β-BL) with CDs is βCD (67%) > αCD (55%) > γCD (7%); see Figure 12.13b. The cavity sizes of the CDs and monomer structure synergistically affect the polymer, indicating that the reaction occurs via the inclusion of lactones in the CD cavity.[40] These results suggest that CDs selectively initiate ring-opening polymerizations of lactones to generate high yields of polyesters without a solvent and co-catalyst.

As described above, CDs selectively include some lactones and initiate polymerization to give polyesters having single CD molecules at the ends of the polyester

FIGURE 12.13 Dependence of yield on time for polymerization of lactones initiated by various cyclodextrins in bulk at 100°C. (a) [δ-VL]/[CDs] = 5. (b) [β-BL]/[CDs] = 5.

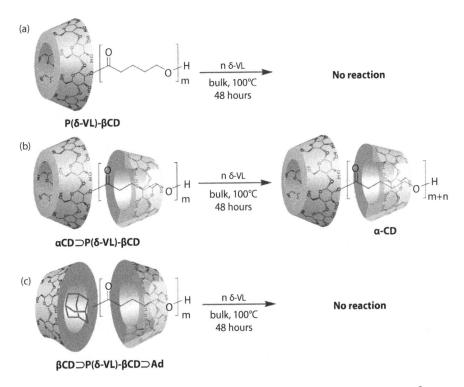

FIGURE 12.14 Reinitiating polymerization of δ-VL by P(δ-VL)-βCD (a), αCD⊃P(δ-VL)-βCD (b), and βCD⊃P(δ-VL)-βCD with adamantane (c) in bulk at 100°C.

chains simply by mixing the CD and lactones. Although the βCD functionalized poly(δ-VL) (P(δ-VL)-βCD) was still expected to have inclusion and polymerization abilities for lactones, the addition of δ-VL to P(δ-VL)-βCD did not change the molecular weight of P(δ-VL)-βCD, indicating that P(δ-VL)-βCD was inactive for polymerization of δ-VL (Figure 12.14a).

The authors estimated that the topological structure of the P(δ-VL)-βCD with intact CDs plays an important role in initiating polymerization again to produce longer polymers. The mixing of αCD and P(δ-VL)-βCD gave poly-*pseudo*-rotaxane, αCD⊃P(δ-VL)-βCD. βCD⊃P(δ-VL)-βCD was also obtained by a similar method. αCD⊃P(δ-VL)-βCD showed polymerization activity for δ-VL. The M_n of αCD⊃P(δ-VL)-βCD increased from 2,300 to 6,000 (Figure 12.14b). Moreover, the addition of adamantane (Ad) as a competitive guest for βCD⊃P(δ-VL)-βCD showed no polymerization activity for δ-VL (Figure 12.14c).

The conclusion was that δ-VL was not included in the βCD at the end of βCD⊃P(δ-VL)-βCD. These observations suggest that the βCD at the end of CDs⊃P(δ-VL)-βCD is the active site for the polymerization of δ-VL and that the intact αCD and βCD threaded onto the P(δ-VL)-βCD exerted no polymerization activity. The intact CDs threaded onto the P(δ-VL)-βCD cannot include δ-VL because they already included polymer chains in their cavities.

These threaded CDs are supposed to play a supplementary role to control the structure of poly(δ-VL) during polymerization. The CDs threaded by the poly(δ-VL) prevented the polymer chain from entangling with itself or with another polymer chain so as to maintain the propagating state of the polyester. These processes are similar to those of chaperone proteins in biological systems that aid protein folding and allow the formation of functional proteins. CDs showed the activation and transformation of monomers analogous to enzymes and also protein-like refolding activity as artificial chaperones.[41]

To take advantage of the polymerization activities of βCDs for lactones, a supramolecular architecture demonstrating polymerization activity was prepared by the reaction of βCDs thiolated at the narrow rims and a Au nanocolloid (HAuCl$_4$). A βCD nanosphere was synthesized by using the Au colloid as a template.[42] After removing the core of the Au colloid with iodine as the oxidant, the βCD units were linked to each other via disulfide bonds to form βCD nanospheres.

The βCD nanosphere yielded 70% oligo(δ-VL) by reaction with δ-VL in bulk at 100°C for 48 hr. The number of oligo(δ-VL) molecules ($M_n = 1,400$) propagated from the surface of the βCD nanosphere was five chains as determined by ^{1}H NMR spectroscopic measurement. On the other hand, the addition of Ad inhibited the polymerization of δ-VL. These results indicate that the βCD cavities on the surface of the sphere are the active sites and the oligomers propagate from the βCD cavities.

To obtain polymers with higher molecular weights, the formation of poly-*pseudo*-rotaxane by the addition of αCD to the βCD nanosphere was resolved. The resulting spherical supramolecule possessed the βCD spherical molecule as a core and was functionalized with the poly-*pseudo*-rotaxane consisting of αCD on the surface. Mixing the poly-*pseudo*-rotaxane modified nanosphere with δ-VL resulted in a re-initiation of the polymerization of δ-VL in 36% yield to obtain nanospheres with longer poly(δ-VL) chains ($M_n = 2,100$).[43]

12.6   ARTIFICIAL MOLECULAR CLAMP CATALYSIS

Selective molecular recognition and activation of substrates are considered essential functions in the basic design of supramolecular catalysts. However, to the best of our knowledge, attempts to improve the inhibition of further substrate recognition have not been reported, probably due to the greatly increased complexities of supramolecular catalytst design.

Recently, sliding clamps with ring-shaped proteins were reported to encircle DNA chains and provide high processivity on DNA polymerases in all living organisms.[44–46] Interestingly, the DNA clamp–clamp loader complex fulfills a function in DNA replication, whereas the sliding clamp only assists in the process.[47] The molecular clamps bind peptides or segments of a polypeptide chain in an extended conformation. Introducing an artificial molecular clamp to supramolecular catalysts should dramatically improve catalytic activity and turnover.

As a novel concept for supramolecular catalysis, an artificial molecular clamp was attached to the activation site. Synthetic polymerases with artificial molecular clamps yielded high-molecular-weight polymers without solvents or co-catalysts.

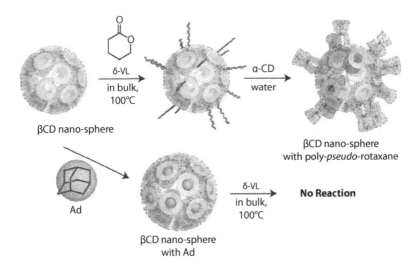

FIGURE 12.15 Polymerization of δ-VL by βCD nanosphere that forms poly-pseudo-rotax-ane on surface by addition of αCD. Addition of adamantane (Ad) inhibited polymerization of δ-VL.

The CD dimers behave as polymerases for cyclic esters in which one CD moiety initiates ring-opening polymerization, while the other propagates polymerization by serving as a molecular clamp. The polymerization activities of α,β-TPA-dimers linked with terephthalamide between αCD and βCD for δ-valerolactone (δ-VL) were studied. Polymerization of δ-VL was carried out by stirring and heating bulk mixtures of the CD dimers and δ-VL ([δ-VL]/[CD unit] = 50) at 100°C. Although the α,α-TPA-dimer linked with terephthalamide between αCDs showed much lower polymerization activity because the activity of αCD for δ-VL is very low (Figure 12.15b), the α,β-TPA-dimer displayed significantly higher polymerization activity to yield poly(δ-VL) with M_n = 11,000 (Figure 12.15a).

Although the molecular clamp CD does not exhibit polymerization activity for δ-VL, it assists polymerization by anchoring the polymer chain and securing the active site.[48] On the other hand, when 50 equivalents of adamantane (Ad) molecules as competitive guests were mixed with the α,β-TPA-dimer and 50 equivalents of δ-VL in a reaction tube in the solid state, the α,β-TPA-dimer with Ad did not show polymerization activity for δ-VL (Figure 12.15c) due to the inhibition of monomer recognition by including Ad in the βCD cavity.

The polymerization activities of the CD dimers depend on linker length. Linkers with the appropriate lengths displayed high levels of polymerization. If a linker was too short, the CD dimer suppressed monomer recognition. In contrast, if the linker was too long, the CD dimer could not guide the growing polymer chain.

To investigate the differences in polymerization activities corresponding to the supramolecular structure of a CD dimer with an oligomer chain, an α,β-TPA-dimer possessing oligo(δ-VL) threaded into a molecular clamp CD was prepared and compared with an α,β-TPA-dimer with the oligo(δ-VL) dethreaded. The α,β-TPA-dimer

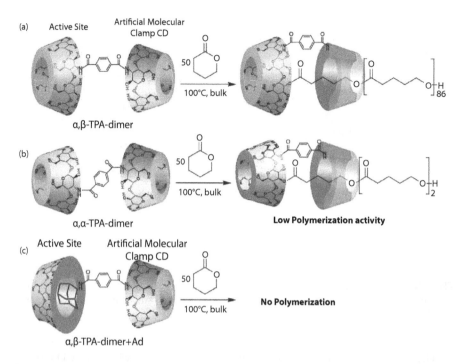

FIGURE 12.16 Reinitiating polymerization of δ-VL by α,β-TPA-dimer (a), α,α-TPA-dimer (b), and α,β-TPA-dimer with adamantane (Ad) (c) in bulk at 100°C.

possessing oligo(δ-VL) threaded into the molecular clamp CD initiated post-polymerization and gave a molecular weight of M_n = 16,500 in 95% yield (Figure 12.16). However, the α,β-TPA-dimer possessing oligo(δ-VL) dethreaded from the molecular clamp CD showed significantly lower polymerization activity (3.5% yield). See Figure 12.17. This behavior is similar to a sliding DNA clamp–clamp loader complex that plays an important role in propagating DNA in biological systems. These CD dimers efficiently initiate polymerization to produce high-molecular-weight polyesters.

12.7 CONCLUSION AND OUTLOOK

This chapter introduced several examples of supramolecular catalysts in recent works. Natural enzymes that show high selectivities and efficiently generate products from substrates at ambient temperature even from their flexible supramolecular structures attracted the interest of chemists in many fields. DNA and RNA polymerases show advanced processes for the synthesis of biomacromolecules. They recognize monomeric nucleotides with their binding sites and catalyze their polymerizations to produce polynucleotides. It is extremely difficult to mimic similar catalytic behaviors even with modern chemical techniques. Conventional molecular catalysts cannot perform these catalytic functions with such high selectivity and efficiency.

Inspired by nature's enzymatic catalysts such as cytochrome P450 and processive enzymes such as cellulases and DNA polymerases, many designed supramolecular

(a) Polymerization activity of α,β-TPA-dimer with threaded oligo(δ-VL)

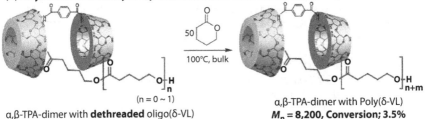

α,β-TPA-dimer with **threaded** oligo(δ-VL)

$(n = 0 \sim 1)$

α,β-TPA-dimer with poly(δ-VL)
$M_n = 16,500$, Conversion; 95%

(b) Polymerization activity of α,β-TPA-dimer with dethreaded oligo(δ-VL)

α,β-TPA-dimer with **dethreaded** oligo(δ-VL)

$(n = 0 \sim 1)$

α,β-TPA-dimer with Poly(δ-VL)
$M_n = 8,200$, Conversion; 3.5%

FIGURE 12.17 (a) Polymerization of δ-VL initiated by α,β-TPA-dimer with oligo(δ-VL) threaded into molecular clamp CD. (b) Polymerization of δ-VL initiated by α,β-TPA-dimer with oligo(δ-VL) dethreaded into molecular clamp CD.

catalysts have been successful in mimicking the catalytic abilities of natural enzymes and surpassing them. Recent supramolecular catalysts have precision substrate-recognition sites and can control reactions precisely via non-covalent interactions with the flexible structures at the reactive sites. The selective molecular recognition and activation processes of substrates are considered essential basic functions in the design of supramolecular catalysts.

Moreover, highly sophisticated supramolecular catalysts improved the inhibition of further substrate recognition. Sliding clamps utilizing ring-shaped proteins shown to encircle DNA chains provide high processivity for DNA polymerases in all living organisms. Artificial molecular clamps attached to activation sites represent an entirely new concept for supramolecular catalysts and should create new paradigms for catalytic reactions of supramolecular catalysts and also provide templates for fine chemical synthesis. Supramolecular catalysts are expected to contribute fundamentally and conceptually different means for achieving chemical transformations. They will also add to our understanding of the science governing biologically catalyzed reactions.

REFERENCES

1. van Leeuwen, P. W. N. 2008. *Supramolecular Catalysis*. Weinheim: Wiley-VCH.
2. Bender, M. L. and Komiyama, M. 1978. *Cyclodextrin Chemistry*. Berlin: Springer.
3. Breslow, R. 1986. In *Advances in Enzymology and Related Areas of Molecular Biology*, Meister, A., Ed. New York: John Wiley & Sons, pp. 1–60.
4. Komiyama, M. and Shigekawa, H. 1996. In *Comprehensive Supramolecular Chemistry*, Szejtli, J. and Osa, T., Eds. Oxford: Pergamon Press, pp. 401–422.

5. Ronald, B. 2005. *Artificial Enzymes*. New York: Wiley-VCH.
6. Bender, M. L. *J. Inclusion Phenom.* 1984, 2, 433–444.
7. D'Souza, V. T. and Bender, M. L. *Acc. Chem. Res.* 1987, 20, 146–152.
8. Komiyama, M., Breaux, E. J., and Bender, M. L. *Bioorg. Chem.* 1977, 6, 127–136.
9. Breslow, R., Doherty, J. B., Guillot, G. et al., *J. Am. Chem. Soc.* 1978, 10, 3227–3229.
10. Breslow, R., Hammond, M., and Lauer, M. 1980. *J. Am. Chem. Soc.* 1980, 102, 421–422.
11. Breslow, R., Canary, J. W., and Varney M. et al. 1990. *J. Am. Chem. Soc.* 112, 5212–5219.
12. Tabushi, I., Kuroda, Y., Yamada, M. et al. *J. Am. Chem. Soc.* 1985, 107, 5545–5546.
13. Fasella, E., Dong, S. D., and Breslow, R. *Bioorg. Med. Chem.* 1999, 7, 709–714.
14. Hilvert, D. and Breslow. R. *Bioorg. Chem.* 1984, 12, 206–220.
15. Breslow, R. and Kool. E. *Tetrahedron Lett.* 1988, 29, 1635–1638.
16. Breslow, R., Groves, K., and Mayer, M. U. *Pure Appl. Chem.* 1998, 70, 1933–1938.
17. Breslow, R., Groves, K., and Mayer, M. U. *J. Am. Chem. Soc.* 2002, 124, 3622–3635.
18. Breslow, R. and Overman, L. E. *J. Am. Chem. Soc.* 1970, 92, 1075–1077.
19. Breslow, R. and Katz, I. *J. Am. Chem. Soc.* 1968, 90, 7376–7377.
20. Dong, S. D. and Breslow, R. 1998, 39, 9343–9346.
21. Rebek, J., Jr. *Acc. Chem. Res.* 1999, 32, 278–286.
22. Richeter, S. and Rebek, J., Jr. *J. Am. Chem. Soc.* 2004, 126, 16280–16281.
23. Purse, B. W. and Rebek, J. Jr. *Proc. Natl. Acad. Sci. USA* 2005, 102, 10777–10782.
24. Wilson, M. E. and Whitesides, G. M. *J. Am. Chem. Soc.* 1978, 100, 306–307.
25. Skander, M., Humbert, N., Collot, J. et al. *J. Am. Chem. Soc.* 2004, 126, 14411–14418.
26. Yamaguchi, H., Hirano, T., Kiminami, H. et al. *Org. Biomol. Chem.* 2006, 4, 3571–3573.
27. Breyer, W. A. and Matthews, B.W. *Protein Sci.* 2001, 10, 1699–1711.
28. Trakselis, M. A., Alley, S. C., Abel-Santos, E. et al. *Proc. Natl. Acad. Sci. USA* 2001, 98, 8368–8375.
29. Benkovic, S. J., Valentine, A. M., and Salinas, F. *Annu. Rev. Biochem.* 2001, 70, 181–208.
30. Stryer, L., Ed. 1996. *Biochemistry,* 4th Rev. Ed. Heidelberg: Spektrum.
31. Szejtli, J. and Osa, T., Eds. 1996. *Comprehensive Supramolecular Chemistry: Cyclodextrins,* Vol. 3. New York: Pergamon.
32. Cyclodextrins: Special Issue. 1998. *Chem. Rev.* 98 (5).
33. Harada, A. and Kamachi, M. *Macromolecules* 1990, 23, 2821–2823.
34. Harada, A. and Kamachi, M. *J. Chem. Soc., Chem. Commun.* 1990, 19, 1322–1323.
35. Harada, A., Li, J., and Kamachi, M. *Nature* 1992, 356, 325–327.
36. Thordarson, P., Bijsterveld, E. J. A., Rowan, A. E. et al. *Nature* 2003, 424, 915–918.
37. Jonsson, S., Odille, F. G. J., Norrby, P.-O. et al. *Chem. Commun.* 2005, 4, 549–551.
38. Jonsson, S., Arribas, C. S., Wendt, O. F. et al. *Org. Biomol. Chem.* 2005, 3, 996–1001.
39. Takashima, Y., Kawaguchi, Y., Nakagawa, S. et al. *Chem. Lett.* 2003, *32*, 1122–1123.
40. Takashima, Y., Osaki, M., and Harada, A. *J. Am. Chem. Soc.* 2004, 126, 13588–13589.
41. Osaki, M., Takashima, Y., Yamaguchi, H. et al. *J. Am. Chem. Soc.* 2007, 129, 14452–14457.
42. (a) Sun, L., Crooksa, R. M., and Chechik, V. *Chem. Commun.* 2001, 4, 359–360. (b) Liu, J., Ong, W., Román, E. et al. *Langmuir* 2000, 16, 3000–3002. (c) Rojas, M. T., Königer, R., Stoddart, S. F. et al. *J. Am. Chem. Soc.* 1995, 117, 336–343.
43. Osaki, M., Takashima, Y., Yamaguchi, H. et al. *J. Org. Chem.* 2009, 74, 1858–1863.
44. Huang, C. C., Hearst, J. E., and Alberts, B. M. *J. Biol. Chem.* 1981, 256, 4087–4094.
45. Prelich, G., Kostura, M., Marshak, D. R. et al. *Nature* 1987, 326, 471–475.
46. Stukenberg, P. T., Studwell-Vaughan, P. S., and O'Donnell, M. *J. Biol. Chem.* 1991, 266, 11328–11334.
47. Bowman, G. D., O'Donnell, M., and Kuriyan, J. *Nature* 2004, 429, 724–730.
48. Takashima, Y., Osaki, M., Ishimaru, Y. et al. *Angew. Chem. Int. Ed.* 2011, 50, 7524–7528.

13 Supramolecular Functions of Cyclodextrin Complex Sensors for Sugar Recognition in Water

Hiroyuki Kobayashi, Takeshi Hashimoto, and Takashi Hayashita

CONTENTS

13.1 INTRODUCTION

Sugars are attracting much attention not only because of their functions as energy sources in living organisms and also due to their involvement in various biological processes such as development, differentiation, carcinogenesis, and immunity. In the body, starch and glycogen are the main energy sources along with glucose and sucrose, whereas cellulose, hemicellulose, chitin, and polysaccharides in cell walls help support organisms and provide structural framework. Similarly, glycoproteins and mucopolysaccharides on the cell surface or between cells are involved in cell identification interactions, cell differentiation, and immune response.[1] Thus, the development of a simple, rapid, and selective method for the recognition of sugars in water is a key issue, given the importance of sugars in life processes.

Blood glucose monitoring is essential in diabetic patients and enzyme-based glucose sensor kits catering to this need are used widely. However, the sensor has several drawbacks, including the sensitivity of the enzymes to heat and pH and limitations in the types of sugars that can be detected. Moreover, the sensor cannot be used as an imaging probe to study sugar distribution *in vivo*. Therefore, the development of chemosensors that are not based on enzymatic reactions but rather utilize artificial host compounds has been pursued.[2]

Artificial sugar receptors have attracted much interest recently. One example is an artificial sugar receptor that uses phenylboronic acid. It is known that phenylboronic acid forms a cyclic ester with a *cis*-diol group on sugar and other molecules and that the formation is accompanied by acid dissociation. Thus, phenylboronic acid is an important functional group in the designs of sensors for sugar recognition.[3] Studies have shown that the use of supramolecular chemosensors in combination with a nanospace reaction field such as that offered by cyclodextrins (CDs) enhances sugar recognition.

Supramolecular chemistry is concerned with an assembly of molecules bound by multiple weak non-covalent forces such as hydrogen bonds, aromatic π-stacking, and van der Waals interactions.[4] These supramolecular structures are expected to produce novel synergistic functions that differ from those found in simple molecules. CDs are water-soluble host compounds having nanosize hydrophobic cavities that enable them to incorporate various organic molecules in water. In addition, optically inert CDs can be combined efficiently with various types of chromo- and fluoroionophores. In this review, phenylboronic acid-based approaches toward sugar recognition using the synergistic functions of supramolecular CD complexes will be discussed with a focus on the authors' studies.[3,5]

13.2 SUGAR RECOGNITION WITH PHENYLBORONIC ACID

Lorand and Edwards were the first to report sugar recognition by phenylboronic acid in 1959.[6] Phenylboronic acid has a simple structure, and the researchers used it to determine the binding constants for 16 compounds having two or three hydroxyl groups each. The results showed that boronic acid binds 1,2- and 1,3-diols, forming 5- and 6-membered cyclic esters, and that even more stable cyclic esters can be formed with *cis*-diols adjacent to sugars than with non-cyclic diols such as ethylene glycol.

Among these, the binding constants for sugars have been reported; the highest were for fructose ($4370 \ M^{-1}$), followed by those for galactose ($276 \ M^{-1}$) and glucose ($110 \ M^{-1}$). This selectivity, as demonstrated by subsequent studies, almost corresponds to the binding order of monophenylboronic acid. Moreover, studies on the structure of phenylboronic acid in water revealed that at neutral pH, boronic acid assumes a trigonal planar geometry with a phenyl group in which the boron atom is sp^2-hybridized ($C_6H_5B(OH)_2$). On the other hand, in the anionic form, it reportedly assumes a tetrahedral geometry in which the boron atom is sp^3-hybridized ($C_6H_5B(OH)_3^-$).

Phenylboronic acid is primarily present in the anionic form in dilute aqueous solution, and on reacting with a diol, it undergoes esterification while maintaining its tetrahedral geometry (Figure 13.1a). It has been demonstrated that with esterification, a shift in acid–base equilibrium toward lower pH occurs. We should note that

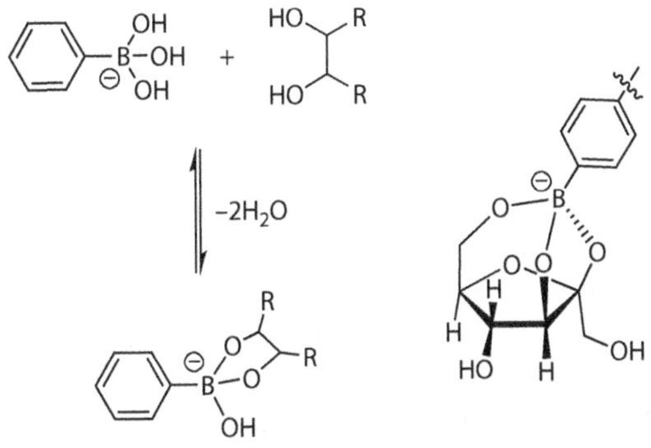

FIGURE 13.1   (a) Reaction of diol with tetrahedral anion to form anionic acid. (*Source*: J. P. Lorand et al., J. *Org. Chem.* 24, 769–774 (1959). Copyright (1959), American Chemical Society. With permission.) (b) Ester formation with fructose (*Source:* S. Arimori et al., *J. Chem. Soc. Perkin Trans.* 1, 803–808 (2002). With permission of Royal Society of Chemistry.)

the high selectivity for fructose is due to the fact that three-point recognition is possible, as shown in Figure 13.1b.

13.3   GLUCOSE-SELECTIVE FLUORESCENT SENSORS

Fluorescent sensors based on monoboronic acid often exhibit selectivity for fructose. The development of glucose-selective chemosensors in the medical field is being pursued vigorously because of the need to measure blood glucose levels for the management of diabetes.

The design of a molecular recognition site that allows the capture of one glucose molecule by two boronic acid residues is expected to increase the selectivity for glucose.[7] As it is difficult for other monosaccharides such as fructose to form esters in a 1:2 ratio, glucose selectivity can be enhanced. James et al. developed a boronic acid fluorescent probe **1** (Figure 13.2) that exhibits glucose selectivity via a two-point

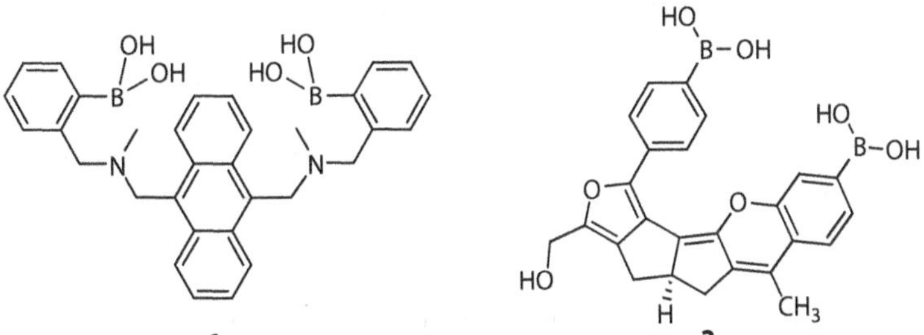

FIGURE 13.2   Molecular structures of diboronic acids for glucose sensing. (*Sources:* **1:** T. D. James et al., *J. Am. Chem. Soc.*, 117, 8982–8987 (1995). Copyright (1995), American Chemical Society. With permission. **2:** W. Yang et al., *Angew. Chem., Intl. Ed.*, 40, 1714–1718 (2001). Copyright (2001), John Wiley and Sons. With permission.)

interaction.[8] Probe **1** has two boronic acid moieties in a single molecule. The binding constant for glucose was 4000 M^{-1} (33.3% methanol/water), compared to 320 M^{-1} for fructose. Since then, various diboronic acid sensors that can form 1:2 type esters have been developed.[9]

Glucose in aqueous solution exists as an equilibrium mixture of a five-membered furanose ring and a six-membered pyranose ring. Norrild and Eggert reported that glucose is bound to boronic acid mainly in the furanose form rather than in the pyranose form, based on ^{1}H NMR analysis.[10] Similarly, when fluorescent probe **1** developed by Shinkai et al. was investigated, bound glucose molecules were found to be in the pyranose form, but it was concluded that the diboronic acid group first binds glucose in the pyranose form, but rearranges over the course of time to bind glucose in the furanose form.[11]

Drueckhammer and coworkers designed fluorescent sensor **2** using the same two-point interaction concept, the objective of which was to bind glucose in the pyranose form (Figure 13.2).[12] As expected, sensor **2** had high glucose selectivity and its binding constant for glucose reached 4000 M^{-1} (30% methanol/water), whereas almost no interaction was observed with fructose.

13.4   DEVELOPMENT OF SUPRAMOLECULAR FLUORESCENT SENSORS

The supramolecular chemistry of molecular complexes in biological systems that are formed by weak interactions (e.g., hydrogen bonding) of a number of molecules is pursued with an eye to developing new functional capabilities differing from those of individual molecules. By introducing the concept of supramolecular sensors that combine molecular self-organization and accompanying optic information conversion function into the design of sugar sensors, various response functions not observed with conventional chemosensors based on 1:1-type interactions can be expected.[13] In the following sections, unique sugar recognition functions demonstrated by CD-based supramolecular complexes will be introduced.

13.4.1   SUGAR RECOGNITION WITH CD COMPLEX

CDs are cyclic oligosaccharides having three known types, α-CD (six-membered), β-CD (seven-membered), and γ-CD (eight-membered) that differ in the number of glucopyranose groups that form the ring structure (Figure 13.3). Water-soluble CD is structured like a bucket from which the bottom is knocked out, and its cavity size varies. The inclusion of various organic molecules in the CD cavity enhances the solubility of the organic molecules in water. We reported that various molecular recognition reactions occur in water by using a supramolecular CD complex with a poorly water-soluble pyrene-based fluorescent sensor included in the CD cavity.

We designed boronic acid fluorescent sensors **3** ($n = 1, 4$) and **4** as supramolecular CD complexes with a sugar recognition functions in which phenylboronic acid and a pyrene fluorophore are connected via a suitable alkyl chain (Figure 13.4).[14,15] Sensor **3** alone shows weak fluorescence as it rearranges in water, but by forming an inclusion

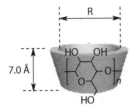

	Cavity diameter	
Cyclodextrin	n	R/Å
α-CD	6	4.5
β-CD	7	7.0
γ-CD	8	8.5

FIGURE 13.3 Structures and cavity diameters of cyclodextrins. (*Source:* T. Hayashita et al., *J. Incl. Phenom. Macrocycl. Chem.*, 50, 87–94 (2004), Springer Science and Business Media, with permission.)

complex with β-CD, pyrene monomer fluorescence is significantly increased even in water. This demonstrates that inclusion complex formation suppresses quenching by water and restricts the motion of a fluorescent sensor, thereby diminishing the non-radiative deactivation process.

Changes in the fluorescence emission spectra and the pH profile of the **3b/**–β-CD complex are shown in Figure 13.5. Whereas the pK_a value is 7.95 in the absence of sugar, the apparent pK_a shifts to 6.06 in the presence of 30 mM fructose. It is clear that this decrease in pK_a leads to an increase in fluorescence intensity when sugar is added under neutral conditions, allowing for the detection of sugar.

The mechanism by which the addition of sugar enhances fluorescence response is shown in Figure 13.4. In this fluorescent sensor, the pyrene fluorophore serves as an electron donor and the phenylboronic acid moiety serves as an electron acceptor. In other words, photo-induced electron transfer (PET) from the donor (pyrene) to the acceptor (acidic or neutral form of phenylboronic acid) causes quenching in the absence of sugar.

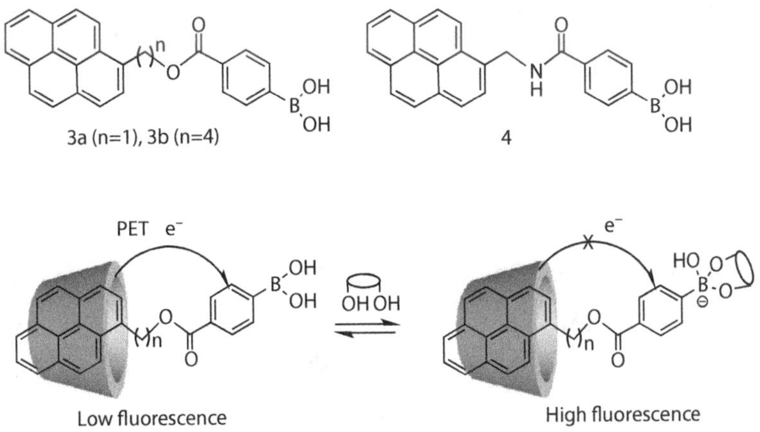

3a (n=1), 3b (n=4)   4

PET e⁻   e⁻

Low fluorescence   High fluorescence

FIGURE 13.4 Fluorescent boronic acid–γ-CD complex sensor. (*Sources:* A. J. Tong et al., *Anal. Chem.*, 73, 1530–1536 (2001). Copyright (2001), American Chemical Society. With permission. M. Kumai et al., *Anal. Sci.*, 28, 121–126 (2012), Copyright (2012), The Japan Society of Analytical Chemistry. With permission.)

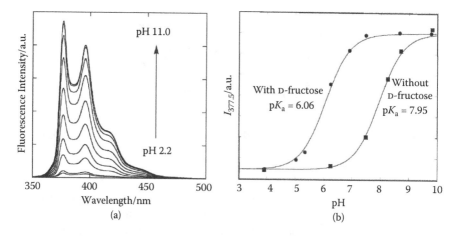

FIGURE 13.5 (a) pH dependence of fluorescence spectra for **3b**/β-CD. (b) $I_{377.5}$ versus pH profile for **3b**/β-CD with and without D-fructose; [**3b**] = 1.0 µM in 2% DMSO-98% water (v/v) containing 5.0 mM β-CD; pH adjusted by 0.015 M phosphate buffer; excitation wavelength = 328 nm. (*Source:* A. J. Tong et al., *Anal. Chem.*, 73, 1530–1536 (2001). Copyright (2001), American Chemical Society. With permission.)

However, this PET-based quenching is inhibited by the formation of an ester by the reaction of sugar and boronic acid molecules that weakens the electron-accepting property. In fact, based on the PET mechanism, the response efficiency depends on the spacer length, and when comparing **3a** and **3b**, **3a**, which has a shorter spacer, shows greater fluorescence recovery in terms of the fluorescence quantum yield under basic conditions. The selectivity for sugar follows the order of fructose > arabinose > galactose > glucose, consistent with the general binding order of sugar and phenylboronic acid. The pK_a changes for fructose, the fluorescence emission recovery rates, and the binding constants for the complexes of sensors **3** and **4** and β-CD are summarized in Table 13.1.

As for glucose recognition, one method is to use the dimerization reaction of a boronic acid probe within γ-CD, which has a large cavity.[16] Recently, Jiang and

TABLE 13.1

Comparison of Characteristics of Boronic Acid Probe–β-CD Complex[a]

Probe	pK_a	DpK_a	ϕ_L/ϕ_{HL}	K_{LS}/M⁻¹
		with 30 mM Fructose		
3a	6.30	1.96	57.4	2800
3b	6.06	1.89	15.2	2500
4	6.03	2.01	34.5	3900

[a] Apparent pK_a was obtained in presence of 30 mM D-fructose. ϕ_L/ϕ_{HL} is ratio of quantum yields of L⁻ and HL forms. K_{LS} is binding constant of probe/β-CD complex for D-fructose.

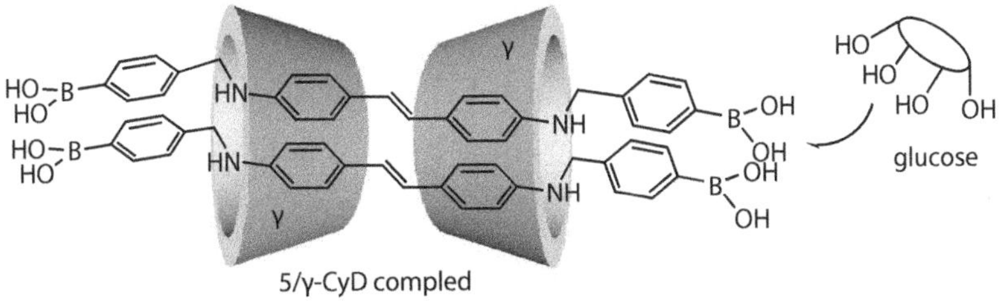

FIGURE 13.6   Supramolecular 2:2 complex for glucose recognition. (*Source:* X. Wu et al., *Chem. Commun.*, 48, 4362–4364 (2012). Royal Society of Chemistry. With permission.)

coworkers designed bisboronic acid **5** with styryl groups in the main chain backbone and evaluated the sugar recognition function of the **5**/CD complex. Fluorescence quenching occurs when sugar is added to the **5**/β-CD complex, and the selectivity for fructose is similar to those of other monophenylboronic acids. On the other hand, in the **5**/γ-CD complex, the formation of a 2:2 inclusion complex for glucose occurs, resulting in high glucose selectivity with fluorescence emission (Figure 13.6)[17].

13.4.2   SUGAR RECOGNITION BY MODIFIED CD

We have been examining a variety of chemically modified CDs for use with fluorescent sensor **4**, which exhibits excellent fluorescence response.[18] For example, we investigated the sugar recognition function of **4**/NH$_2$-CD complexes for both β-CD and γ-CD, which differ in inner cavity size, using 6-NH$_2$-CD (amino-substituted CD at the primary hydroxyl sites) and 3-NH$_2$-CD (amino-substituted CD at the secondary hydroxyl sites). We found almost no change in the fluorescence response of the 6-NH$_2$/CD complex compared to that of the non-amino-substituted CD complex. This suggests that amination at the primary hydroxyl sites does not affect sugar recognition; in other words, the boronic acid recognition site is on the secondary hydroxyl site of the CD complex.

On the other hand, using 3-NH$_2$-CD, we found that the inclusion function of 3-NH$_2$-β-CD for **4** was lost due to a steric hindrance of NH$_2$ group in 3-NH$_2$-β-CD, whereas 3-NH$_2$-γ-CD formed a stable inclusion complex with **4** and the **4**/3-NH$_2$–γ-CD complex exhibited high glucose selectivity. The glucose recognition functions of various **4**/CD complexes are summarized in Table 13.2.

It is evident that multi-point interaction, including electrostatic interaction, facilitates recognition by the **4**/3-NH$_2$–γ-CD complex (Figure 13.7a). These findings demonstrate that the sugar recognition function greatly changes with the type of CD used in complex formation.

The design of supramolecular complexes based on chemically modified CDs incorporating boronic acid has been reported. Suzuki et al. reported that an inclusion complex of receptor **6** (Figure 13.8) composed of β-CD bearing a phenylboronic acid moiety and styryl derivative **7**, a cationic fluorescent probe, can serve as a unique supramolecular sensor with selectivity for glucose.[19,20] The phenylboronic acid residue on **6** acts as the sugar receptor, whereas fluorescent probe **7** emits fluorescence that fluctuates depending on the environment.

TABLE 13.2

D-Glucose Recognition Function of 4–CD Complexes[a]

CD	pK_a	ΔpK_a	K_{LS}/M^{-1}
β–CD	7.60	0.58	93
6-NH$_2$- β–CD	7.54	0.53	80
3-NH$_2$- β–CD	7.50	0.61	1.0×10^2
γ–CD	7.36	0.49	70
6-NH$_2$- γ–CD	6.91	0.37	19
3-NH$_2$- γ–CD	6.95	1.07	3.6×10^2

[a] [4] = 1.0×10^{-6} M in 2% DMSO-98% water (v/v). [CD] = 5 mM, pH at 25°C adjusted with 0.01 M phosphate buffer, $I = 0.1$ M with NaCl. Apparent pK_a was obtained in presence of 30 mM D-glucose. K_{LS} is binding constant of 4/CD complex for D-glucose.

Probe **7** forms a 1:2 inclusion complex with **6**, which is a pseudo-rotaxane-type complex. Interestingly, this sensor yields a stronger fluorescence response in glucose than in fructose. In this supramolecular system, the binding of fructose by **6** decreases the ability of fluorescent probe **7** to form an inclusion complex with **6** due to steric hindrance. However, the binding of glucose by **6** increases its inclusion capability for fluorescent probe **7** compared to that of unbound **6**, which is facilitated by electrostatic interaction without the effect of steric hindrance. This in turn leads to an increase in fluorescence intensity with selectivity for glucose. This system represents a novel synergistic response that takes advantage of the dynamic changes in inclusion equilibrium based on sugar recognition.

We also revealed that when phenylboronic acid is introduced to the secondary hydroxyl sites of γ-CD (receptor **6**), the complex of receptor **6** with fluorescent sensor **4** exhibits glucose selectivity in water due to multi-point interaction (Figure 13.7b).[21] Fluorescent sensor **4**, when used alone, is a selective fructose sensor. As illustrated by these examples, supramolecular CD complexes have great appeal in that the

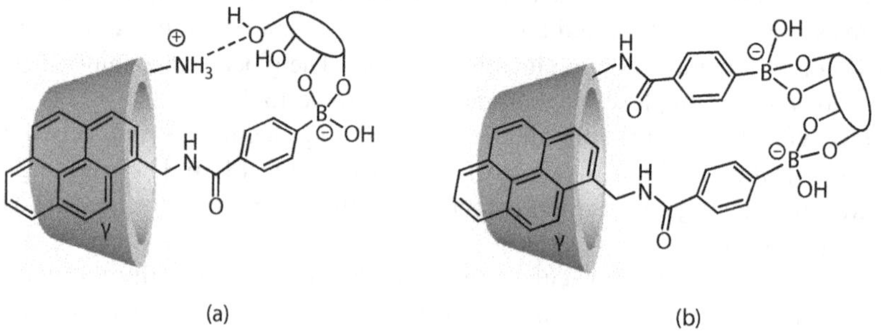

(a)   (b)

FIGURE 13.7 D-Glucose recognition by (a) **4**/3-NH$_2$-γ-CD complex [Reproduced from M. Kumai et al., *J. Ion Exchange*, *21*(3), 249–253 (2010) with permission of the Japan Society of Ion Exchange (JSIE)] and (b) **4**/phenylbronic acid-modified γ-CD(**6**) complex [Reproduced from M. Kumai et al., *Anal. Sci.*, **28**, 121-126 (2012), copyright (2012) The Japan Society of Analytical Chemistry].

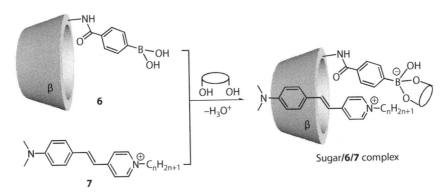

FIGURE 13.8 Sugar recognition by styrylpyridinium fluorophore (**7**)/phenylboronic acid-modified γ-CD/(**6**) complex. (*Source:* I. Suzuki et al., *Anal. Sci.*, 23, 1167–1171 (2007). Copyright (2007), The Japan Society of Analytical Chemistry. With permission.)

diversity of combinations is expected to contribute to the emergence of novel responses that cannot be observed with individual fluorescent sensors alone.

13.4.3 Sugar Recognition by Supramolecular CD Gel

Our studies revealed that conformational changes of the probes inside CD cavities contributed to the large difference in selectivity for saccharides. This supramolecular recognition function is expected to act as a novel sugar separation system in which the hydrophobic phenylboronic acid ligands are dynamically fixed inside a polymeric CD gel.[22]

We synthesized 4-(4′-alkoxyphenylazo)-phenylboronic acid (**8**) and introduced it into a γ-CD to form a gel using ethylene glycol digrycyl ether (EGDE)[23] in the presence

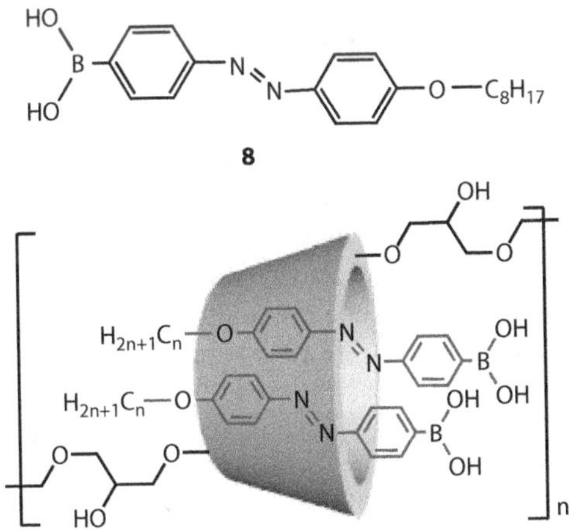

FIGURE 13.9 Structures of azoprobes and supramolecular gels. (a) **8** and (b) **8′(B-Azo-C*n*)**-γ-CD gel. (*Source:* T. Hashimoto et al., *Chem. Lett.* 43(2), 228–230 (2014), the Chemical Society of Japan. With permission.)

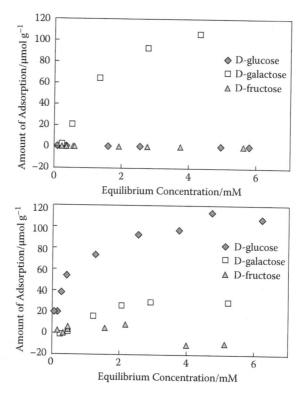

FIGURE 13.10 Adsorption isotherm of D-glucose, D-fructose, and D-galactose by azoprobes/γ-CD gels in water. (a) **8**/γ-CD gel without template sugar. (b) **8**/γ-CD gel with D-glucose template. NaCl = 0.1 M, Na_2CO_3 = 10 mM, sugar = 0 to 6 mM at 30°C. (*Source:* T. Hashimoto et al., *Chem. Lett.* 43(2), 228–230 (2014), the Chemical Society of Japan. With permission.)

or absence of sugar for the selective adsorption of sugars based on the template effect (Figure 13.9). This system has several merits: (a) the selective concentration of sugars is feasible *via* specific adsorption; (b) the template effect can be activated efficiently by using an adsorption system; and (c) the repositioning of the binding sites for target sugar recognition by repeating sugar adsorption (self-learning ability[24] of CD gel) is expected.

The gel without template sugar exhibited selective adsorption for galactose, and no adsorption was noted for glucose and fructose (Figure 13.10a). It is evident that **8** is positioned at appropriate recognition sites for galactose inside the γ-CD of template-free gel. In contrast, the glucose template gel showed selective and high adsorption ability for glucose, as expected (Figure 13.10b). This difference was caused by changes in the positions of the two azoprobes inside γ-CD, as confirmed by the induced circular dichromism spectra of sugar/**8**–γ-CD complexes in water.[22]

13.5 CONCLUSIONS

In this review, we discussed sugar recognition by boronic acid-based fluorescent sensors and azoprobes in combination with CD, with focus on our research. Among the boronic acid–based fluorescent sensors, those that utilize the PET mechanism allow for molecular design aimed to achieve high sugar selectivity by pre-orienting molecular recognition sites by using suitable spacers to position various fluorophores and boronic acids. We also demonstrated that unique sugar selectivity could be achieved with supramolecular chemosensors such as CD complexes by taking advantage of their diverse combinations and synergism.

Other molecular complex-based sugar sensors have been reported,[25–27] including those that form complexes in the presence of polycations[28,29] and those that utilize polymers modified with boronic acid functional groups,[30–34] hydrogels,[35–37] and nanoparticles.[38,39]

The principle of sugar recognition introduced in this review can be applied to chemosensors and also to materials used to isolate various carbohydrate moieties such as glycoproteins and glycated hemoglobin. The future goal will be to develop supramolecules that are even closer to biological systems by shifting the focus from simple molecular chemistry based on 1:1 recognition and responses of molecules to the construction of multi-molecular and macro-molecular systems that possess even more complex molecular recognition functions. The development of artificial boronolectin, which makes use of the boronic acid recognition moiety to achieve high sugar recognition ability as demonstrated by biological systems such as lectins, represents one such objective.

ACKNOWLEDGMENTS

We would like to express our deepest condolences over the passing of Dr. Hiroshi Tsukube (Osaka City University), who had been on the frontlines of molecular recognition chemistry and supramolecular chemistry. This study was supported by a Grant-in-Aid for Scientific Research (No. 22350039) from the Ministry of Education, Culture, Sports, Science and Technology of Japan.

REFERENCES

1. H. Iwase, M. Onishi, M. Kiso et al., *Introduction to Sugar Chain Chemistry*. Baifukan: Tokyo (1999).
2. T. D. James, *Boronic Acids*. Wiley-VCH: Weinheim, pp. 441–479 (2005).
3. A. Yamauchi, I. Suzuki, and T. Hayashita, *Glucose Sensing, Topics in Fluorescence Spectroscopy*, Vol. 11. Springer: New York, pp. 237–258 (2006).
4. J. M. Lehn, *Supramolecular Chemistry: Concepts and Perspectives*, VCH: Weinheim (1995).
5. R. Ozawa and T. Hayashita, J. Ion Exchange, 18, 26–35 (2007).
6. J. P. Lorand and J. O. Edwards, J. Org. Chem., 24, 769–774 (1959).
7. S. Arimori, M. L. Bell, C. S. Oh et al., *J. Chem. Soc., Perkin Trans.* 1, 803–808 (2002).
8. T. D. James, K. R. A. S. Sandanayake, R. Iguchi et al., *J. Am. Chem. Soc.*, 117, 8982–8987 (1995).
9. J. D. Larkin, K. A. Frimat, T. M. Fyles et al., *New J. Chem.*, 34, 2922–2931 (2010).
10. J. C. Norrild and H. Eggert, *J. Am. Chem. Soc.*, 117, 1479–1484 (1995).
11. M. Bielecki, H. Eggert, and J. C. Norrild, *J. Chem. Soc., Perkin Trans.* 2, 449–455 (1999).
12. W. Yang, H. He, and D. G. Drueckhammer, *Angew. Chem., Intl. Ed.*, 40, 1714–1718 (2001).

13. T. Hayashita and H. Tsukube, *Molecular Recognition and Supramolecules*, Sankyoshuppan (2007).
14. A. J. Tong, A. Yamauchi, T. Hayashita et al., *Anal. Chem.*, 73, 1530–1536 (2001).
15. T. Hayashita, A. Yamauchi, A. J. Tong et al., *J. Incl. Phenom. Macrocycl. Chem.*, 50, 87–94 (2004).
16. C. Shimpuku, R. Ozawa, A. Sasaki et al., *Chem. Commun.*, 1709–1711 (2009).
17. X. Wu, L. R. Lin, Y. J. Huang et al., *Chem. Commun.*, 48, 4362–4364 (2012).
18. R. Ozawa, T. Hashimoto, A. Yamauchi et al., *Anal. Sci.*, 24, 207–212 (2008).
19. A. Yamauchi, Y. Sakashita, K. Hirose et al., *Chem. Commun.*, 4312–4314 (2006).
20. I. Suzuki, A. Yamauchi, Y. Sakashita et al., *Anal. Sci.*, 23, 1167–1171 (2007).
21. M. Kumai, S. Kozuka, M. Samizo et al., *Anal. Sci.*, 28, 121–126 (2012).
22. T. Hashimoto, M. Yamasaki, H. Ishii et al., *Chem. Lett.*, 43(2), 228–230 (2014).
23. C. Rodorigeus-Tenreiro, L. Diez-Buero, A. Concheiro et al., *J. Control. Release* 123, 56–66 (2007).
24. A. Cazacu, Y. M. Legrand, A. Pasc et al., *Proc. Natl. Acad. Sci. USA* 106, 8117–8122 (2009).
25. X. Wu, Z. Li, X. X. Chen et al., *Chem. Soc. Rev.*, 42, 8032–8048 (2013).
26. R. Nishiyabu, Y. Kubo, T. D. James et al., *Chem. Commun.*, 47, 1106–1123 (2012).
27. J. S. Hansen, J. B. Christensen, J. F. Petersen et al., *Sensors Actuators B*, 161, 45–79 (2012).
28. Y. Kanekiyo and H. Tao, *Chem. Lett.*, 34, 196–197 (2005).
29. W. Takayoshi, M. Imajo, M. Iijima et al., *Sensors Actuators B*, 192, 776–781 (2014).
30. C. Yu and V. W. W. Yam, *Chem. Commun.*, 1347–1349 (2009).
31. F. Cheng and F. Jäkle, *Polym. Chem.*, 2, 2122–2132 (2011).
32. H. Kim, Y. J. Kang, S. Kang et al., *J. Am. Chem. Soc.*, 134, 4030–4033 (2012).
33. J. N. Cambre and B. S. Sumerlin, *Polymer*, 52, 4631–4643 (2011).
34. C. Y. S. Chung, K. H. Y. Chan, and V. W. W. Yam, *Chem. Commun.*, 47, 2000–2002 (2011).
35. C. Anda, V. Lapeyre, I. Gosse et al., *Langmuir*, 27, 12693–12701 (2011).
36. Y. Guan and Y. Zhang, *Chem. Soc. Rev.*, 42, 8106–8121 (2013).
37. C. Zhang, M. D. Losego, and P. V. Braun, *Chem. Mater.*, 25, 3239–3250 (2013).
38. W. K. Oh, Y. Seon, K. J. Lee et al., *Anal. Methods*, 4, 913–918 (2012).
39. W. Wu, T. Zhou, A. Berliner et al., *Angew. Chem. Intl. Ed.*, 49, 6554–6558 (2010).

14 Design of Chirality-Sensing Systems Based on Supramolecular Transmission of Chirality

Ryo Katoono

CONTENTS

14.1 INTRODUCTION

Chirality-transferring events between molecules are important in wide areas of supramolecular chemistry dealing with the design of asymmetric catalysts,[1,2] memory devices,[3–5] and sensors.[6–20] Chirality generates diversity in molecular structures when it interacts with other chiralities to yield two structures with different energies (diastereomers). In a case where one of two chiral units is a racemic pair of enantiomers that exist in equilibrium (dynamic chirality), the other chiral unit perturbs the equilibrium to prefer a particular enantiomer through the formation of a complex.

For such events, we consider that chirality is transferred. The chirality sensing that we describe below is based on the supramolecular transmission of point chirality to dynamic chirality. Most chiral substances, including naturally occurring ones show low activity levels in circular dichroism (CD) spectroscopy[6–8] due to lack of distinct chromophores. Less CD-active chirality becomes detectable as an enhanced readout if it is transferred to the dynamic chiralities of chromophoric molecules.

In this chapter, we focus on the supramolecular transmission of chirality in host–guest complexes and present a design concept of chirality-sensing systems using CD spectroscopy by taking a terephthalamide skeleton[17] as an example and also discuss representative successful designs of a chirality-sensing host.[6–16] We first demonstrate a simple system based on the supramolecular transmission of guest chirality to dynamic helicity of a host and then demonstrate a lock-and-key system through the cooperative transmission of chiralities on a host and guest. We do not deal with oligomers and polymers that adopt helix structures[16] or other sensing systems using UV absorption[18,19] or fluorescence[20] as readouts.

14.2 DESIGN OF CHIRALITY-SENSING HOSTS

Three requisites for the design of a chirality-sensing host based on supramolecular transmission of chirality are (1) a binding site for capturing a guest, (2) a chromophoric part for displaying guest chirality, and (3) adoption of racemic forms that interconvert between R/S- or P/M-enantiomeric conformations in a complexed state in which the host prefers a certain dynamic chirality (R/S) or helicity (P/M) (one of the two diastereomeric complexes) and therefore provides chiroptical outputs as a readout in the absorption region of the chromophoric host (Scheme 14.1).

The host is not required to adopt chiral forms in the absence of a chiral guest. In a complexed state, (a) the chiral guest biases the equilibrium between enantiomeric conformations of the host and (b) the chiral guest induces the host to change in structure from achiral to chiral upon complexation and biases the newly generated equilibrium through the supramolecular transmission of chirality.

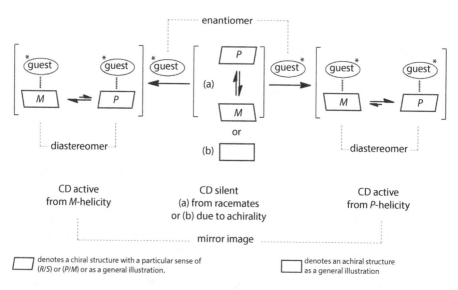

SCHEME 14.1 Supramolecular transmission of chirality in a guest to dynamic helicity of a chirality-sensing host: (a) adopting helical forms and (b) adopting a non-helical form in the absence of a guest.

14.3 TEREPHTHALAMIDES

14.3.1 STRUCTURAL FEATURES

N,N′- or N,N,N′,N′-substituted terephthalamides provide suitable platforms for the design of dynamic host molecules. The amide carbonyl groups act as binding sites through the formation of hydrogen bonds, and chemical modification is available on the amide nitrogens and/or on the central benzene ring, leading to large dihedral angles between amide planes and the central benzene ring. The twisting motion of the amide groups with respect to the central benzene ring allows the molecule to adopt many conformations that interconvert dynamically by rotating about the $C_{C=O}$-$C_{central}$ single bonds, the ease of which is dependent on the bulkiness of substituents (Scheme 14.2).

Regarding the relative direction of the two amide groups, anti and syn forms are possible. The anti form is commonly more stable than the syn form due to cancellation of the dipole moments of the two amide groups. The syn form seems to be suitable for binding with a ditopic guest to form a 1:1 complex through a two-fold hydrogen bond at the two amide carbonyls. These two structural characteristics lead us to expect conformational switching from anti to syn form upon complexation with a ditopic guest. Conformational switching is not a requisite for chirality sensing but is important when designing advanced systems based on the dynamic features of terephthalamides (Section 14.3.4).

We can expect that for both anti and syn forms twisting of the two amide groups would minimize or reduce the dipole–dipole interaction. Disrotatory twisting of the two amide groups leads to a non-helical conformation with a mirror plane or a center of symmetry in the unit. Conrotatory twisting allows a unit to adopt helical conformations that interconvert between P and M helicities about the two-fold axis of symmetry (dynamic helicity; Scheme 14.2). In some cases, conformational switching from a non-helical to a helical conformation can be induced by complexation with a hydrogen-bonding ditopic guest at the two amide carbonyls, resulting in the generation of dynamic helicity in the host upon complexation (case (b) in Scheme 14.1).

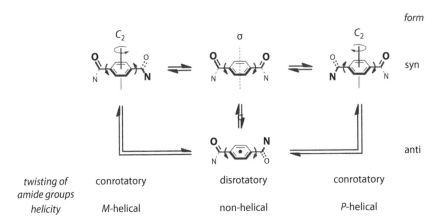

SCHEME 14.2 Dynamic interconversion among conformations of a terephthalamide skeleton based on the rotation and twisting of the two amide groups.

14.3.2 CHIRALITY SENSING THROUGH SUPRAMOLECULAR TRANSMISSION OF CHIRALITY

Helicity biasing can be attained by complexation with a chiral ditopic guest through supramolecular transmission of chirality to dynamic helicity (Scheme 14.1). Examples of cases (a) and (b) are given below.

One is a propeller-shaped tertiary terephthalamide derivative **1** with four aryl blades on the central benzene ring and N,N-dialkyl substitutions on each amide nitrogen [CH$_2$(cHex) and Me] (Figure 14.1).[21,22] Two forms of anti and syn were separated and isolated as a pair of atropisomers (anti-**1** and syn-**1**) due to no rotation about the $C_{C=O}$-$C_{central}$ single bonds. In such a sterically congested situation, all six blades are almost perpendicular to the central benzene ring and allowed to twist in both directions to give numerous conformations.

For the syn atropisomer (syn-**1**), propeller-shaped conformations are the most stable among numerous conformations. All the blades are twisted in a conrotatory manner and two energetically equivalent propellers with P or M helicity intercon-vert to each other (CD silent). We used an N,N'-dialkyl-xylylenediammonium salt (R,R)-**2** as a hydrogen-bonding chiral ditopic guest [(R)-C*HMe(cHex)] that formed a 1:1 complex with a syn-formed terephthalamide host at the two amide carbonyls.

Positive Cotton effects emerged in the absorption region of syn-**1** on addition of (R,R)-**2**, which is considered a result of biasing the equilibrium to a particular sense in the complex through supramolecular transmission of guest chirality to propeller-shaped dynamic helicity. The structure of the propeller-shaped host was unchanged but the ratio of P:M enantiomers changed during complexation (case (a) in Scheme 14.1).

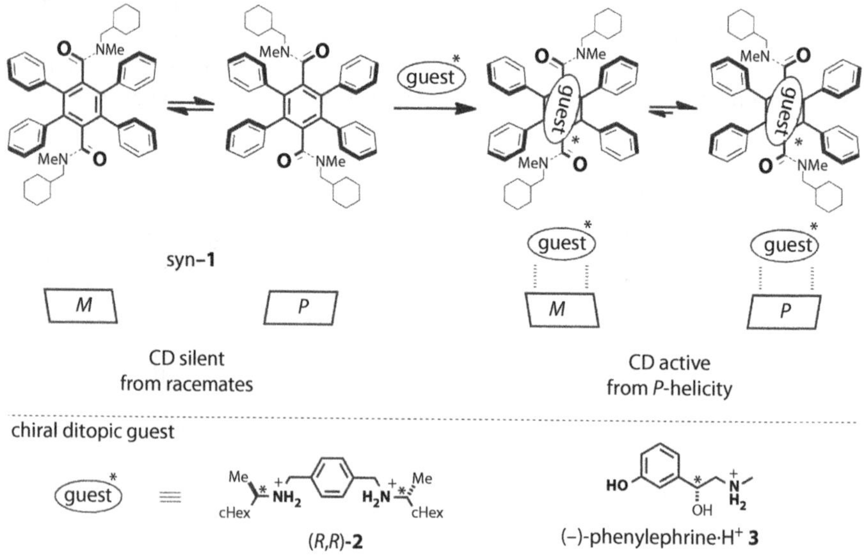

FIGURE 14.1 Dynamic propeller-shaped helicity based on tetraarylterephthalamide derivative syn-**1**, and chemical structures of chiral ditopic guests **2** and **3**.

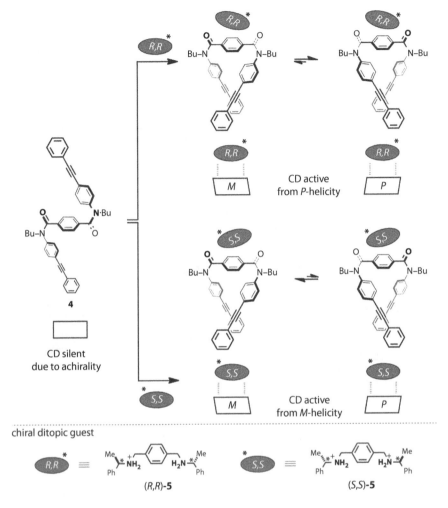

FIGURE 14.2 Dynamic double-arm-crossing helicity based on tertiary terephthalamide derivative **4**, and chemical structures of chiral ditopic guests **5**.

The small CD activity of the guest ($|\Delta\varepsilon| < 0.2$) may be enhanced successfully to be detected easily as a readout in the CD spectrum ($\Delta\varepsilon$ +8.8) (chiroptical enhancement). This protocol was applied to the detection of a (−)-phenylephrine·H⁺ salt **3** ($|\Delta\varepsilon| < 0.2$) to give an enhanced readout ($\Delta\varepsilon$ +9.6). For the anti-atropisomer (anti-**1**), no chiroptical signal was induced on the addition of (R,R)-**2** or (−)-phenylephrine·H⁺ salt **3** due to a non-propeller-shaped structure.

The example of case (b) is a double-armed tertiary terephthalamide derivative **4** with an N-aryl [$C_6H_4CCC_6H_5$] and N-alkyl [nBu] group on each amide nitrogen, which was isolated as a mixture of anti and syn forms that interconvert to each other due to free rotation about the $C_{C=O}-C_{central}$ single bonds (Figure 14.2).[23] The anti form adopts a non-helical conformation with a center of symmetry in which the dipoles

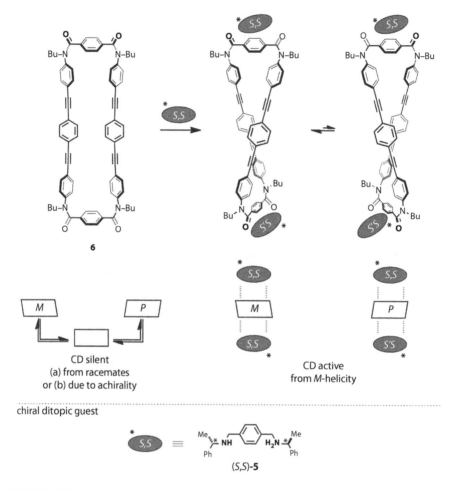

FIGURE 14.3 Dynamic figure-of-eight-shaped helicity based on cyclic bis(terephthalamide) derivative **6**, and chemical structure of chiral ditopic guest **5**.

are well arranged to cancel polarity. This form is therefore more stable than syn forms in the absence of a guest.

According to the "*cis*-preference of aromatic N-methylamides" reported by Azumaya et al.,[24,25] the aryl group occupies the vertical position that produces a racemic pair of helical conformations with P or M helicity in the syn form. As a chiral guest, we again used N,N'-dialkyl-xylylenediammonium salts (R,R)-**5** and (S,S)-**5** [C*HMePh]. A positive couplet emerged in the absorption region of the diphenyl-acetylenic chromophore on the addition of (R,R)-**5**.

The mirror image was induced by the addition of (S,S)-**5**. The observation is considered a result of conformational switching from a non-helical anti form to a helical syn form, followed by biasing the newly generated equilibrium to a particular sense in the syn-formed complex through supramolecular transmission of guest chirality to double-arm-crossing dynamic helicity (case (b) in Scheme 14.1).

A derivative can no longer adopt an anti form when forced to form a rigid cycle, as shown by the following example of a cyclic tertiary bis(terephthalamide) derivative **6** with an alkyl substitution [nBu] on each nitrogen (Figure 14.3).[26] Even in such a situation, the two amide groups can twist in both directions to avoid being strictly perpendicular to the central benzene ring. Disrotatory twisting of the two amide groups in a terephthalamide unit leads to a non-helical rectangular form. Conrotatory twisting allows the macrocycle to adopt helical figure of eight-shaped forms (dynamic helicity; Scheme 14.2). A negative couplet was induced in the absorption region of the bis(phenylethynyl)benzene chromophore when complexed with two molecules of (S,S)-5 through supramolecular transmission of guest chirality to figure of eight-shaped dynamic helicity (case (a) or (b) in Scheme 14.1).

14.3.3 INTRAMOLECULAR TRANSMISSION OF CHIRALITY

A racemic host in a complex with a chiral guest prefers a particular sense of dynamic helicity through supramolecular transmission of chirality (Section 14.3.2; Scheme 14.1). Similar helicity biasing can also be attained by the attachment of point chirality to the nitrogen through intramolecular transmission of chirality (Scheme 14.3 and below).

In (a), two helical forms with P or M helicity are a diastereomeric pair due to a stereogenic center on the nitrogen. They are energetically non-equivalent, and therefore an easily interconverting mixture of diastereomers is CD active due to a preferred helicity. In (b), a stereogenic center on the nitrogen provides a local chiral geometry, but in some cases, it cannot be transferred to the whole molecule due to a non-helical structure. Also in this case, CD signals are not silent. They come from the local chiral geometry but not from dynamic helicity. Two examples of (a) and (b) are shown below.

In a tertiary terephthalamide derivative (R,R)-syn-**7** with four aryl blades on the central benzene ring and N,N-dialkyl substitution on each amide nitrogen [(R)-C*HMe(cHex) and Me], two helical propellers with P or M helicity exist in equilibrium, and one of

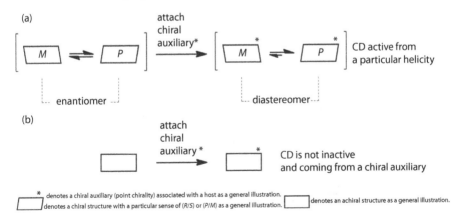

SCHEME 14.3   (a) Intramolecular transmission of chirality associated with a host to dynamic helicity; (b) A chiral auxiliary does not lead to a preferred sense due to the achiral structure of a host.

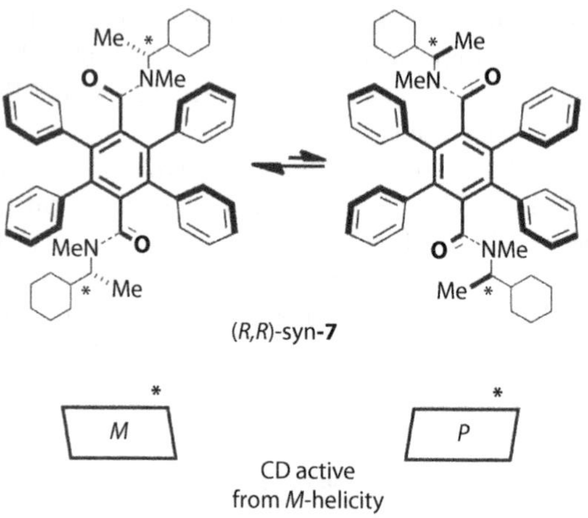

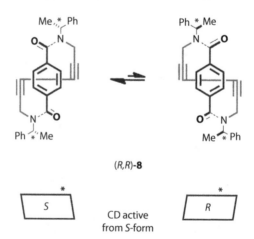

FIGURE 14.4 An equilibrium between diastereomers with (M)- or (P)-helicity, and a preferred sense of dynamic helicity due to the intramolecular transmission of chirality in propeller-shaped tertiary derivative (R,R)-syn-**7** leading to a chiroptical readout.

them forms in preference to the other in solution (Figure 14.4; Scheme 14.3a).[22] Two diastereomeric conformations are available using dynamic planar chirality, as shown by a cyclic tertiary terephthalamide derivative (R,R)-**8** [(R)-C*HMePh and $(CH_2CC)_2$], in which two planar-chiral conformations with R or S form exist in equilibrium but in a different ratio in solution (Figure 14.5; Scheme 14.3a).[27]

A secondary terephthalamide derivative (R,R)-**9** with four aryl blades on the central benzene ring and N-alkyl substitution on each nitrogen [(R)-C*HMe(cHex)], which is the synthetic precursor of the above-mentioned propeller-shaped tertiary derivative (R,R)-syn-7 (Figure 14.4), exists as a mixture of anti and syn forms due

FIGURE 14.5 An equilibrium between diastereomers with (S)- or (R)-planar chirality, and a preferred sense of dynamic planar chirality due to the intramolecular transmission of chirality in cyclic tertiary derivative (R,R)-**8** leading to a chiroptical readout.

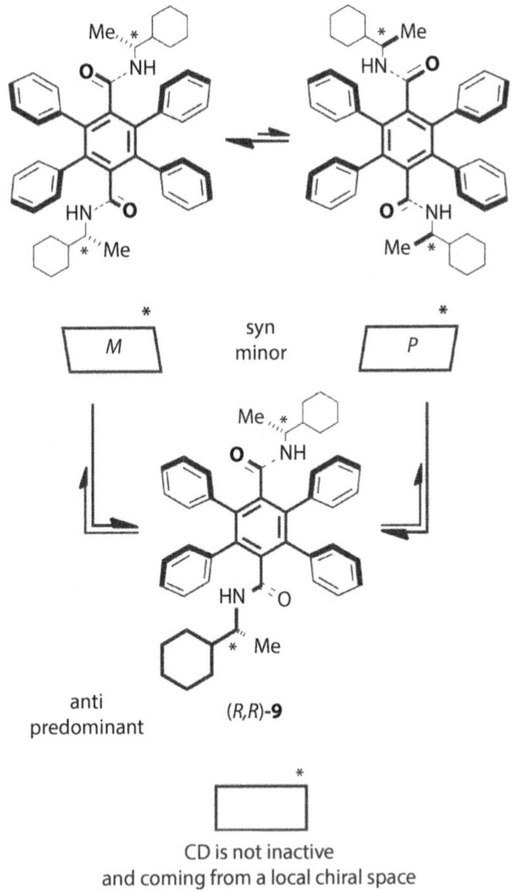

FIGURE 14.6 Equilibriums between a non-propeller form and a propeller-form with (*M*)- or (*P*)-helicity, and a preferred non-propeller form of secondary derivative (*R*,*R*)-**9** leading to a weak readout even with the attachment of chirality to each nitrogen.

to free rotation about the $C_{C=O}-C_{central}$ single bonds. A non-propeller-shaped anti form with a pseudocentrosymmetric center is predominant due to dipole cancellation (Figure 14.6; Scheme 14.3b).[21]

A double-armed tertiary terephthalamide derivative (R,R)-**10** with an N-aryl $[C_6H_4CCC_6H_5]$ and N-alkyl $[(R)-C*HMePh]$ group on each amide nitrogen prefers a non-helical anti form in the absence of a guest (Figure 14.7; Scheme 14.3b).[23] In these cases, chirality of a stereogenic center on the nitrogen is not transferred to the skeletal asymmetry due to a non-helical structure of the anti (predominant) form. However, it could act as a chiral handle to control the helical preference in a syn (minor) form. It was verified that a stereogenic center acted as a chiral handle to prefer a particular sense of dynamic helicity in the syn form when complexed with an achiral ditopic guest such as a dopamine·H$^+$ salt **11**, which has no directing preference (Scheme 14.4).[23]

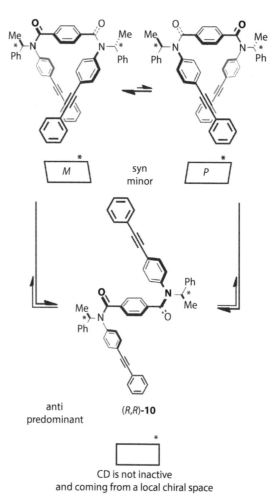

FIGURE 14.7 Equilibriums between a non-helical form and a helical-form with (*M*)- or (*P*)-helicity, and a preferred non-helical form of tertiary derivative (*R*,*R*)-**10** leading to a weak readout even with the attachment of chirality to each nitrogen.

14.3.4 SYNERGY IN TRANSMISSION OF CHIRALITY

We have demonstrated that chirality has a directing or potentially directing preference in biasing the equilibrium between enantiomeric forms through supramolecular trans-mission of guest chirality (Section 14.3.2) and through intramolecular transmission of internal chirality associated with a host (Section 14.3.3). In this section, we demonstrate a variety of transmissions of chirality during diastereomeric complexation of a host with internal chirality and a chiral guest. The host is not required to adopt helical forms in the absence of a chiral ditopic guest. Some guests may induce conformational switching of a host from a non-helical to a helical form during complexation, as mentioned above.

In Case (a), when both guest chirality and internal chirality have the same direct-ing preference (matched pair), they are synergistically transmitted to dynamic

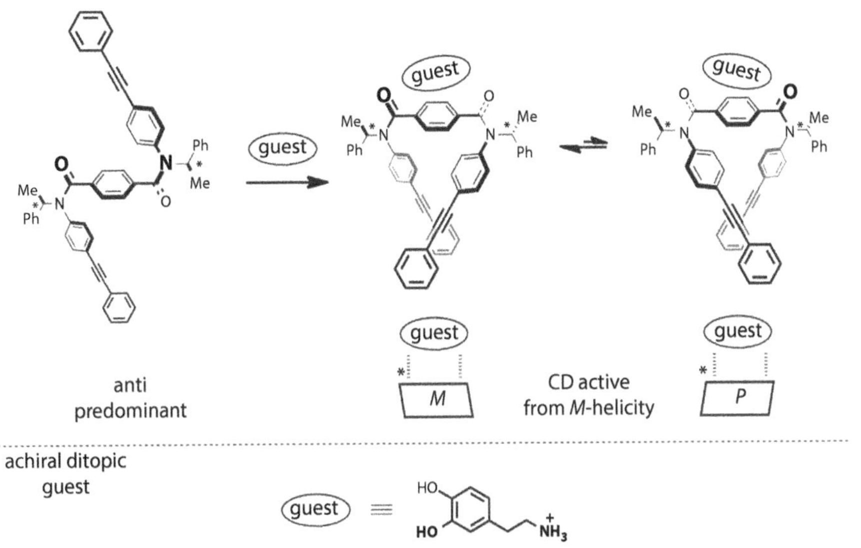

SCHEME 14.4 Intramolecular transmission of chirality associated with a host to dynamic helicity that generated in a complex with an achiral ditopic guest, and chemical structure of dopamine' H+ salt 11.

helicity and produce an enhanced readout from a helically biased syn-formed complex. When each directing preference is opposite (mismatched pair), (b) a preferred helicity in a syn-formed complex obeys the directing preference of guest chirality, (c) a preferred helicity in a syn-formed complex obeys the directing preference of internal chirality, or (d) they work competitively and a helical structure is not induced in a syn-formed complex. Results of diastereomeric complexation depicted in Figure 14.8 represent Cases (a) through (d).

Cases (a) and (d) are especially interesting because the host adopts a non-helical structure due to failure in intramolecular transmission of internal chirality in the absence of a guest (Figure 14.8). In some cases, such a host can undergo a stereospecific change in structure upon complexation with a pair of enantiomeric guests. A helical structure that prefers a particular sense of dynamic helicity is induced only when the directing preference is matched. Alternatively, a non-helical structure is maintained during complexation with a mismatched guest (Scheme 14.5).

A tris(terephthalamide) derivative (R,R,R)-**12** with a threefold cyclic structure presents an example of the stereospecific conformational switching in response to an enantiomeric pair of chiral guests (R,R)-**2** and (S,S)-**2** (Figure 14.9).[28] The host prefers a non-helical structure even with the attachment of a stereogenic center to one of the two amide nitrogens [(R)-C*HMe(cHex)]. Upon complexation with (R,R)-**2**, it undergoes a change in conformation to a helical form in each cyclic structure through the cooperative transmission of chiralities on the host and guest, resulting in a helically biased C_3-symmetric propeller (a). Alternatively, chirality was neither intramolecularly nor intermolecularly transferred when the directing preference was

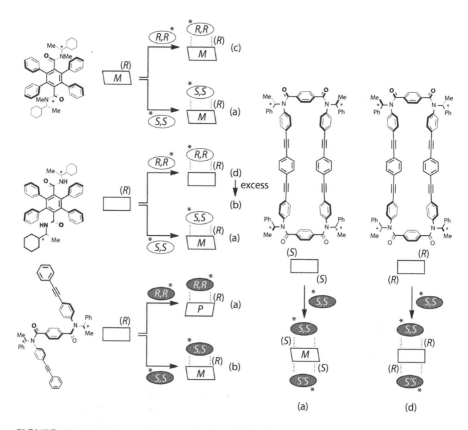

FIGURE 14.8 Summary of a transmission of chirality in cases (a) through (d).

mismatched, even though a complex was formed. During complexation with (S,S)-**2**, the non-helical structure of the host remained unchanged (d).

We can consider such chiral recognition based on stereospecific conformational switching as a lock-and-key system in which only a proper key (matched guest) can turn in a particular direction and open the lock, while an improper key (mismatched guest)

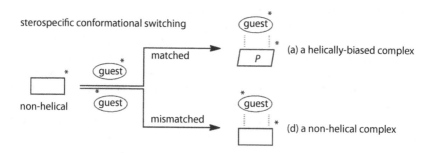

SCHEME 14.5 Molecular lock-and-key system based on the stereospecific change in conformation of a dynamic host.

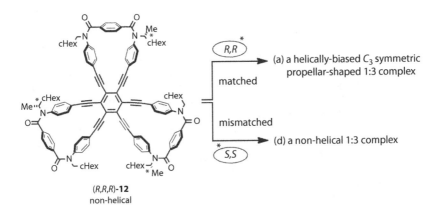

FIGURE 14.9 Stereospecific change in conformation of tris(terephthalamide) host (*R*,*R*,*R*)-**12** upon complexation with a chiral ditopic guest (molecular lock-and-key system).

cannot do so even though a complex like a competitive antagonist forms. This is a less recognized aspect of chiral discrimination[21,26,28] and is different from the well-documented cases, in which a chiral host prefers one of the enantiomeric guests to the other in terms of binding affinity.[29–34] It will be an important event when we develop an advanced system in which a stereospecific change in structure acts as a trigger to transduce information sequentially.

REFERENCES

1. Shirakawa, S. and Maruoka, K. 2013. *Angew. Chem. Int. Ed.* 52: 4312–4348.
2. Sandoval, C. A. and Noyori, R. 2012. In *Organic Chemistry: Breakthroughs and Perspectives*, Ding, K. and Dai, L. X., Eds. Wiley-VCH, Weimheim, pp. 335–366.
3. Rosaria, L., D'urso, A., Mammana, A. et al. 2011. *Top. Curr. Chem.* 298: 143–188.
4. Rosaria, L., D'urso, A., Mammana, A. et al. 2008. *Chirality* 20: 411–419.
5. Purrello, R. 2003. *Nat. Mater.* 2: 216–217.
6. Harada, N. and Berova, N. *Comp. Chirality* 8: 449–477.
7. Wolf, C. and Bentley, K. W. 2013. *Chem. Soc. Rev.* 42: 5408–5424.
8. Giovanni, G., Stefano, L., Stefano, M. et al., 2008. *Chirality* 20: 471–485.
9. Canary, J. W., Dai, Z., and Mortezaei, S. 2012. *Comp. Chirality* 8: 600–624.
10. Clayden, J. 2009. *Chem. Soc. Rev.* 38: 817–829.
11. Hembury, G. A., Borovkov V. V., and Inoue, Y. 2008. *Chem. Rev.* 108: 1–73.
12. Pijper, D. and Feringa, B. L. 2008. *Soft. Matter.* 4: 1349–1372.
13. Calama, M., and Reinhoudt, D. N. 2006. *Top. Curr. Chem.* 265.
14. Luo, J. and Zheng, Y. S. 2012. *Curr. Org. Chem.* 16: 483–506.
15. Cragg, P. J. and Sharma, K. 2012. *Chem. Soc. Rev.* 41: 597–607.
16. Yashima, E., Maeda, K., Iida, H. et al. 2009. *Chem. Rev.* 109: 6102–6211.
17. Katoono, R., Kawai, H., Fujiwara, K. et al. *J. Syn. Org. Chem. Jpn.* 70: 640–650.
18. Kubo, Y., Maeda, S., Tokita, S. et al. *Nature* 382: 522–524.
19. Iwaniuk, D. P., Spangler, K., and Wolf, C. 2012. *J. Org. Chem.* 77: 5203–5208.
20. Accetta, A., Corradini, R., and Marchelli, R. 2011. *Top. Curr. Chem.* 300: 175–216.
21. Katoono, R., Kawai, H., Fujiwara, K. et al. 2009. *J. Am. Chem. Soc.* 131: 16896–16904.

22. Katoono, R., Kawai, H., Fujiwara, K. et al. 2008. *Chem. Commun.* 4906–4908.
23. Katoono, R., Kawai, H., Fujiwara, K. et al. 2006. *Tetrahedron Lett.* 47: 1513–1518.
24. Azumaya, I., Kagechika, H., Fujiwara, Y. et al. 1991. *J. Am. Chem. Soc.* 113: 2833–2838.
25. Itai, A., Toriumi, Y., Tomioka, N. et al. 1989. *Tetrahedron Lett.* 30: 6177–6180.
26. Katoono, R., Kawai, H., Fujiwara, K. et al. 2005. *Chem. Commun.* 5154–5156.
27. Katoono, R., Kawai, H., Fujiwara, K. et al. 2004. *Tetrahedron Lett.* 45: 8455–8459.
28. Katoono, R., Kawai, H., Fujiwara, K. et al. *Chem. Commun.* DOI:10.1039/C3CC43571G.
29. Tanaka. T. 2012. *Comp. Chirality* 8: 63–90.
30. Wang, B. Y., Stojanović, S., Turner, D. A. et al. 2013. *Chem. Eur. J.* 19: 4767–4775.
31. Bois, J., Bonnamour, I., Duchamp, C., et al. 2009. *New J. Chem.* 33: 2128–2135.
32. Demirtas, H. N., Bozkurt, S., Durmaz, M. et al. 2009. *Tetrahedron* 65: 3014–3018.
33. Rekharsky, M. V., Yamamura, H., Inoue, C. et al., 2006. *J. Am. Chem. Soc.* 128: 14871–14880.
34. Rekharsky, M. V., Yamamura, H., Kawai, M. et al. 2003. *J. Org. Chem.* 68: 5228–5235.

15 Fundamental Aspects of Host–Guest Complexation from 1:1 to Synergistic Binding

Keiji Hirose

CONTENTS

15.1   INTRODUCTION

The formation of a complex of a host and a guest is a fundamental and important process in supramolecular chemistry. A well-accepted quantitative measure of complex formation is the binding constant. Selective complexations of hosts with specific guests can be seen on many occasions both in artificial systems and those in nature.

A ratio of binding constants of corresponding complexations is usually treated as a measure of selectivity. Consequently, binding constants have been used as fundamental criteria for evaluation of host–guest complexation processes.[1] In addition to the ratio of binding constants, the temperature dependence of selectivity also provides important insights into the origins of supramolecular functions.

Indeed, the selectivities of some supramolecular systems were reported to be significantly influenced by temperature,[2] including temperature-dependent inversion of enantioselectivity.[3] To express molecular recognition abilities by taking such temperature dependencies into account, thermodynamic parameters (enthalpy (ΔH) and entropy (ΔS)) are suitable criteria. Since a determination of binding constants at different temperatures offers these thermodynamic parameters from the slope and intercept in a van't Hoff plot (Figure 15.1), the important point in quantitative analyses of host–guest complexations is therefore how to determine binding constants with high reliability.

In this chapter, general methods for the determination of binding constants are discussed from theoretical and practical aspects. In particular 1:1 binding, 1:2 binding, and 1:n binding host–guest systems will be treated as fundamental examples that allow us to consider issues concerned with synergistic effects of bindings (Figure 15.2).

Publisher's note: There are further data and information for other appendices (E1–E4) available from the author at the following website (http://hdl.handle.next/11094/39419)

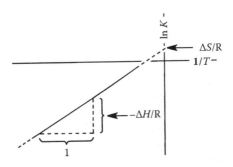

FIGURE 15.1 Correlation of ΔH, ΔS, K and temperature according to the van't Hoff equation.

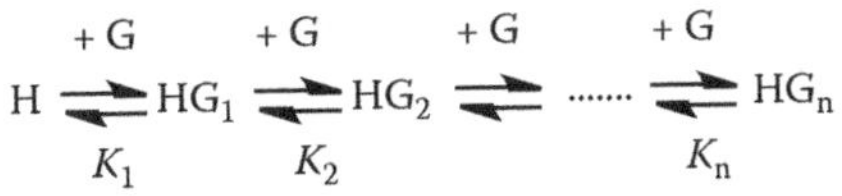

$$H \underset{K_1}{\overset{+G}{\rightleftharpoons}} HG_1 \underset{K_2}{\overset{+G}{\rightleftharpoons}} HG_2 \overset{+G}{\rightleftharpoons} \cdots \cdots \underset{K_n}{\overset{+G}{\rightleftharpoons}} HG_n$$

FIGURE 15.2 General representation of host-guest complexations including 1:1, 1:2 and 1:n bindings.

15.2 FUNDAMENTAL BINDING PROCESSES AND PRACTICAL COURSE FOR DETERMINATION OF BINDING CONSTANTS

Rate constants for complexation (k_1) and decomplexation (k_{-1}) are defined as shown in Equations (15.1) through (15.4).

Complexation Process (Association):

$$a \cdot H + b \cdot G \xrightarrow{k_1} C \tag{15.1}$$

$$\frac{d[C]_{association}}{dt} = k_1[H]^a[G]^b \tag{15.2}$$

Decomplexation Process (Dissociation):

$$a \cdot H + b \cdot G \xleftarrow[k_{-1}]{} C \tag{15.3}$$

$$\frac{d[C]_{dissociation}}{dt} = -k_{-1}[C] \tag{15.4}$$

The complexation and corresponding decomplexation rates are the same at equilibrium,

$$\frac{d[C]}{dt} = k_1[H]^a[G]^b - K_{-1}[C] = 0 \tag{15.5}$$

$$\frac{k_1}{k_{-1}} = \frac{[C]}{[H]^a \cdot [G]^b} \left(= K = K_d^{-1} \right) \tag{15.6}$$

The binding constant ($=K$), association constant, equilibrium constant, stability constant, and formation constant terms are synonymous with one another. Dissociation constant K_d is defined as K^{-1}. The binding constant K, the ratio of rate constants k_1 over k_{-1}, is expressed as shown in Equation (15.6). The binding constant can be obtained from concentrations [C], [H], and [G]. The activity coefficients are generally unknown, and the binding constant K based on concentrations is usually employed. Based on this situation, the question of the activity coefficients of the solutes is disregarded in order to simplify the discussion. Nevertheless, it should be remembered that this point is not always insignificant. The generally accepted way to determine the binding constant is based on a simple binding equilibrium model, i.e., Equation (15.7).

$$a \cdot H + b \cdot G \rightleftharpoons G \tag{15.7}$$

The basic equations for the host–guest complexation are (15.8) through (15.10).

$$K = \frac{[C]}{[H]^a \cdot [G]^b} \tag{15.8}$$

$$[H]_0 = [H] + a \cdot [C] \tag{15.9}$$

$$[G]_0 = [G] + b \cdot [C] \tag{15.10}$$

H = host; G = guest; C = complex ($H_a \cdot G_b$); a, b = stoichiometry as shown in Equation (15.7); $[H]_0$ = initial (total) concentration of host molecule; $[G]_0$ = initial (total) concentration of guest molecule; and [H], [G], and [C] = concentrations of host, guest, and complex, respectively, at equilibrium.

Equation (15.11) is derived from Equations (15.8) through (15.10).

$$K = \frac{[C]}{([H]_0 - a \cdot [C])^a \cdot ([G]_0 - b \cdot [C])^b} \tag{15.11}$$

Classification of parameters: K, a, b = constants (a and b are integers larger than or equal to 1); $[H]_0$, $[G]_0$ = variables that can be set up as experimental conditions; [H], [G], [C] = variables dependent on equilibrium.

The following guidelines can be deduced for experiments to determine the binding constant from Equation (15.11) and the classification of its parameters. When [C] is obtained at equilibrium in which a and b are known, K can be derived directly according to Equation (15.11) from the experimental conditions $[H]_0$ and $[G]_0$. Consequently, the following four steps must be carried out to determine the binding constants:

- Determination of stoichiometry (a and b)
- Evaluation of complex concentration [C]
- Setting up initial concentration conditions $[H]_0$ and $[G]_0$
- Data treatment

The following material deals with the principles and the practical issues necessary for understanding and performing the above four points.

In a system whose equilibrium can be expressed simply with one parameter, K (binding constant) is the most useful quantitative measure of the host. For systems that can be interpreted by combination of two or more equilibria, that is, in addition to the descriptions for the fundamental binding process, ambiguity remains. However, these systems can be handled theoretically with a set of binding constants (e.g., K_1, K_2, ...). The use of cooperative interactions at each equilibrium in a multi-step binding system makes it possible to realize very high selectivity (synergistic binding) that cannot be achieved in a single-step equilibrium system.

In the following sections, a method of determining a binding constant (K) precisely with practical precautions for general systems is described, followed by descriptions for 1:2 binding systems utilizing a spontaneous binding model. More complex equilibria will also be examined using 1:2 host–guest systems with a stepwise binding model having a synergistic binding process. Then, 1:n host–guest systems with a spontaneous binding model followed by the systems with stepwise binding models with and without synergistic binding processes will be discussed.

15.3 DETERMINATION OF STOICHIOMETRY

The different methods to determine the stoichiometry of a chemical reaction include continuous variation methods,[4] the slope ratio method,[5,6] the mole ratio method,[7] and others. Here we adopt the continuous variation method to determine the stoichiometry since it is the most widely employed method by the community. To determine the stoichiometry of a reaction by the continuous variation method, the following four points must be considered and carried out:

- Keeping the sum of $[H]_0$ and $[G]_0$ constant (α)
- Changing $[H]_0$ from 0 to α
- Measuring $[C]$
- Data treatment (Job's plot)

As shown in Figure 15.3, the stoichiometry ($\frac{a}{a+b}$) is obtained from the x-coordinate at the maximum in a Job's curve where the y-axis is $[C]$ and x-axis is $\frac{[H]_0}{([H]_0+[G]_0)}$.

For the comprehension of the theoretical background of the continuous variation method, the required equations are (15.7) through (15.10) and (15.12) through (15.14).

$$\alpha = [H]_0 + [G]_0 \tag{15.12}$$

$$x = \frac{[H]_0}{([H]_0 + [G]_0)} \tag{15.13}$$

$$y = [C] \tag{15.14}$$

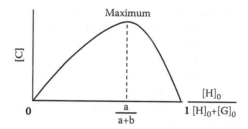

FIGURE 15.3 Correlation between stoichiometry (a, b) and x-coordinate at the maximum of the curve in Job's Plot.

$[H]_0$ and $[G]_0$ will be substituted by the function of x and α from Equations (15.12) and (15.13).

$$[H]_0 = \alpha \cdot x \qquad\qquad (15.15)$$

$$[G]_0 = \alpha - \alpha \cdot x \qquad\qquad (15.16)$$

From Equations (15.7) through (15.10) and (15.14) through (15.16):

$$K = \frac{y}{\{(\alpha \cdot x - a \cdot y)^a \cdot (\alpha - b \cdot y - \alpha \cdot x)^b\}}$$

$$\therefore K \cdot (\alpha - b \cdot y - \alpha \cdot x)^b \cdot (\alpha \cdot x - a \cdot y)^a = y \qquad (15.17)$$

Equation (15.17) is then differentiated, and the $\frac{dy}{dx}$ is substituted by zero. This allows the x-coordinate at the maximum in the curve to be obtained.

$$K \cdot [(\alpha - b \cdot y - \alpha \cdot x)^b \cdot \{(\alpha \cdot x - a \cdot y)^a\}' + \{(\alpha - b \cdot y - \alpha \cdot x)\}^b]' \cdot (\alpha \cdot x - a \cdot y)^a] = \frac{dy}{dx}$$

$$K \cdot \left[(\alpha - b \cdot y - \alpha \cdot x)^b \cdot a \cdot (\alpha \cdot x - a \cdot y)^{a-1} \cdot \left(\alpha - a \cdot \frac{dy}{dx} \right) + \right.$$

$$\left. b \cdot (\alpha - b \cdot y - \alpha \cdot x)^{b-1} \cdot \left(-b \cdot \frac{dy}{dx} - \alpha \right) \cdot (\alpha \cdot x - a \cdot y)^a \right] = \frac{dy}{dx}$$

Substitution of $\frac{dy}{dx}$ by zero yields:

$$K \cdot [(\alpha - b \cdot y - \alpha \cdot x)^b \cdot a \cdot (\alpha \cdot x - a \cdot y)^{a-1} \cdot \alpha + b \cdot (\alpha - b \cdot y - \alpha \cdot x)^{b-1}$$

$$\cdot (-\alpha) \cdot (\alpha \cdot x - a \cdot y)^a] = 0$$

Division by $K \cdot (\alpha - b \cdot y - \alpha \cdot x)^{b-1} \cdot (\alpha \cdot x - a \cdot y)^{a-1} \cdot \alpha$ produces

$$a \cdot (\alpha - b \cdot y - \alpha \cdot x) - b \cdot (\alpha \cdot x - a \cdot y) = 0$$

$$a \cdot \alpha - a \cdot b \cdot y - a \cdot \alpha \cdot x - b \cdot \alpha \cdot x + b \cdot a \cdot y = 0$$

$$a \cdot \alpha - a \cdot \alpha \cdot x - b \cdot \alpha \cdot x = 0$$

Division by α

$$a - ax - bx = 0$$

$$\therefore x = \frac{a}{a+b} \tag{15.18}$$

Equation (15.18) implies that $\frac{a}{a+b}$ is the x-coordinate at the maximum $\left(\frac{dy}{dx} = 0\right)$ in the curve of Equation (15.17), thus providing the correlation between the stoichiometry and the x-coordinate at the maximum in Job's plot. For example, when 1:1 complexation is predominant at equilibrium, the maximum appears $x = 0.5$ ($a = b = 1$). In the case of 1:2 complexation, the maximum is at $x = 0.333$.

From a practical view, the following point is important. Even if the concentration of a complex [C] cannot be measured directly, the [C] (y-axis) can be replaced with a parameter proportional to [C]. Then the same x-coordinate and with it the same stoichiometry is obtained at the maximum in Job's plot. In other words, the stoichiometry can be determined even when [C] cannot be obtained directly. The practical question for each individual experiment is then how to modify the y-coordinate.

Depending on experiments, a suitable observable property should be selected and it should replace the complex concentration [C] in Job's plot. In the following section, the UV/vis spectroscopic response is discussed as a representative example.

In the case of investigation by means of UV/vis spectroscopy, the concentrations and absorbances of each species are related by Equations (15.19) through (15.21). The observed absorbance is expressed as Equation (15.22) and Figure 15.4. The length of the optical cell is fixed here to 1 cm as a premise. The definitions of the abbreviations

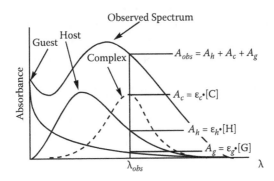

FIGURE 15.4 Representative UV-visible spectra to show correlation of observed spectra and each component.

are given below. The definitions of other abbreviations (a, b, $[H]_0$, $[G]_0$, $[H]$, $[G]$, $[C]$) are the same as described earlier.

$$A_h = \varepsilon_h \cdot [H] = \varepsilon_h \cdot ([H]_0 - a \cdot [C]) \tag{15.19}$$

$$A_g = \varepsilon_g \cdot [G] = \varepsilon_g \cdot ([G]_0 - b \cdot [C]) \tag{15.20}$$

$$A_c = \varepsilon_c \cdot [C] \tag{15.21}$$

$$A_{obs} = A_h + A_g + A_c \tag{15.22}$$

A_{obs} = observed absorbance; A_h, A_g, A_c = absorbances of host, guest, and complex, respectively; and ε_h, ε_g, ε_c = molar absorptivities of host, guest, and complex, respectively. Equation (15.22) is transformed to Equation (15.23) by using Equations (15.19) through (15.21).

$$A_{obs} = \varepsilon_h \cdot ([H]_0 - a \cdot [C]) + \varepsilon_g \cdot ([G]_0 - b \cdot [C]) + \varepsilon_c \cdot [C]$$
$$\therefore A_{obs} - \varepsilon_h \cdot [H]_0 - \varepsilon_g \cdot [G]_0 = (\varepsilon_c - a \cdot \varepsilon_h - b \cdot \varepsilon_g) \cdot [C] \tag{15.23}$$

Equation (15.23) shows that $A_{obs} - \varepsilon_h \cdot [H]_0 - \varepsilon_g \cdot [G]_0$ is proportional to $[C]$ because $(\varepsilon_c - a \cdot \varepsilon_h - b \cdot \varepsilon_g)$ is constant. The molar absorptivities ε_h, ε_g can be determined from independent measurements using a pure host and a pure guest, respectively. The concentrations $[H]_0$ and $[G]_0$ are known because they are experimental conditions set up by the researcher. Consequently, $(A_{obs} - \varepsilon_h \cdot [H]_0 - \varepsilon_g \cdot [G]_0)$ is determined from the experiments by means of UV/vis spectroscopy. The stoichiometry is determined from the x-coordinate at the maximum in the curve that might be called a modified Job's plot where $(A_{obs} - \varepsilon_h \cdot [H]_0 - \varepsilon_g \cdot [G]_0)$ is plotted as the y-coordinate instead of $[C]$. To get a better feeling of the practical experiment, a spreadsheet for the continuous variation method is available on the author's website.

15.4 EVALUATION OF COMPLEX CONCENTRATION

When the observed property is the complex concentration ($[C]$) at equilibrium, no further explanation is required. However, the actual complex concentration cannot be directly observed in most cases. Thus, the question of how to evaluate $[C]$ is an important practical issue and naturally depends on the properties that may be observed in each experiment. In this section, two typical cases for the evaluation of the complex concentration at equilibrium by UV/vis spectroscopic methods are discussed.

15.4.1 CASE 1: ABSORPTION BANDS OF HOST, GUEST, AND COMPLEX OVERLAP

From Equation (15.23), the following Equation (15.24) is derived.

$$[C] = \frac{A_{obs} - \varepsilon_h \cdot [H]_0 - \varepsilon_g \cdot [G]_0}{\varepsilon_c - a \cdot \varepsilon_h - b \cdot \varepsilon_g} \tag{15.24}$$

$[C]$ can be determined using A_{obs} and the experimental conditions $[H]_0$ and $[G]_0$ when all constants (a, b, ε_h, ε_g and ε_c) are known. Since the molar absorptivity of the complex ε_c is not measurable directly, a titration experiment and regression analysis are necessary for the evaluation of complex concentration.

This case is the most complicated host–guest complexation systems detected by means of UV/vis spectroscopy because all absorption bands of components, host, guest, and complex overlap. However, the situation can be simplified very often by choosing a detection wavelength at which one component (e.g., guest) has an $\varepsilon = 0$. This scenario is discussed as Case 2.

15.4.2   CASE 2: ABSORPTION BANDS OF ONLY TWO COMPONENTS OVERLAP

As a typical example, the complexation of a chromophoric chiral crown ether and an amino alcohol in chloroform is shown in Figure 15.5.[2e] In the visible region, 2-amino-1-propanol (**2**) has no absorption. However, both host **1** and complex **3** show clear absorption bands that overlap. Figure 15.5 shows UV/vis spectra of a chloroform solution of pure **1** and its mixture with **2**. To avoid the complexities arising from overlapping absorption bands and maximize spectral change, a detection wavelength of 555 nm was chosen for the measurements.

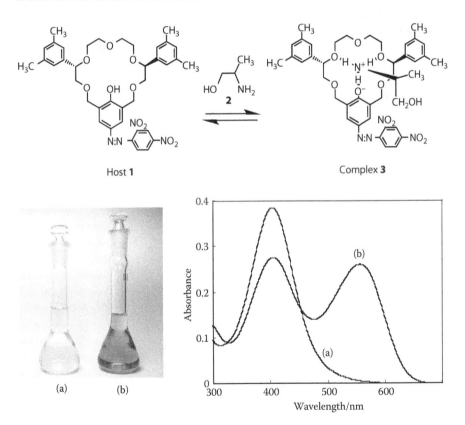

FIGURE 15.5   Typical example of color change caused by host-guest complexation: pictures and corresponding UV-VIS spectra: (a) A chloroform solution of Host **1** (1.85×10^{-5} M) at 25°C (solid line); (b) The same solution as (a) containing 2 equivalent of (R)-$_2$ (dotted line).

In this example, ε_g is zero, ε_h is small but not negligible, and ε_c is large at λ_{max} of complex **3** (555 nm). Equation (15.24) can be simplified and yields Equation (15.25) just by substitution of ε_g by zero.

$$[C] = \frac{A_{obs} - \varepsilon_h \cdot [H]_0}{\varepsilon_c - \alpha \cdot \varepsilon_h} \qquad (15.25)$$

Because three parameters (b, ε_g, and $[G]_0$) disappear from Equation (15.24), data treatment will be much simpler. If all constants (a, ε_h and ε_c) are obtained, [C] may be determined using A_{obs} and the experimental condition $[H]_0$. Nevertheless, since the molar absorptivity of the complex (ε_c) is not directly measurable, a titration experiment and regression are necessary for the evaluation of complex concentration in this case.

15.5   PRECAUTIONS FOR SETTING CONCENTRATION CONDITIONS OF TITRATION EXPERIMENT

Each method for binding analysis has limitations. Sources of systematic errors are encountered often in host–guest complexation, for example, the danger of carrying out titrations at concentrations unsuitable for the equilibrium to be measured. The origin of this error will be discussed and methods for avoiding such problems will be presented in the next three sections.

15.5.1   CORRELATION OF $[H]_0$, $[G]_0$, x, AND K

The experimental conditions that can be set up are $[H]_0$ and $[G]_0$ (see Equation (15.11) and classification of variables). How should the experimental conditions, $[H]_0$ and $[G]_0$, be changed for the titration? Figure 15.6 depicts possibilities including a dilution experiment (a), a continuous variation experiment (b), an experiment utilizing constant $[H]_0$ with different $[G]_0$ (c), and two more possible examples. The criteria for deciding condition changes include:

- Ease of experimentation and calculation
- Acceptability

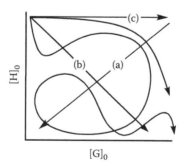

FIGURE 15.6   Graphical expression showing possible ways to change $[H]_0$ and $[G]_0$ for titration experiment.

- Applicability
- Reliability

Let us consider how to choose experimental conditions $[H]_0$ and $[G]_0$ using a 1:1 host–guest complexation stoichiometry for simplicity.

$$H + G \rightleftharpoons C \tag{15.26}$$

From Equation (15.11)

$$K = \frac{[C]}{([H]_0 - [C]) \cdot ([G]_0 - [C])} \tag{15.27}$$

Let us calculate the correlation between K and complexation ratio (x).

$$y = K, x = \frac{[C]}{[H]_0} \tag{15.28}$$

Equation (15.27) is transformed to

$$y = \frac{x}{(1 - x) \cdot ([G]_0 - [H]_0 \cdot x)} \tag{15.29}$$

Figure 15.7 is a graph of Equation (15.29) in which the x (x-coordinate) is 0 to 1, the K (y-coordinate) is 10 to 100,000,000, $[H]_0$ is 0.01 to 0.000001, and $[H]_0 = [G]_0$ is a premise. In general, caution is expressed as: *measurements below 20 and above 80% complexation ratio (x) yield uncertain values*. This caution is interpreted with Figure 15.7 as follows.

The steep rises of K at a complexation ratio below 20% and exceeding 80% cause the transfer of magnified errors from the complexation ratio into K. When K is determined based on the measurement of a property directly connected to or proportional to the complexation ratio, the obtained errors in K are magnified compared with those for the observed property. In the case where $[H]_0 = 0.0001$ M as an example, the complexation ratio is between 0.2 and 0.8 when K is between 3000 and 200 000 M^{-1}, so an accurate experiment is carried out. This brief discussion shows how the accuracy of K is governed by the choice of concentrations of $[H]_0$, $[G]_0$, and K.

15.5.2 SET-UP OF $[H]_0$

Setting up the concentration of host $[H]_0$ is limited by the measured properties, the apparatus, and other features of the experiment. For example, $[H]_0$ for UV/vis spectroscopy, which depends greatly on molar absorptivity, is roughly in the range of 0.0001 M because of absorbance limitation from noise level to <2. $[H]_0$ for NMR spectroscopy is roughly around 0.01 M because of signal-to-noise ratio

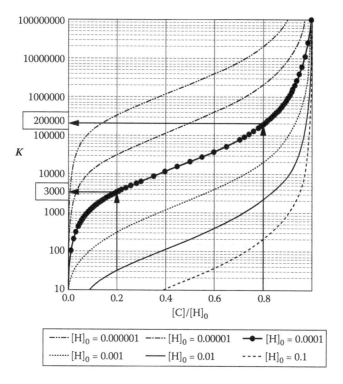

FIGURE 15.7   The correlation between complexation ratio ($[C]/[H]_0$) and binding constant (K).

(SNR) and accumulation time of limitation. Therefore, $[G]_0$ is often the only variable that can be set up in a wide range because $[H]_0$ is usually governed by the experimental method.

15.5.3   Set-Up of $[G]_0$

We now consider how to set up the concentration of $[G]_0$ using Figure 15.8, which is based on Equation (15.29) in which $[H]_0 = 0.0001$ and $\frac{[G]_0}{[H]_0}$ is changed from 0.1 to 1000. The correlation between complexation ratio x and the accurately obtainable K range by changing $[G]_0$ with constant $[H]_0$ (= 0.0001 M) is clear when based on Figure 15.8. Considering the suitable x range (0.2 < x < 0.8) for reliable measurement in the figure, the combination of $[H]_0$, $[G]_0$, and K is determined. For example, when $[G]_0 = 0.001$ M and $[H]_0 = 0.0001$ M, then $[G]_0/[H]_0 = 10$, and consequently, a reliable range of K of 250 to 4000 M^{-1} is obtained by following the arrows in Figure 15.8.

The obtained K ranges are summarized in Figure 15.9 by repeating the above procedures for several combinations of $[H]_0$ and $[G]_0$. This figure is useful for a preliminary check of the experimental concentration conditions and for choosing a suitable experimental method.

In most cases K is determined with a titration experiment followed by regression of the obtained data based on the above-mentioned theoretical equations. In the commonly

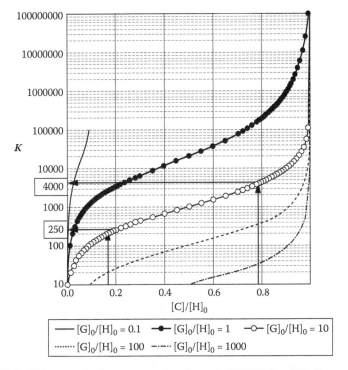

FIGURE 15.8 The correlation between complexation ratio (($[C]/[H]_0$)) and binding constant (K).

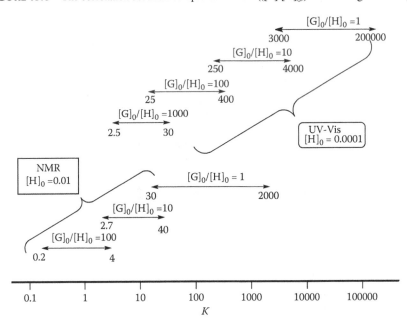

FIGURE 15.9 Reliable regions of $[H]_0$ and $[G]_0$ for K determination shown for representative concentration of UV-visible and NMR spectroscopies.

used titration experiment, $[G]_0$ is changed while the range of $[H]_0$ change is limited by the detection method. Therefore, the important point is how to set up the range of $[G]_0$.

A representative answer to this question is obtained by considering the correlation of $[G]_0/[H]_0$ and the complexation ratio based on Figure 15.10 in which the x-coordinate is the concentration ratio of guest over host and the y-coordinate is the complexation ratio. The graph in Figure 15.10 is based on Equation (15.32) which is derived from Equation (15.27) by multiplying both sides of the equation by $[H]_0$, dividing the denominator and numerator by $[H]_0^2$, then substituting with y and x according to Equation (15.30).

$$y = \frac{[C]}{[H]_0}, x = \frac{[G]_0}{[H]_0} \tag{15.30}$$

$$[H]_0 \cdot K = \frac{y}{(1-y)\cdot(x-y)} \tag{15.31}$$

Displacement using equation $\alpha = [H]_0 \cdot K$ and transformation produces

$$\alpha \cdot y^2 - (\alpha + \alpha \cdot x + 1) \cdot y + \alpha \cdot x = 0 \tag{15.32}$$

Figure 15.10 is obtained by changing α from 0.0001 to 1000, which corresponds to the change of K from $K = \frac{0.0001}{[H]_0}$ to $K = \frac{1000}{[H]_0}$. While tracing from the bottom to the top of the S-curve in Figure 15.10 is necessary for complete identification of each

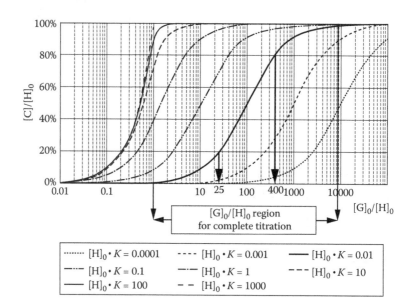

FIGURE 15.10 The calculated curves plotted between $[G]_0/[H]_0$ and $[C]/[H]_0$ for $[G]_0$ range determination of the titration experiment.

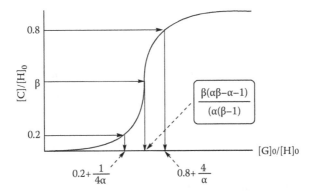

FIGURE 15.11 Useful graph for $[G]_0$ range determination of titration experiment ($\alpha = [H]_0 \cdot$ K, $\beta = [C]/[H]_0$).

equilibrium, it is possible to determine the binding constant by plotting the data $\frac{[C]}{[H]_0}$, $\frac{[G]_0}{[H]_0}$, as in Figure 15.10 obtained from experimentation, followed by curve fitting using Equation (15.32).

When $[H]_0 \cdot K = 0.01$ is selected as an example, the range of $[G]_0$ for a complete titration is $1 \cdot [H]_0$ to $10\,000 \cdot [H]_0$ M as indicated in Figure 15.10. To reduce error, the $[G]_0$ area should be avoided where lines are close together. When the experimental conditions are in an area of the figure crowded with graphs, a small error in $[G]_0$ causes a plot on a different S-curve whose K is much different. Therefore, this unsuitable concentration setting results in low reliability.

Based on this consideration, the range of the complexation ratio between 0.2 to 0.8 is again suitable for a reliable measurement. The suitable range of $[G]_0$ can be obtained from Figure 15.10. For this example, the suitable range of $[G]_0$ is from $25 \cdot [H]_0$ to $400 \cdot [H]_0$ expressed relative to $[H]_0$. Figure 15.11 depicts this suitable range of $[G]_0$. It is possible to express the x-coordinate using α and β as follows based on the Equation (15.32)

$$x = \frac{\beta \cdot (\alpha \cdot \beta - \alpha - 1)}{\alpha \cdot (\beta - 1)}$$

$$x = \frac{[G]_0}{[H]_0}, \alpha = [H]_0 \cdot K, \beta = \frac{[C]}{[H]_0}$$

(15.33)

The complexation ratio here is β. With Equation (15.33), the $[G]_0$ range for the titration experiment in which the complexation ratio is 0.2 to 0.8 is obtained as a function of α by just entering $\beta = 0.2$ or $\beta = 0.8$ into Equation (15.33). The result is summarized in Figure 15.11. On entering $\alpha = 0.01$, the suitable x range is obtained easily from Figure 15.11 as

$$25.2 = 0.2 + \frac{1}{4 \cdot 0.01} \leq \frac{[G]_0}{[H]_0} \leq 0.8 + \frac{4}{0.01} = 400.8$$

One more consideration for $[G]_0$ should be mentioned here, referring back to Figure 15.10. When $[H]_0 \cdot K$ is larger than 1, the curves are close together even if the complexation ratio is between 0.2 and 0.8. Consequently, as a premise for a reliable experiment, $[H]_0 \cdot K$ should be *smaller than 1*. When $[H]_0 \cdot K$ is larger than 1, $[H]_0$ should be reduced. When $[H]_0$ cannot be reduced, other observable parameters should be chosen along with an appropriate spectroscopic method. Based on Figure 15.11, the $[G]_0$ range is limited as described below.

$$0.2 + \frac{1}{4 \cdot \alpha} \leq \frac{[G]_0}{[H]_0} \leq 0.8 + \frac{4}{\alpha} \quad \text{where} \quad \alpha = [H]_0 \cdot K \tag{15.34}$$

Multiplying by $[H]_0$, followed by transformation results in

$$(0.2 \cdot [H]_0 \cdot K + 0.25) \cdot \frac{1}{K} \leq [G]_0 \leq (0.8 \cdot [H]_0 \cdot K + 4) \cdot \frac{1}{K} \tag{15.35}$$

When $[H]_0 \cdot K$ is set smaller than 1, the range of $[H]_0 \cdot K$ is from 0 to 1 and

$$0.25 \leq (0.2 \cdot [H]_0 \cdot K + 0.25) \leq 0.45$$

$$4 \leq (0.8 \cdot [H]_0 \cdot K + 4) \leq 4.8 \tag{15.36}$$

$$\therefore 0.25 \cdot K_{diss} \leq [G]_0 \leq 4.8 \cdot K_{diss}$$

In summary, we can conclude with respect to the concentration ranges of host and guest that for a reliable experiment, the magnitude of K should be predicted with an educated guess. The method selected (NMR spectroscopy,[8] UV/vis or fluorometry spectroscopy,[9] mass spectroscopy,[10] capillary electrophoresis,[10,11] or other method) approximately decides the range of $[H]_0$ and finally decides the range of $[G]_0$ using Figure 15.11 and/or Equation (15.35).

15.6 DATA TREATMENT

15.6.1 General View

The next step is treatment of data collected to obtain the K values after titration experiments. Some data treatment methods are general, some rely on approximations and thus are subject to some cautions, and some are only regression methods. Typical examples of the approximate methods are the Benesi-Hildebrand,[12] Ketelaar,[13] Nagakura-Baba,[14,15] Scott,[16] and Hammond[17] methods that approximate [G] by $[G]_0$. From Equations (15.8) and (15.10) and a = b = 1, we derive

$$[G]_0 = [G] + K \cdot [H] \cdot [G]$$

$$\therefore [G]_0 = [G](1 + K \cdot [H]) \tag{15.37}$$

If $K \cdot [H] \ll 1$, it can safely be assumed that $[G]_0 = [G]$. This condition is encountered frequently for weak complexation where K is small. Furthermore, $[G]_0 \gg [H]_0$

is usually employed for practical titration, which is thought to be essential. However, all systems cannot be investigated under condition $[G]_0 \gg [H]_0$ ($K \cdot [H] \ll 1$).

When the assumption $[G]_0 = [G]$ cannot be applied, other approximation or regression methods must be employed. Typical examples of the regression methods are the Rose–Drago,[15a,18] Nakano,[19] and Creswell-Allred[20] methods. Because of its wide applicability, a practical guide based on the Rose-Drago method is presented as an example employing UV/vis spectroscopy.

Originally, the Rose-Drago method was used for UV/vis spectroscopy for evaluating acid–base equilibria and complexation of iodine. The only assumption for the original method is that there are at most two species observed that obey Beer's law in the concentration range employed. Since no other assumptions are involved, the Rose-Drago method is widely applicable. The results are presented graphically and by close inspection one can determine the precision quantitatively. The following section describes an example case.

15.6.2 ROSE-DRAGO METHOD FOR UV/VIS SPECTROSCOPY

The Rose-Drago method will be described using a 1:1 host–guest complexation stoichiometry detected by UV/vis spectroscopy. The observed property of a supramolecular system is absorbance determined in a titration experiment. For details of data treatment with the Rose-Drago method, a spreadsheet program is available on the author's website.[21]

We substitute a = b = 1 into Equation (15.11). Then the reciprocal is

$$\frac{1}{K} = [C] - ([H]_0 + [G]_0) + \frac{[H]_0 \cdot [G]_0}{[C]} \tag{15.38}$$

Combining Equation (15.24) with Equation (15.38) gives

$$\begin{aligned}
\frac{1}{K} &= \frac{A_{obs} - \varepsilon_h \cdot [H]_0 - \varepsilon_g \cdot [G]_0}{\varepsilon_c - \varepsilon_h - \varepsilon_g} - ([H]_0 + [G]_0) \\
&+ \frac{\varepsilon_c - \varepsilon_h - \varepsilon_g}{A_{obs} - \varepsilon_h \cdot [H]_0 - \varepsilon_g \cdot [G]_0} \cdot [H]_0 \cdot [G]_0
\end{aligned} \tag{15.39}$$

For an evaluation of K values, the constants ε_h, and ε_g in Equation (15.39) must be obtained from independent measurements because they are molar absorptivities of the pure host and pure guest, respectively. A_{obs} then must be measured at different combinations of $[H]_0$ and $[G]_0$ followed by regression of the obtained data using Equation (15.39). Theoretically, A_{obs} values at more than two different combinations of $[H]_0$ and $[G]_0$ give two unknowns K and ε_c. Measurement of absorbance at different combinations of $[H]_0$ and $[G]_0$ supplies a matrix $\{A_{obsn}, [H]_{0n}, [G]_{0n}\}$ consisting of three elements:

$A_{obsn} =$ observed absorbance of nth measurement
$[H]_{0n} =$ concentration of host molecule at initial stage for nth measurement
$[G]_{0n} =$ concentration of guest molecule at initial stage for nth measurement

Combining Equation (15.39) and the definitions in Equations (15.40) through (15.44), we have Equation (15.45).

$$Y = \frac{1}{K} \tag{15.40}$$

$$X = \varepsilon_c - \varepsilon_h - \varepsilon_g \tag{15.41}$$

$$a_n = A_{obsn} - \varepsilon_h \cdot [H]_{0n} - \varepsilon_g \cdot [G]_{0n} \tag{15.42}$$

$$b_n = [H]_{0n} + [G]_{0n} \tag{15.43}$$

$$c_n = \frac{[H]_{0n} \cdot [G]_{0n}}{A_{obsn} - \varepsilon_h \cdot [H]_{0n} - \varepsilon_g \cdot [G]_{0n}} \tag{15.44}$$

Then

$$Y = \frac{a_n}{X} - b_n + c_n \cdot X \tag{15.45}$$

One combination of data (e.g., $\{A_{obs1}, [H]_{01}, [G]_{01}\}$ and $\{A_{obs2}, [H]_{02}, [G]_{02}\}$) supplies a matrix of answers $\{X, Y\}$ according to Equation (15.45). A representative solution is shown below. As an example, one combination of data where n = 1 and n = 2 (e.g., $\{A_{obs1}, [H]_{01}, [G]_{01}\}$ and $\{A_{obs2}, [H]_{02}, [G]_{02}\}$) is used.

$$Y = \frac{a_1}{X} - b_1 + c_1 \cdot X \tag{15.46}$$

$$Y = \frac{a_2}{X} - b_2 + c_2 \cdot X \tag{15.47}$$

Subtraction of both sides, followed by multiplication by X results in

$$(c_1 - c_2) \cdot X^2 + (b_1 - b_2) \cdot X + (a_1 - a_2) = 0 \tag{15.48}$$

$$\therefore X = \frac{-(b_1 - b_2) \pm \sqrt{(b_1 - b_2)^2 - 4 \cdot (c_1 - c_2) \cdot (a_1 - a_2)}}{2 \cdot (c_1 - c_2)} \tag{15.49}$$

Substituting Equation (15.46) with Equation (15.49) yields an expression for Y. The obtained $\{X, Y\}$ is merely an answer that satisfies both Equations (15.46) and (15.47), but it is not necessarily a chemically correct answer. For example, a chemically reasonable Y value should have a positive sign. Based on such chemical limitation, correct sets of answers should be collected.

The maximum number of obtainable answer pairs $\{X, Y\}$ is $_nC_2$ for n combinations of concentration conditions. For example, five pairs of $\{[H]_{0n}, [G]_{0n}\}$ yield ten ($=_5C_2$) pairs of $\{X, Y\}$. *These $\{X, Y\}$ values are obtained under the premise where 1:1 complexation is in effect. No approximation is introduced into this solution.* The reciprocal of the obtained Ys is the binding constant K. The number of obtained K at this stage is $_nC_2$.

15.6.3 ESTIMATION OF ERROR

Statistics teaches that the deviation of data based on fewer than 30 measurements is not a normal distribution but rather a Student's t-distribution. It is thus suitable to express the binding constant K with a 95% confidence interval calculated by applying Student's t-distribution. When the number of measurements is more than 30, Student's t-distribution and the normal distribution are practically the same. The actual function of Student's t-distribution is complicated, and it is rarely used directly.

A conventional way to apply Student's t-distribution is to pick up data from the critical value table of Student's t-distribution and considering degrees of freedom, levels of significance, and measured data. It is troublesome to repeat this conventional method many times. Most spreadsheet software even for personal computers includes a Student's t-distribution function. Without any tedious work, namely, picking up data from the table, statistical treatment can be applied to experimental results based on Student's t-distribution with the aid of a computer. Figure 15.12 displays an

A1	B	C	D	E	F	G
2				How to use this sheet		
3		Data	34. 72	1.Input data at D3 to D9		
4			36. 84	2.Input level of significance (α)		
5			33. 25	3.Find answer in D18 and D19		
6			37. 78			
7			39. 16			
8			35. 08			
9			34 54			
10		Average	35. 91	=AVERAGE(D3:D9)		
11		s	1 93	=STDEVP(D3:D9)		
12						
13		α	0. 05		0.05	
14		Degree of freedom	6	=COUNTA(D3:D9)-1		
15		$t_{\alpha/2}$	2. 447	=TINV(D13, D14)		
16						
17		95% confidence interval		Confidence interval		
18		$K=$	35.91	=AVERAGE(D3:D9)		
19		$\pm$	1.78	=D15*D11/COUNTA(D3:D9)$^{0.5}$		
20						
21						

FIGURE 15.12   Spreadsheet for statistical data-treatment based on Student's t-distribution.

example. When the measurement data are input into the gray cells, answers can be obtained in cells D18 and D19 instantaneously.

When the confidence interval obtained after statistical treatment is very wide, the probability is high that a precise experiment has not been conducted. If this is the case, the experimental conditions and all procedures should be checked.

15.6.4 CONCLUSIONS FOR UV/VIS SPECTROSCOPIC METHODS

The method described here includes no approximation with regard to data treatment and may be used generally. The required level of mathematical knowledge is not high. Only a formula for polynomials of degree 2 are involved and therefore the logical basis can be understood easily. Moreover, statistical treatment of the obtained data is understandable with primary statistics background.

When the stoichiometry of the complex is not 1:1 or if other premises are not satisfied, the method of data treatment should be changed or modified. Nonlinear least square data treatment is one of the best approximations. By using Equations (15.11) and (15.24) for UV/vis spectroscopy, other complexations may be applied even if a, b do not equal 1.

15.7 GUIDELINES FOR DETERMINING BINDING CONSTANTS IN A 1:n COMPLEXATION SYSTEM WITH AND WITHOUT SYNERGISTIC EFFECT OF GUEST BINDING ON BINDING PROPERTIES OF HOST

One fundamental example of synergistic binding of host with guest is a 1:n complexation. Figure 15.13(a) describes this process in a spontaneous sense. Generalization of 1:n binding with stepwise binding involves two typical methods, with apparent binding constants or binding constants at each binding site. The analyses are summarized in Figure 15.13(b); relevant abbreviations, parameters, and explanations are covered in text.

15.7.1 TYPICAL ANALYSIS FOR 1:n COMPLEXATION WITH SPONTANEOUS BINDING MODEL

The basic equations for 1:n host–guest complexation of spontaneous binding model are the following.

$$H + n \cdot G \rightleftarrows HG_n \qquad (15.50)$$

$$[H]_0 = [H] + [HG_n] \qquad (15.51)$$

H = host; G = guest; C = complex ($H \cdot G_n$); n = stoichiometry; $[H]_0$ = initial (total) concentration of host molecule; $[G]_0$ = initial (total) concentration of guest molecule; and [H], [G], [C] = equilibrium concentrations of host, guest, and complex, respectively.

$$a\,H + b\,G \underset{k_{-1}}{\overset{k_1}{\rightleftharpoons}} C\ (=Ha\,Gb) \qquad K = \frac{k_1}{k_{-1}} = \frac{[C]}{[H]^a\,[G]^b}$$

(a) Spontaneous Binding Model

$$H + nG \underset{k_{-1}}{\overset{k_1}{\rightleftharpoons}} HG_n \qquad K = \frac{k_1}{k_{-1}} = \frac{[HG_n]}{[H]\,[G]^n}$$

(b) Stepwise Binding Model

apparent binding constant   genuine binding constant
(binding constant at each binding site)

$$H_0 + G \underset{k_{-1}}{\overset{k_1}{\rightleftharpoons}} H_1 \qquad K_1 = \frac{k_1}{k_{-1}} = \frac{[H_1]}{[H_0]\,[G]} \qquad L_1 = \frac{[H_1]}{n[H_0]\,[G]}$$

$$H_1 + G \underset{k_{-2}}{\overset{k_2}{\rightleftharpoons}} H_2 \qquad K_2 = \frac{k_2}{k_{-2}} = \frac{[H_2]}{[H_1]\,[G]} \qquad L_2 = \frac{2[H_2]}{(n-1)[H_1]\,[G]}$$

$$\vdots \qquad\qquad\qquad \vdots \qquad\qquad\qquad \vdots$$

$$H_{i-1} + G \underset{k_{-i}}{\overset{k_i}{\rightleftharpoons}} H_i \qquad K_i = \frac{k_i}{k_{-i}} = \frac{[H_i]}{[H_{i-1}]\,[G]} \qquad L_i = \frac{i[H_i]}{(n-i+1)[H_{i-1}]\,[G]}$$

$$\vdots \qquad\qquad\qquad \vdots \qquad\qquad\qquad \vdots$$

$$H_{n-1} + G \underset{k_{-n}}{\overset{k_n}{\rightleftharpoons}} H_n \qquad K_n = \frac{k_n}{k_{-n}} = \frac{[H_n]}{n[H_{n-1}]\,[G]} \qquad L_n = \frac{n[H_n]}{[H_{n-1}]\,[G]}$$

FIGURE 15.13 Representative expression of models for 1:n complexation and their parameters.

Rate constants for complexation (k_1) and decomplexation (k_{-1}) are defined as shown in Equations (15.52) through (15.55).
Complexation Process (Association):

$$H + n \cdot G \xrightarrow{k_1} HG_n \tag{15.52}$$

$$\frac{d[HG_n]_{association}}{dt} = k_1[H][G]^n \tag{15.53}$$

Decomplexation Process (Dissociation):

$$H + n \cdot G \xleftarrow[k_{-1}]{} HG_n \tag{15.54}$$

$$\frac{d[HG_n]_{dissociation}}{dt} = -K_{-1}[HG_n] \tag{15.55}$$

The complexation and corresponding decomplexation rates must be the same at equilibrium; then

$$\frac{d[HG_n]}{dt} = k_1[H][G]^n - k_{-1}[HG_n] = 0 \tag{15.56}$$

$$\therefore \frac{k_1}{k_{-1}} = \frac{[HG_n]}{[H][G]^n}(= K) \tag{15.57}$$

The complexation ratio (θ = fraction of occupied binding sites) is derived as follows:

$$K = \frac{[HG_n]}{[H][G]^n} = \frac{[HG_n]}{([H_0] - [HG_n])[G]^n}$$

$$K_d = K^{-1} = \frac{([H]_0 - [HG_n])[G]^n}{[HG_n]}$$

$$K_d[HG_n] = ([H]_0 - [HG_n])[G]^n$$

$$(K_d + [G]^n)[HG_n] = [H]_0[G]^n$$

$$\therefore \theta = \frac{[HG_n]}{[H]_0} = \frac{[G]^n}{K_d + [G]^n} \quad \text{(Hill equation)} \tag{15.58}$$

Equation (15.58) is called the Hill equation and the n is the Hill coefficient. Recall that complexation ratios are proportional to absorptions in UV/vis spectra, chemical shift differences in NMR spectra, and other experimentally obtainable values. Therefore, complexation ratios can be useful parameters for practical analyses. Based on Equation (15.58),

$$\theta(K_d + [G]^n) = [G]^n$$

$$(1 - \theta)[G]^n = \theta K_d \tag{15.59}$$

$$\frac{[G]^n}{K_d} = \frac{\theta}{1 - \theta}$$

$$\log\left(\frac{[G]^n}{K_d}\right) = \log\left(\frac{\theta}{1 - \theta}\right) \tag{15.60}$$

$$\therefore \log\left(\frac{\theta}{1 - \theta}\right) = n\log[G] - \log K_d$$

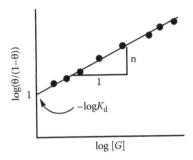

FIGURE 15.14 Representative expression of the Hill Plot, dissociation constant (K_d), and the Hill coefficient (*n*).

By taking $\log\left(\frac{\theta}{1-\theta}\right)$ as the y-axis and log [G] as the x-axis (Hill plot), the Hill coefficient (n) and K_d can be obtained from the slope and the intercepts of the Hill plot, respectively (Figure 15.14).

15.7.2 Typical Prognosis for Synergistic Binding System Using Hill Plots and Hill Coefficients (Spontaneous Binding Model)

Typical changes in the shapes of Hill Plots (n = 0.5, 1, 2, ..., 9) are shown in Figure 15.15. In Figure 15.16, corresponding saturation plots (y-axis, which is $\frac{[HG_n]}{[H]_0}$ degree of saturation, complexation ratio, or fraction of occupied hosts)[22] are shown. To model 1:n binding equilibrium, n must be a natural number (positive integer) scientifically. As shown in Figure 15.16, for n > 1, the Hill plots take sigmoidal-shaped curves. When observed saturation plots can be fitted with the Hill equation, the obtained Hill

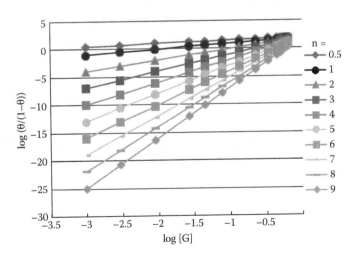

FIGURE 15.15 Typical shape of the Hill Plots for complexation of host and guest with different Hill coefficient (n).

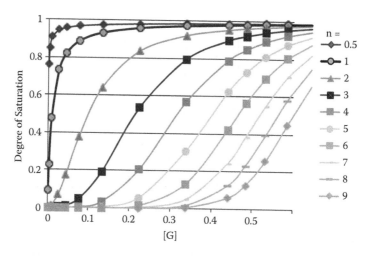

FIGURE 15.16 Change of saturation degree curves in host-guest complexation from saturation to sigmoidal curve with different Hill coefficients (n = 0.5 to 9) with the curve of the Hill equation (n = 2).

coefficient n is treated as a useful parameter to show synergistic effects. Complexation systems can be divided into three categories based on the Hill coefficient n.

Systems having n < 1 are treated as negative synergistic complexations.
Systems having n = 1 are treated as containing no synergistic complexations.
Systems having n > 1 are treated as containing positive synergistic complexations.

15.7.3 General Considerations of Analysis for 1:n Complexation with Stepwise Binding Model

15.7.3.1 General Considerations Using Apparent Binding Constants for Stepwise Binding Model

In 1:n complexation systems that the Hill equation dealt with, n guest molecules bind to the host molecule spontaneously with one binding constant K. However, for a step-by-step complexation process, a set of binding constants (K_1, K_2, ...) is required. The following explanation for generalized 1:n complexation systems with stepwise binding models takes these considerations into account. The fundamental equations needed to describe this process are Equations (15.61) through (15.64).

$$H_0 + G \underset{k_{-1}}{\overset{k_1}{\rightleftarrows}} H_1 \quad \text{where} \quad H_1 \equiv H_0G_1 \quad K_1 = \frac{k_1}{k_{-1}} \tag{15.61}$$

.....

$$H_{i-2} + G \underset{k_{-(i-1)}}{\overset{k_{i-1}}{\rightleftarrows}} H_{i-1} \quad \text{where} \quad H_{i-1} \equiv H_{i-2}G_1 \quad K_{i-1} = \frac{k_{i-1}}{k_{-(i-1)}} \tag{15.62}$$

$$H_{i-1} + G \underset{k_{-i}}{\overset{k_i}{\rightleftarrows}} H_i \quad \text{where} \quad H_i \equiv H_{i-1}G_1 \quad K_i = \frac{k_i}{k_{-i}} \tag{15.63}$$

.....

$$H_{n-1} + G \underset{k_{-n}}{\overset{k_n}{\rightleftarrows}} H_n \quad \text{where} \quad H_n \equiv H_{n-1}G_1 \quad K_n = \frac{k_n}{k_{-n}} \tag{15.64}$$

where i = running index expressing a step number; n = maximum number of guest molecules and steps of the complexation equilibria (number of binding sites on host); k_i = forward rate constant of ith step; k_{-i} = backward rate constant of ith step; K_i = apparent binding constant at ith step; $[H]_0$ = concentration of host molecule at initial stage; $[G]_0$ = concentration of guest molecule at initial stage; $[H]_n$ = concentration of host molecule after nth complexation step; and $[G]_n$ = concentration of guest molecule after nth complexation step. The total concentration of guests is calculated as follows:

$$[G]_0 = [G] + \sum_1^n i[H_i] \tag{15.65}$$

The average number (r) of guests bound to hosts is calculated as a ratio of total concentration of guests bound to hosts over total concentration of hosts.

$$r = \frac{\sum_1^n i[H_i]}{\sum_0^n [H_i]} \tag{15.66}$$

The average number (r) is between 0 and n (0 < r < n; ratio $\frac{r}{n}$ is the fraction of occupied binding sites). For each step, Equations (15.67) through (15.70) can be formulated as follows.

First step:

$$\frac{d[H_0]}{dt} = -k_1[G][H_0] + k_{-1}[H_1] \tag{15.67}$$

Second step:

$$\frac{d[H_1]}{dt} = -k_2[G][H_1] + k_{-2}[H_2] + k_1[G][H_0] - k_{-1}[H_1] \tag{15.68}$$

....

ith step:

$$\frac{d[H_{i-1}]}{dt} = -k_i[G][H_{i-1}] + k_{-i}[H_i] + k_{i-1}[G][H_{i-2}] - k_{-(i-1)}[H_{i-1}] \tag{15.69}$$

....

nth step:

$$\frac{d[H_n]}{dt} = +k_n[G][H_{n-1}] - k_{-n}[H_n] \tag{15.70}$$

At equilibrium state, all species in the system are constant.

$$\frac{d[H_0]}{dt} = \frac{d[H_1]}{dt} = \frac{d[H_2]}{dt} = \cdots\cdots = \frac{d[H_{i-1}]}{dt} = \cdots\cdots = \frac{d[H_n]}{dt}(=0) \quad (15.71)$$

The following equations are derived (see Appendix 15).

First step:

$$[H_1] = K_1[G][H_0] \quad (15.72)$$

Second step:

$$[H_2] = K_2[G][H_1] = K_1K_2[G]^2[H_0] \quad (15.73)$$

nth step:

$$[H_n] = K_n[G][H_{n-1}] = \left(\prod_{i=1}^{n} K_i\right)[G]^n[H_0] \quad (15.74)$$

Replacements of $[H_n]$s in Equation (15.66) according to Equation (15.74) produce the Adair equation:

$$r = \frac{K_1[G]^1 + 2K_1K_2[G]^2 + \cdots\cdots + nK_1K_2K_3\ldots K_n[G]^n}{1 + K_1[G]^1 + K_1K_2[G]^2 + \cdots\cdots + K_1K_2K_3\ldots K_n[G]^n} \quad \text{(The Adair equation)}$$

$$15.75)$$

15.7.3.2 General Considerations for Use of Genuine Binding Constants at Binding Sites for Stepwise Binding Model

Because binding constants K_i ($i = 1 - n$) are apparent binding constants for 1:n complexation at each step, it is usually not possible to obtain genuine binding constants (L_i ($i = 1 - n$):binding constants at each binding site of ith step). L_i values are considered here as follows. As a premise, the host has n binding sites. Then,

$$H_0 + G \underset{k_{-1}}{\overset{k_1}{\rightleftharpoons}} H_1 \quad \text{where} \quad H_1 \equiv H_0G \quad K_1 = \frac{k_1}{k_{-1}} \quad (15.76)$$

$$H_1 + G \underset{k_{-2}}{\overset{k_2}{\rightleftharpoons}} H_2 \quad \text{where} \quad H_2 \equiv H_1G \quad K_2 = \frac{k_2}{k_{-2}} \quad (15.77)$$

$$\ldots\ldots$$

$$H_{n-1} + G \underset{k_n}{\overset{k_n}{\rightleftharpoons}} H_n \quad \text{where} \quad H_n \equiv H_{n-1}G \quad K_n = \frac{k_n}{k_{-n}} \quad (15.78)$$

Because occupied binding sites of H_i is i and unoccupied sites are $(n - i)$, concentration of each unoccupied binding site of H_i ($[H_i]$) is $(n-i)[H_i]$. Then,

$$\frac{d[H_0]}{dt} = -k_1[G] \cdot n[H_0] - k_{-1}[H_1] \tag{15.79}$$

$$\frac{d[H_1]}{dt} = -k_2[G] \cdot (n-1)[H_1] + k_{-2}2[H_2] + k_1[G] \cdot n[H_0] - k_{-1}[H_1] \tag{15.80}$$

....

$$\frac{d[H_{i-1}]}{dt} = -k_i[G](n-i+1)[H_{i-1}] + k_{-i}i[H_i] + k_{i-1}[G](n-i+2)[H_{i-2}]$$
$$- k_{-(i-1)}(i-1)[H_{i-1}] \tag{15.81}$$

....

At equilibrium state, all species in the system are constant.

$$\frac{d[H_0]}{dt} = \frac{d[H_1]}{dt} = \frac{d[H_2]}{dt} = ...= \frac{d[H_{i-1}]}{dt} = ...(= 0) \tag{15.82}$$

Then, the following equations are derived (see Appendix 2). Therefore, for the first step:

$$L_1 = \frac{[H_1]}{[G]n[H_0]} \tag{15.83}$$

....

For the ith step:

$$L_i = \frac{i[H_i]}{[G](n-i+1)[H_{n-1}]} \tag{15.84}$$

Finally, the relation between the two binding constants is formulated as follows.

$$L_i = \frac{i}{(n-i+1)}K_i \tag{15.85}$$

15.7.3.2.1 Complexation without Synergistic Binding ($L_1 = L_2 = ...$)

In cases without synergistic binding, the binding constants at any step are the same.

$$L_1 = L_2 = ... = L_n = L$$

The relation between K_i and L_i (Equation (15.85)) gives the following:

$$K_1 = n \cdot L$$

$$K_2 = \frac{(n-1)}{2} \cdot L$$

$$\cdots\cdots\cdots$$ (15.86)

$$K_n = \frac{1}{n} \cdot L$$

The fraction of occupied binding sites on host (Y), which is the average number of binding sites associated with guests is formulated as follows from Equation (15.85) and the "Adair Equation" Equation (15.75). Details are shown in Appendix 3.

$$Y = \frac{r}{n} = \frac{L[G]}{1 + L[G]} = \frac{[G]}{\frac{1}{L} + [G]}$$ (15.87)

Because Y is the fraction of occupied binding sites on the host, Equation (15.87) corresponds to Equation (15.58) in the Hill equations with $n = 1$ (one-step complexation) and $K_d (= L^{-1})$. Apparently binding constant K_1 is the same as L_1 from Equation (15.85), ($n = 1$, $i = 1$).

The important point here is that if L_i is constant at any step, the relation between fraction of occupied binding sites and concentrations of guest is the same as that of 1:1 binding with apparent binding constants $K_1 (= L)$ regardless of the number of binding sites (n) on the host. Therefore, in a host–guest system *without* synergistic binding (all L_i are constant $(= L)$) the slope of the Hill plot (Hill coefficient) of *any* titration result must be 1 regardless of the number of binding sites on the host.

15.7.3.2.2 Complexation with Synergistic Binding ($L_1 \neq L_2,...$)

Situations in cases with synergistic binding (binding constants are *not* always the same: $L_1 \neq L_2 ... L_n$) become much more complicated. At a typical but simple example of complexation with different binding constants, a 1:2 binding complexation system is considered here to explain how to analyze synergistic binding phenomena. At the first step,

$$K_1 = 2L_1 \text{ (for a first binding site)}$$ (15.88)

At the second step:

$$K_2 = \frac{1}{2}L_2$$

$$\left(\because K_i = \frac{(n-i+1)}{i} L_i, i = 2, n = 2 \right)$$ (15.89)

Therefore, Equation (15.75) is

$$r = \frac{K_1[G] + 2K_1K_2[G]^2}{1 + K_1[G] + K_1K_2[G]^2}$$

$$= \frac{2L_1[G] + 2 \cdot 2L_1 \dfrac{L_2}{2}[G]^2}{1 + 2L_1[G] + 2L_1 \dfrac{L_2}{2}[G]^2} \tag{15.90}$$

$$= \frac{2L_1[G] + 2L_1L_2[G]^2}{1 + 2L_1[G] + L_1L_2[G]^2}$$

Then,

$$Y = \frac{r}{n}$$

$$= \frac{1}{2}\frac{2L_1[G] + L_1L_2[G]^2}{1 + 2L_1[G] + L_1L_2[G]^2} \quad (\because n = 2) \tag{15.91}$$

$$= \frac{L_1[G] + L_1L_2[G]^2}{1 + 2L_1[G] + L_1L_2[G]^2}$$

Equation (15.91) is the general equation of regression for synergistic complexation systems.

For a non-synergistic binding system, $L_1 = L_2 (= L)$, then,

$$Y = \frac{L[G] + L^2[G]^2}{1 + 2L[G]^1 + L^2[G]^2}$$

$$= \frac{L[G](1 + L[G])}{(1 + L[G])^2} \tag{15.92}$$

$$= \frac{L[G]}{1 + L[G]}$$

$$= \frac{[G]}{\frac{1}{L} + [G]}$$

The resulting Equation (15.92) is in the same form as the Hill equation with Hill coefficient (n) = 1 and $K_d = 1/L$. Namely, if binding constants (L_i) of all steps are the same (non-synergistic binding cases), the Hill coefficient n is always 1 regardless of complexation step number. (In this exemplified case with two-step complexation, step n is 2.) Next, for consideration of the synergistic binding cases ($L_1 \neq L_2$), L_2 is expressed in terms of L_1 using Equation (15.93).

$$L_2 = j \cdot L_1 \tag{15.93}$$

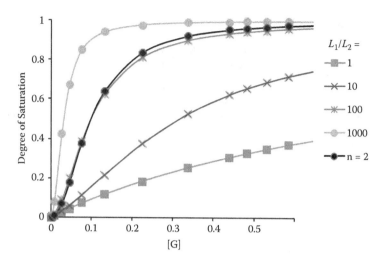

FIGURE 15.17 Change of saturation degree curves in 1:2 host–guest complexation with different ratio of binding constants (L_2/L_1= 0.1 to 1000) with the curve of Hill equation (n=2).

The parameter (j) can be used to classify the cases as follows. In the case of positive binding (j > 1), graphs of saturation degree in 1:2 complexation obtained from Equations (15.91, 15.93, 15.94) (L_1 = 1, j = 1, 10, 100, and 1000) are shown in Figure 15.17. In the case of negative binding (j<1), graphs of saturation degree in 1:2 complexation obtained from Equation (15.91, 15.93, 15.94) (L_1 = 100, j = 0.001, 0.01, 0.1, and 1) are shown in Figure 15.18.

$$Y_S = \frac{L_1[G]+ j \cdot L_1^2[G]^2}{1+2L_1[G]+ j \cdot L_1^2[G]^2} \tag{15.94}$$

When the first binding of a guest to a binding site on the host enhances the second binding event (positive synergistic binding), the binding constant at the second step (L_2) is larger than that of first step (L_1) (L_2> L_1). Examples of positive binding titration curves are shown in Figure 15.17 in which L_1= 1 and L_2= 1 to 1000. The curve for L_2= 100 (L_2/L_1 = 100) is very similar to the curve of the Hill plot (n = 2) having the same value of binding constant K (= $L_2 \cdot L_1$). The similarity of these two curves is always observed when the synergistic effect is large enough ($L_2 \gg L_1$). These similarities create difficulties in analysis of titration results with positive synergistic bindings having more than two parameters.

When the first binding of a guest to a binding site on a host suppresses the second binding event (negative synergistic binding), the binding constant at the second step (L_2) is smaller than that of first step (L_1). Examples of negative synergistic titration curves are shown in Figure 15.18 in which L_1 = 100 and L_2= 0.001 to 1. All the curves are clearly unlike those of the corresponding Hill plot (n = 2) for the 1:2 complexation model. To determine the binding constants of each step, a precise experimental set-up dependent on the physical properties observed is required. The next section

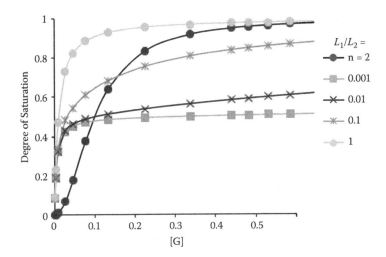

FIGURE 15.18 Change of saturation degree curves in 1:2 host-guest complexation with different ratio of binding constants ($L_2/L_1 = 0.1$ to 1000).

details a practical example for determining binding constants of a 1:2 host–guest binding system.

15.7.4 Practical Data Treatment for 1:2 Complexation System

15.7.4.1 Typical Case Study of Spontaneous Binding Model for 1:2 Complexation System Using ¹H NMR Spectroscopy

Equation (15.95) is a model scheme of 1:2 complexation of host (H) and guest (G):

Scheme 15.1:

$$H + 2G \rightleftharpoons C \tag{15.95}$$

$$K = \frac{[C]}{[H]\cdot[G]^2} \tag{15.96}$$

$$[H] = [H]_0 - [C] \tag{15.97}$$

$$[G] = [G]_0 - 2\cdot[C] \tag{15.98}$$

$$\delta_{calcd} = \delta_{host}\cdot\frac{[H]}{[H]_0} + \delta_{couplex}\cdot\frac{[C]}{[H]_0} \tag{15.99}$$

G = guest; C = complex (H· G_2); $[H]_0$ = initial (total) concentration of host molecule; $[G]_0$ = initial (total) concentration of guest molecule; [H], [G], [C] = equilibrium concentrations of host, guest, and complex, respectively; δ_{calcd} = calculated chemical shift assigned to host probe proton at equilibrium; δ_{host} = chemical shift assigned to host probe proton of free host; and $\delta_{complex}$ = chemical shift assigned to host probe proton of complex. From Equations (15.96) and (15.97):

$$K \cdot [H] \cdot [G]^2 = [C]$$

$$K \cdot \{[H]_0 - [C]\} \cdot [G]^2 = [C]$$

Then

$$K \cdot \left\{ [H]_0 - \frac{[G]_0 - [G]}{2} \right\} \cdot [G]^2 = \frac{[G]_0 - [G]}{2}$$

(15.100)

$$\because [C] = \frac{[G]_0 - [G]}{2} \quad \text{(obtained from Equation (15.98))}$$

Multiply both sides of Equation (15.100) by 2:

$$K \cdot \{2 \cdot [H]_0 - [G]_0 + [G]\} \cdot [G]^2 = [G]_0 - [G]$$

$$\therefore K \cdot [G]^3 + K \cdot \{2 \cdot [H]_0 - [G]_0\} \cdot [G]^2 + [G] - [G]_0 = 0$$

(15.101)

The parameters can be classified as K = constants (larger than 0); $[H]_0$, $[G]_0$ = variables that may be set up as experimental conditions; and $[H]$, $[G]$, $[C]$ = variables dependent on equilibrium.

From Equation (15.101) and the classification of its parameters, we can deduce the following guideline for experiments to determine the binding constant. When K is assumed, $[G]$ at equilibrium can be derived according to Equation (15.101) from the experimental conditions $[H]_0$ and $[G]_0$. Then, $[H]$ and $[G]$ are derived from Equations (15.97) and (15.98). With these data obtained, calculated chemical shift (δ_{calcd}) at that complexation equilibrium can be derived according to Equation (15.99).

Following regression by minimization of the sum of square deviation ($\Sigma(\Delta\delta)^2 : \Delta\delta = \delta_{calcd} - \delta_{obs}$) affords binding constant K and chemical shift of host probe proton in complex ($\delta_{complex}$). Therefore, the following six steps must be followed to determine the binding constants:

1. Assume K and $\delta_{complex}$.
2. $[G]$ can be obtained from Equation (15.101).
3. $[C]$ can be obtained from $[G]_0$ and Equation (15.98).
4. $[H]$ can be obtained from $[C]$, $[H]_0$, and Equation (15.97).
5. $\Delta\delta$ (= $\delta_{calcd} - \delta_{obs}$) is calculated from Equation (15.99) and experimentally obtained δ_{obs}.
6. Minimization of the sum of square deviation ($\Sigma(\Delta\delta)^2 : \Delta\delta = \delta_{calcd} - \delta_{obs}$).

Appendix 15.1 is a concrete example of a spreadsheet applicable for this system. The three candidates of the solutions ([G]s) of third degree Equation (15.101) appear in the spreadsheet as 1-[G], 2-[G], and 3-[G]. With the solutions, ([G]s), [C]s, and [H]s are obtained according to Equations (15.98) and (15.97). At this step, chemically correct solutions must be selected. Only real numbers can be the solutions. (Only the real numbers of [G] are shown in the spreadsheet. Imaginary numbers are omitted.) Moreover, the solutions must be positive numbers because the solutions correspond to concentration [G]. With the selected solution(s), subsequent data treatment is carried out, then, K and δ_{calcd} can be obtained.

15.7.4.2 Estimation of Error for General Functions

To obtain a standard deviation to express an experimental error with a confidence area, procedures to obtain standard deviations of parameters of general functions are indispensable. These are, however, quite complex. With the aid of spreadsheet software, the standard deviations can be obtained relatively easily. In this case, a convenient way to obtain the standard deviation is to apply the "Solvstat" macro to the spreadsheet. Doing so is easy and provides reliable results. First, a spreadsheet to obtain regression coefficients by using "Solver" must be prepared. Then the "Solvstat" add-in can be used to elucidate standard deviations of regression coefficients.[23]

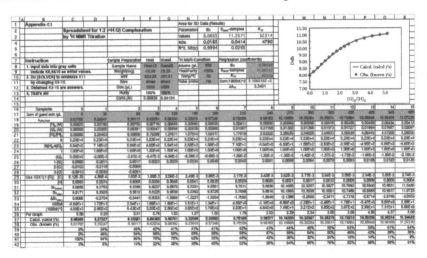

15.7.5 MODIFICATION FOR 1:2 COMPLEXATIONS WITH GUEST PROBE PROTON MONITORING

When one cannot observe a chemical shift change of a host probe proton, a guest probe proton may be used instead. In this case, Equation (15.99) must be modified as follows and all other calculations can be carried out in the same way:

$$\delta_{calcd} = \delta_{guest} \cdot \frac{[G]}{[G]_0} + \delta_{complex} \cdot \frac{2 \cdot [C]}{[G]_0} \qquad (15.99)$$

15.7.6 MODIFICATION FOR 2:1 COMPLEXATIONS

$$2H + G \rightleftharpoons C \tag{15.102}$$

Scheme 15.2 Model scheme of 2:1 complexation of host (H) and guest (G) .

The spreadsheet for 1:2 complexation can be applied easily to 2:1 complexation shown in Scheme 15.2. If the spreadsheet for 1:2 complexation monitored by a host probe proton is modified simply by interchanging host and guest, the resulting spreadsheet is applicable to 2:1 complexation observing the chemical shift change of a guest probe proton.

15.7.7 TYPICAL CASE STUDY OF STEPWISE BINDING MODEL FOR SYSTEMS HAVING 1:1 AND 1:2 COMPLEXATIONS

A stepwise binding model for host–guest complexation systems having 1:1 and 1:2 complexation equilibria shown in Scheme 15.3 is discussed in this section.

$$H + G \rightleftharpoons HG \tag{15.103}$$

Scheme 15.3 Model scheme of 1:1 and 1:2 complexation equilibria of host and guest (G).

$$HG + G \rightleftharpoons GHG \tag{15.104}$$

Equations (15.103) and (15.104) constitute Scheme 15.3 (model scheme of 1:1 and 1:2 complexation equilibria of host (H) and guest (G)). The fundamental equations are (15.105) and (15.106).

$$K_{11} = \frac{[HG]}{[H] \cdot [G]} \tag{15.105}$$

$$K_{12} = \frac{[GHG]}{[HG] \cdot [G]} \tag{15.106}$$

$$[H]_0 = [H] + [HG] + [GHG] \tag{15.107}$$

$$[G]_0 = [G] + [HG] + 2 \cdot [GHG] \tag{15.108}$$

H = host; G = guest; C = complex (H· G_2); $[H]_0$ = initial (total) concentration of host molecule; $[G]_0$ = initial (total) concentration of guest molecule; and [H], [G], [HG], and [GHG] = concentrations of host, guest, 1:1 complex, and 1:2 complex at equilibrium, respectively. From Equations (15.105) through (15.108):

$$[HG] = \frac{K_{11} \cdot [H]_0 \cdot [G]}{1 + K_{11} \cdot [G] + K_{11} \cdot K_{12} \cdot [G]^2} \tag{15.109}$$

$$[GHG] = \frac{K_{11} \cdot K_{12} \cdot [H]_0 \cdot [G]^2}{1 + K_{11} \cdot [G] + K_{11} \cdot K_{12} \cdot [G]^2} \tag{15.110}$$

$$K_{11} \cdot K_{12} \cdot [G]^3 + K_{11} \cdot (2 \cdot K_{12} \cdot [H]_0 - K_{12} \cdot [G]_0 + 1) \cdot [G]^2$$
$$+ (1 - K_{11} \cdot [G]_0 + K_{11} \cdot [H]_0) \cdot [G] - [G]_0 = 0 \tag{15.111}$$

Chemical shift change ($\Delta\delta$) of guest probe proton is formulated as a function shown below.

$$\Delta\delta = \frac{[HG]}{[G]_0}\Delta\delta_{11} + \frac{2\cdot[GHG]}{[G]_0}\Delta\delta_{12}$$

$$\Delta\delta = \delta - \delta_{guest}$$

$$\Delta\delta_{11} = \delta_{12} - \delta_{guest}$$

$$\Delta\delta_{12} = \delta_{12} - \delta_{guest} \tag{15.112}$$

Δ = chemical shift assigned to guest probe proton at equilibrium; δ_{guest} = chemical shift assigned to guest probe proton of free host; δ_{11} = chemical shift assigned to guest probe proton of 1:1 complex; and δ_{12} = chemical shift assigned to guest probe proton of 1:2 complex. The parameters can be classified as follows: K_{11}, K_{12}, δ_{guest}, δ_{11}, δ_{12} = constants (K_{11}, K_{12} are larger than 0); $[H]_0$ and $[G]_0$ = variables that can be set up as experimental conditions; and $[H]$, $[G]$, and $[C]$ = variables dependent on the equilibrium.

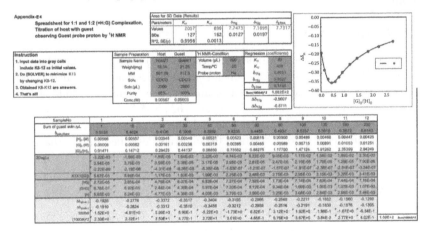

From Equation (15.112) and the classification of its parameters, we can deduce the following guidelines for experiments to determine the binding constant. When K_{11} and K_{12} are assumed, $[G]$ at equilibrium can be derived according to Equation (15.111) from the experimental conditions $[G]_0$ and $[H]_0$. Then $[HG]$ and $[GHG]$ are derived from Equations (15.109) and (15.110), respectively. With these data obtained, calculated chemical shift ($\Delta\delta_{calcd}$) at the complexation equilibria can be derived according to Equation (15.112) with assumed $\Delta\delta_{11}$ and $\Delta\delta_{12}$.

Minimization of the sum of square deviation ($\Sigma(\Delta\Delta\delta)^2$ where $\Delta\Delta\delta = \Delta\delta_{calcd} - \Delta\delta_{obs}$) affords binding constants K_{11}, K_{12}, chemical shift of guest probe proton in the 1:1 complex (δ_{11}), and in the 1:2 complex (δ_{12}). Consequently, the following six steps must be followed to determine the binding constants:

1. Assume K_{11}, K_{12}, δ_{11}, and δ_{12}.
2. $[G]$ can be obtained from Equation (15.111).

3. [HG] can be obtained from [G] and Equation (15.109).
4. [GHG] can be obtained from [G] and Equation (15.110).
5. $\Delta\Delta\delta (= \Delta\delta_{calcd} - \Delta\delta_{obs})$ is calculated from Equation (15.112) and experimentally obtained $\Delta\delta_{obs}$.
6. Minimization of the sum of square deviation ($\Sigma (\Delta\Delta\delta)^2$ where $\Delta\Delta\delta = \Delta\delta_{calcd} - \Delta\delta_{obs}$) is performed.

See author's website for a concrete example of a spreadsheet applicable for this system. Among the three candidates of the solutions [G] of cubic Equation (15.111), chemically correct solutions must be selected. With the selected solution(s), subsequent data treatment may be carried out, then K_{11}, K_{12}, δ_{11}, and δ_{12} will be obtained. Finally, estimations of errors are generated based on the obtained standard deviations and statics.

For the analysis of 1:1 and 1:2 binding systems, four parameters are required. It is not always easy to determine the four parameters practically and precisely. Careful consideration and enough data sets at different conditions (e.g., temperature dependence, etc.) are required generally. If you are required to set up a more complicated system (three, four, or more steps), the reliability of parameters must be determined carefully.

15.8 CONCLUSION

This chapter describes theoretical principles underlying binding constant determinations in detail using basic levels of mathematics, statistics, and spreadsheet software. As concrete examples, the practical measurements and data treatments of UV/vis and NMR titration experiments are discussed. The programs attached as appendices will function with commonly available spreadsheet software on personal computers and provide another way to understand the contents described in this chapter. The appendices are also useful for readers when running experiments.

REFERENCES AND NOTES

1. (a) Izatt, R. M., Bradshaw, J. S., Nielsen, S. A. et al. *Chem. Rev.* 1985, *85* (4), 271–339. (b) Izatt, R. M., Pawlak, K., Bradshaw, J. S. et al. *Chem. Rev.* 1991, *91* (8), 1721–2085. (c) Izatt, R. M., Bradshaw, J. S., Pawlak, K. et al. *Chem. Rev.* 1992, *92* (6), 1261–1354. (d) Zhang, X. X., Bradshaw, J. S., and Izatt, R. M. *Chem. Rev.* 1997, *97* (8), 3313–3361.
2. (a) Ogasahara, K., Hirose, K., Tobe, Y. et al. *J. Chem. Soc. Perkin Trans. 1* (21), 1997, 3227–3236. (b) Naemura, K., Wakebe, T., Hirose, K. et al. *Tetrahedron Asymmet.* 1997, *8* (15), 2585–2595. (c) Naemura, K., Nishioka, K., Ogasahara, K. et al. *Tetrahedron: Asymmet.* 1998, *9* (4), 563–574. (d) Naemura, K., Matsunaga, K., Fuji, J. et al. *Anal. Sci.* 1998, *14* (1), 175–182. (e) Hirose, K., Aksharanandana, P., Suzuki, M. et al. *Heterocycles* 2005, *66*, 405–431.
3. (a) Naemura, K., Fuji, J., Ogasahara, K. et al. *Chem. Commun.* 1996, (24), 2749–2750. (b) Hirose, K., Fuji, J., Kamada, K. et al. *J. Chem. Soc. Perk. Trans. 2* 1997, (9), 1649–1657.
4. (a) Shibata, Y., Inouye, T., and Nakatsuka, Y. *Nippon Kag. Kaishi* 1921, *42*, 983–1005. (b) Job, P. *Compt. Rend. Hebdom. Seances Acad. Sci.* 1925, *180*, 1932–1934. (c) Job, P. *Compt. Rend. Hebdom. Seances Acad. Sci.* 1925, *180*, 1108–1110. (d) Job, P. *Ann. Chim. France* 1928, *9*, 113–203. (e) Tsuchida, R. *Bull. Chem. Soc. Jpn.* 1935, *10*, 27–39.

5. Bent, H. E. and French, C. L. *J. Am. Chem. Soc.* 1941, *63*, 568–572.
6. Harvey, A. E. and Manning, D. L. *J. Am. Chem. Soc.* 1950, *72* (10), 4488–4493.
7. Yoe, J. H. and Jones, A. L. *Ind. Eng. Chem., Anal. Ed.* 1944, *16*, 111–115.
8. Fielding, L. *Progr. Nucl. Magn. Reson. Spectrosc.* 2007, *51* (4), 219–242.
9. Dufour, C. and Dangles, O. *Biochim. Biophys. Acta* 2005, *1721* (1–3), 164–173.
10. (a) Rundlett, K. L. and Armstrong, D. W. *Electrophoresis* 2001, 22 (7), 1419–1427. (b) Kempen, E. C. and Brodbelt, J. S. *Anal. Chem.* 2000, 72 (21), 5411–5416.
11. Busch, M. H. A., Kraak, J. C., and Poppe, H. *J. Chromatogr. A* 1997, *777* (2), 329–353.
12. Benesi, H. A. and Hildebrand, J. H. *J. Am. Chem. Soc.* 1949, *71* (8), 2703–2707.
13. Ketelaar, J. A. A., Vandestolpe, C., Goudsmit, A. et al. *Recueil Travaux Chim. Pays-Bas* 1952, *71* (9–10), 1104–1114.
14. Nagakura, S. *J. Am. Chem. Soc.* 1954, *76* (11), 3070–3073.
15. (a) Nagakura, S. *J. Am. Chem. Soc.* 1958, *80* (3), 520–534. (b) Baba, H. *Bull. Chem. Soc. Jpn.* 1958, *31* (2), 169–172.
16. Scott, R. L. *Recueil Travaux Chim. Pays-Bas* 1956, *75* (6), 787–789.
17. Hammond, P. R. *J. Chem. Soc.* 1964, Jan., 479–484.
18. Rose, N. J. and Drago, R. S. *J. Am. Chem. Soc.* 1959, *81* (23), 6138–6141.
19. Nakano, M., Nakano, N. I., and Higuchi, T. *J. Phys. Chem.* 1967, *71* (12), 3954–3959.
20. (a) Creswell, C. J. and Allred, A. L. *J. Phys. Chem.* 1962, *66* (8), 1469–1472. (b) Stamm, H., Lamberty, W., and Stafe, J. *Tetrahedron* 1976, *32* (16), 2045–2050.
21. Hirose, K. Spreadsheet software with simple instructions.
22. This is not a graph showing relation between "fraction of occupied binding sites" and concentration of guest at equilibrium
23. (a) Billo, E. J. *Excel for Chemists*, 2nd Ed. Wiley, New York, 2001, p. 233. (b) Billo, E. J., *Excel for Scientists and Engineers*, Wiley, New York, 2007, p. 327.

APPENDIX 15.1: DERIVATION OF EQUATIONS (15.71) TO (15.75) IN SECTION 15.7.3.1

At equilibrium state, all species in the system are constant.

$$\frac{d[H_0]}{dt} = \frac{d[H_1]}{dt} = \frac{d[H_2]}{dt} = \cdots\cdots\cdots = \frac{d[H_{i-1}]}{dt} = \cdots\cdots = \frac{d[H_n]}{dt}(=0) \qquad (15.71)$$

Therefore, for the first step:

$$\frac{d[H_0]}{dt} = -k_1[G][H_0] + k_{-1}[H_1] = 0$$

$$\frac{k_1}{k_{-1}} = \frac{[H_1]}{[G][H_0]} = K_1 \qquad (15.72)$$

$$\therefore [H_1] = K_1[G][H_0]$$

For the second step:

There are further data and information for the appendices available from the author at http://hdl.handle.net/11094/39419

$$\frac{d[H_1]}{dt} = -k_2[G][H_1] + k_{-2}[H_2] + k_1[G][H_0] - k_{-1}[H_1]$$

$$= -k_2[G][H_1] + k_{-2}[H_2] + 0$$

$$= 0$$

$$\frac{k_2}{k_{-2}} = \frac{[H_2]}{[G][H_1]} = K_2 \tag{15.73}$$

$$\therefore [H_2] = K_2[G][H_1]$$

$$= K_2[G]K_1[G][H_0]$$

$$= K_1K_2[G]^2[H_0]$$

For the nth step:

$$\frac{k_n}{k_{-n}} = \frac{[H_n]}{[G][H_{n-1}]} = K_n$$

$$\therefore [H_n] = K_n[G][H_{n-1}]$$

$$= K_1K_2K_3.....K_n[G]^n[H_0] \tag{15.74}$$

$$= \left(\prod_{i=1}^{n} K_i \right)[G]^n[H_0]$$

Replacements of $[H_n]$s in Equation (15.66) according to Equation (15.74) produce the Adair Equation (15.75):

$$r = \frac{\sum_{1}^{n} i[H_i]}{\sum_{0}^{n} [H_i]}$$

$$= \frac{K_1[G]^1[H_0] + 2K_1K_2[G]^2[H_0] + \cdots + nK_1K_2K_3....K_n[G]^n[H_0]}{[H_0] + K_1[G]^1[H_0] + K_1K_2[G]^2[H_0] + \cdots + K_1K_2K_3...K_n[G]^n[H_0]} \tag{15.75}$$

$$= \frac{K_1[G] + 2K_1K_2[G]^2 + \cdots + nK_1K_2K_3....K_n[G]^n}{1 + K_1[G] + K_1K_2[G]^2 + \cdots + K_1K_2K_3....K_n[G]^n}$$

APPENDIX 15.2: DERIVATION OF EQUATIONS (15.83) TO (15.85) IN SECTION 15.7.3.2

At equilibrium state, all species in the system are constant.

$$\frac{d[H_0]}{dt} = \frac{d[H_1]}{dt} = \frac{d[H_2]}{dt} = \ldots\ldots = \frac{d[H_{i-1}]}{dt} = \ldots.(=0) \tag{15.82}$$

Therefore, for the first step:

$$\frac{d[H_0]}{dt} = -k_1[G]n[H_0] + k_{-1}[H_1] = 0$$

$$\therefore \frac{k_1}{k_{-1}} = \frac{[H_1]}{[G]n[H_0]} = L_1 \tag{15.83}$$

. . . .

For the ith step:

$$\frac{d[H_{i-1}]}{dt} = -k_i[G](n-i+1)[H_{i-1}] + k_{-1}i[H_i] + k_{i-1}[G](n-i+2)[H_{i-2}]$$

$$-k_{-(i-1)}(i-1)[H_{i-1}]$$

$$= -k_i[G](n-i+1)[H_{i-1}] + k_{-1}i[H_i] \tag{15.84}$$

$$= 0$$

$$\therefore \frac{k_i}{k_{-1}} = \frac{i[H]}{[G](n-i+1)[H_{n-1}]} = L_i$$

Finally, the relation between the two binding constants is formulated as

$$L_i = \frac{k_i}{k_{-1}}$$

$$= \frac{i[H_i]}{[G](n-i+1)[H_{n-1}]} \tag{15.85}$$

$$= \frac{i[H_i]}{(n-i+1)[G][H_{n-1}]}$$

$$= \frac{i}{(n-i+1)}K_i$$

$$\therefore K_i = \frac{(n-i+1)}{i}.L_i$$

APPENDIX 15.3:   DERIVATION OF EQUATION (15.87)

The fraction of occupied binding sites on host (Y), which is the average number of binding sites associated with guests, is formulated as follows from Equation (15.85) and the Adair Equation (15.75).

$$Y = \frac{r}{n}$$

$$= \frac{1}{n} \cdot \frac{K_1[G] + 2K_1K_2[G]^2 + \cdots + nK_1K_2K_3\ldots K_n[G]^n}{1 + K_1[G] + K_1K_2[G]^2 + \cdots + K_1K_2K_3\ldots K_n[G]^n}$$

$$= \frac{1}{n} \cdot \frac{\displaystyle\sum_{i=1}^{n}\left(i \cdot nL \cdot \frac{n-1}{2}L \cdot \frac{n-2}{3}L \cdot \ldots \cdot \frac{n-i+1}{i}L \cdot [G_i]^i\right)}{1 + \displaystyle\sum_{i=1}^{n}\left(nL \cdot \frac{n-1}{2}L \cdot \frac{n-2}{3}L \cdot \ldots \cdot \frac{n-i+1}{i}L \cdot [G_i]^i\right)}$$

$$= \frac{1}{n} \cdot \frac{\displaystyle\sum_{i=1}^{n}\left(nL[G] \cdot \frac{(n-1)!}{(i-1)!(n-i)!}L^{i-1} \cdot [G_i]^{i-1}\right)}{1 + \displaystyle\sum_{i=1}^{n}\left(\frac{n!}{i!(n-i)!}L^i \cdot [G_i]^i\right)}$$

$$= \frac{1}{n} \cdot \frac{nL[G] \cdot \displaystyle\sum_{i=1}^{n}\left(\frac{(n-1)!}{(i-1)!(n-i)!}L^{i-1} \cdot [G_i]^{i-1}\right)}{\displaystyle\sum_{i=0}^{n}\left(\frac{n!}{i!(n-i)!}L^i \cdot [G_i]^i\right)}$$ (15.87)

$$= \frac{1}{n} \cdot \frac{nL[G] \cdot (1 + L[G])^{n-1}}{(1 + L[G])^n}$$

$$= \frac{1}{n} \cdot \frac{nL[G]}{1 + L[G]}$$

$$= \frac{L[G]}{1 + L[G]}$$

$$= \frac{[G]}{\frac{1}{L} + [G]}$$

For the above transformation, the following binomial theorem was used:

$$\because (x+y)^n = \sum_{k=0}^{n}\left(\frac{n!}{k!(n-k)!}x^{n-k} \cdot y^k\right)$$

For (denominator) : $x = 1, y = L[G], k = i$

For (numerator) : $x = 1, y = L[G], k = i - 1, n = n - 1$

16 Supramolecular Chemistry Strategies for Naked-Eye Detection and Sensing

Kamaljit Singh, Paramjit Kaur, Hiroyuki Miyake, and Hiroshi Tsukube

CONTENTS

16.1 INTRODUCTION

Supramolecular chemistry has emerged as a vibrant research field that has provided new approaches to multidisciplinary molecular architecture. An ensemble of molecules organized into higher-order functional structures through weak, non-covalent forces[1–3] offers a wide range of nano- and meso-structured assemblies. These assemblies display useful functional behavior, the ability to integrate electronic and photonic properties, biocatalysis, and other properties.

Based on intermolecular interactions of π-conjugated systems[4,5] and donor–acceptor dyads[6–9] and their propensity to assemble as supramolecular stacks, new applications have been developed. The wide diversity of organic and inorganic receptors[10–12] capable of self-assembly and molecular recognition evoke considerable potential for detection of single or multiple analytes. These receptors also provide insights into chemical reactivity, and supramolecular interactions[13] upon binding with appropriate guests.

Molecular recognition events produce quantifiable changes in electrochemical, photochemical, or optical characteristics to yield practical tools for real-world applications. Receptors that show optical changes are promising candidates for applications in "naked-eye" detection and sensing of targeted guests. The principles of supramolecular recognition described in this chapter represent a selected series of analyte recognition events through exquisite synergistic binding leading to sensing events. The usefulness and challenges of supramolecular strategies for naked-eye detections of various analytes are described. Although covering all types of supramolecular sensors was not possible, in this chapter design strategies involving recognition of analytes through molecular recognition and/or self-assembly leading to quantifiable optical signals are described.

16.2 NAKED-EYE SENSING STRATEGIES THROUGH SUPRAMOLECULAR RECOGNITION OR SELF-ASSEMBLY USING ORGANIC RECEPTORS

Figure 16.1 illustrates water-soluble, benzene-bridged fluorescent receptor 1 encompassing a pyrene excimer transducer and imidazolium cation that selectively and effectively recognizes biologically important adenosine triphosphate (ATP).[14] Recognition occurs through formation of sandwiched pyrene–adenine-pyrene π–π stacks in aqueous solutions at physiological pH (7.4). Upon ATP binding, 1 displayed a large fluorescent quenching effect in its excimer emission, but a selective fluorescent enhancement in its monomeric emission. Other nucleoside triphosphates such as guanosine 5′-triphosphate (GTP) did not affect the pyrene–pyrene dimeric stacking of 1 and interacted only from outside.

The rationally designed supramolecular self-assembled synthetic pores 4 (Figure 16.2) represent a collection of synthetic multifunctional pores[15] of varying thermodynamic and kinetic stabilities. The rigid-rod β-barrel pores 4 were formed in lipid bilayer membranes by self-assembly of the monomeric conjugate 3 consisting of p-octiphenyl backbone with 8 pentapeptides 2 (Figure 16.2). The assembly 4 featured minimal internal counter ion immobilization and enhanced internal charge repulsion

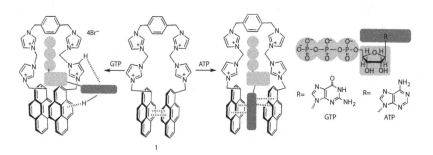

FIGURE 16.1 Proposed binding modes of **1** with ATP and GTP. (*Source:* Copyright (2009), American Chemical Society. With permission.)

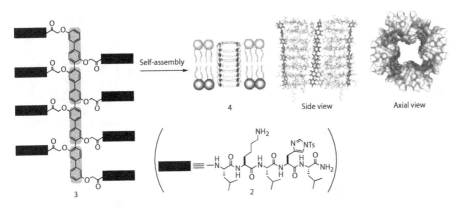

FIGURE 16.2 Peptide-*p*-octiphenyl conjugate **3** with three-dimensional supramolecular assembly **4** with optimized side and axial views without guest. (*Source:* Copyright (2004), American Chemical Society. With permission.)

that also accounted for the large inner diameter (d ≅ 12 Å). Due to the tendency of several appropriately sized blockers to fill the available internal nanospace, pores such as **4** offer opportunities for molecular recognition of single small substrates and supramolecular catalysis. The application of multifunctional synthetic pores has been extended to the naked-eye sensing of sugar in soft drinks, using invertases and kinases as co-sensors and the esterolysis of 8-acetoxy-1,3,6-pyrenetrisulfonate.

Employing a non-empirical approach, synthetic pores with reactive signal amplifiers such as artificial (electronic) tongues that capture analytes selectively after enzymatic signal generation have been reported. These synthetic pores have been applied to detect flavors in food and act as efficient naked-eye sensors.[16,17] Self-assembled pores of the rigid rod **5** having p-octiphenyl backbones with the π-acidic naphthalenediimide pentapeptide **6** (Figure 16.3) are equipped with peptide sequences that produce hydrophobic exteriors and hydrophilic cationic interiors. The sequences bind π-basic anionic guests such as dialkoxynaphthalene amplifier **12** through supramolecular electron donor–acceptor interactions leading to pore blockage.

Such large pores have been used for naked-eye detection of analytes such as L-glutamate **7**, the umami flavor, through reactive amplification (with transaminase, Figure 16.3b) mediated by an active pore-amplifier-analyte complex. Thus, the amplified π-basic "umami signal" **11** was recognized by the π-acidic sites in the pores with excellent selectivity and sensitivity through supramolecular inclusion complex (π-clamping, Figure 16.3a).

Conceptually, **7** and the co-substrate pyruvate **8**, under the influence of transaminase, furnish L-alanine **9** and α-ketoglutarate **10**. The latter upon condensation with π-basic amplifier conjugate **12** (Figure 16.3) furnishes **11** which subsequently forms an inclusion complex within the pore, selectively and without interference from the corresponding conjugate of **8**.

Aptamers, non-natural, synthetic oligonucleotides with defined tertiary structures, like single-stranded nucleic acids, possess specific ligand binding sites and assume a wide variety of folding topologies (3-dimensional [3D] structures) with

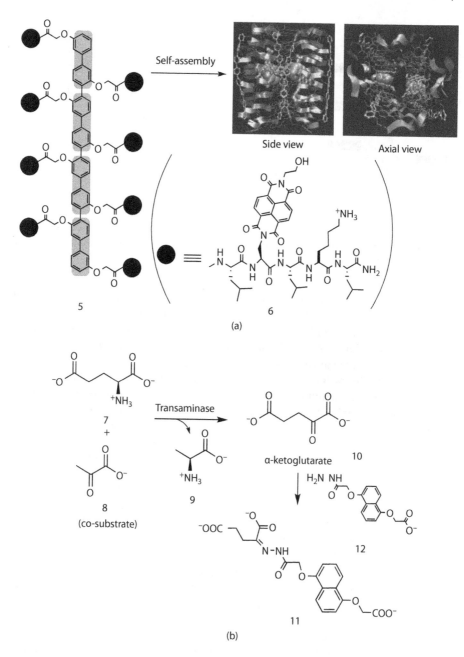

FIGURE 16.3 (a) Peptide *p*-octiphenyl conjugate **5**. Molecular models of self-assembled pores showing simulated geometries (side and axial views) with umami amplifier conjugate π-clamped by adjacent naphthalenediimide residues inside pores. (b) Selective amplification of α-ketoglutarate **10** with transaminase: generation of umami signal **11**. (*Source:* Copyright (2007), Macmillan Publishers Ltd. With permission.)

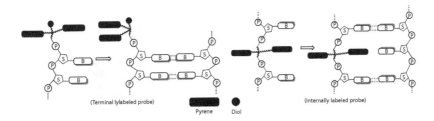

FIGURE 16.4 Supramolecular interactions of DNA with modified oligonucleotides. (*Source:* Copyright (2002), American Chemical Society. With permission.)

structural and functional diversity. In a manner analogous to antibody–antigen binding, aptamers bind a complementary DNA sequence to form a duplex structure and exhibit intricate molecular recognition of small molecules and proteins, including supramolecular complexes such as viruses and cells with high affinity and specificity.

In the folded state, aptamers may act as either ligands for larger receptors or as receptors for small molecules. Their flexible secondary structures along with structural and functional diversity assume conformational topologies conducive for capturing certain analytes, leading to a variety of highly sensitive optical and other signal transduction modes.

A propane diol appended with two pyrene units was incorporated into the 5′-terminal or internal positions of an oligonucleotide using standard phosphoramidate chemistry and desired affinity to complementary DNA (Figure 16.4). Supramolecular stacking interactions between pyrene and DNA bases strongly enhanced excimer fluorescence when a terminally labeled probe was used. The internally labeled oligonucleotides showed comparatively small changes in fluorescence quantum yield as well as excimer:monomer ratios and provided sequence-specific fluorescent probes for DNA.[18] In a similar sensing event, an aptamer enabled highly enantioselective recognition of L-arginine (K_d value of 330 nM) that was bound 12,000 times stronger (7.2 kcal mol^{-1})[19] than D-arginine.

Another example of conformational changes in a small RNA aptamer upon binding involves binding to the bronchodilator theophylline that was selective (100-fold stronger) with a dissociation constant K_d of 0.1 μM.[20] It also possessed a 10,000-fold selectivity over structurally related caffeine.

A partially folded DNA aptamer possessing a three-way junction was engineered to create a ligand-induced binding pocket that recognized cocaine **13** in a micromolar range through supramolecular trapping in the lipophilic cavity (Figure 16.5a)[21] due to excellent complementarity[22] of stems of appropriate length and conformational changes.[23] The close proximity of the two ends of the aptamer labeled with a fluorophore (fluorescein) and a quencher [4-(4′-dimethylaminophenylazo)benzoic acid] (dabcyl), led to quenching of the fluorescence and the consequent signal.[21]

A similar design strategy based on conformational changes in molecular aptamer beacons (MABs) resulted in effective sensing of thrombin **14** upon supramolecular recognition of proteins (K_d 4.87 ± 0.55 nM; Figure 16.5b).[24] In this sensitive (limit of detection: 429 ± 63 pM) recognition event, the change in fluorescence efficiency

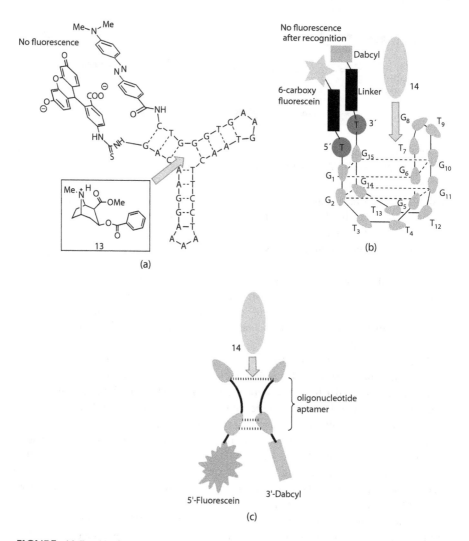

FIGURE 16.5 (a) Cocaine **13** binding DNA aptamer bearing fluorescein and dabcyl units at 5′ and 3′ends.[21] (b) MAB for thrombin **14** binding.[24] (c) thrombin beacon design.[25] (*Sources:* Copyright (2005), American Chemical Society and copyright (2002), Elsevier. With permission.)

through modulation of fluorescence resonance energy transfer (FRET) was used as a signal. A similar assembly of two signaling nucleotides as a consequence of thrombin recognition[25] resulted in FRET quenching (40%; Figure 16.5c) between the fluorophores in complex biological mixtures with picomolar sensitivity.

A non-fluorescent tripartite ensemble consisting of three DNA aptamer units (A, B, and C) was created (Figure 16.6). The 5-fluorescein (F) and quencher (Q, dabcyl), upon recognition of ATP by one of the aptamer units, enhanced fluorescence due to release of a dabcyl-appended aptamer. This "structure-switching" ATP reporter

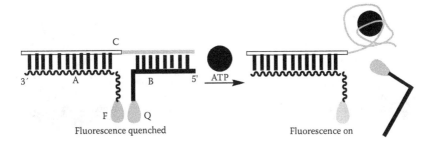

FIGURE 16.6 Design of structure-switching signaling aptamers. (*Source:* Copyright (2003), American Chemical Society. With permission.)

had an apparent K_d of ~600 μM—about 60-fold greater than that reported for the original ATP aptamer.[26,27]

In the examples above, supramolecular recognition of different analytes was described. The process involved recognition of appropriate analytes through binding at the receptor sites, leading to either conformational changes or structure-switching and attendant absorption and/or fluorescence changes that resulted in naked-eye color changes.

Molecular recognition at interfaces has been studied extensively using Langmuir monolayers, self-assembled monolayers (SAMs), and lipid assemblies as recognition media. *Self-assembly* involves the spontaneous assembly of simpler subunits or building blocks such as atoms, molecules, biomolecules, or simple biological structures into discrete nanometer-sized units.[28] Based on natural self-assembled structures such as biological membranes, cellular structures, and viruses, chemists have taken advantage of the structural flexibility and diverse physical and chemical characteristics of SAMs. These monolayers are obtained easily through spontaneous adsorption of molecules (or atoms) on solid surfaces (gold, glass, metal oxides, etc.) from a solution or vapor phase to support different molecular recognition elements (Figure 16.7).

Monolayers can be transferred onto solid supports using the Langmuir-Blodgett (LB) technique. An LB film composed of 4-n-dodecyl-6-(2-thiazolylazo)resorcinol was used[29] for naked-eye detection of submicromolar levels of cadmium ions. The reaction was accompanied by visually perceptible reddish-orange to pinkish-purple color transitions upon recognition over a wide concentration range (0.04 to 44.5 mM), without major interference from other cations or anions. Many examples of sensing using densely packed SAMs revealed sensing of metal ions[30] using metal-specific terminal heads and biosensing of DNA[31] fragments.

Using pyrene as a reporting unit and an oligo(oxyethylene) unit as a hydrophilic spacer, a fluorescent self-assembled monolayer film sensor (Figure 16.7b) was used for selective detection of structurally similar nitro-aromatic compounds.[32] Different aggregation states appeared due to hydrophilic oligo(oxyethylene) units in the sensor and the reduced degree of freedom from immobilization on a glass surface. Quenching of the emission when excited at 350 nm demonstrated that the film responded differently and reversibly due to weak interactions with the nitro-aromatics.

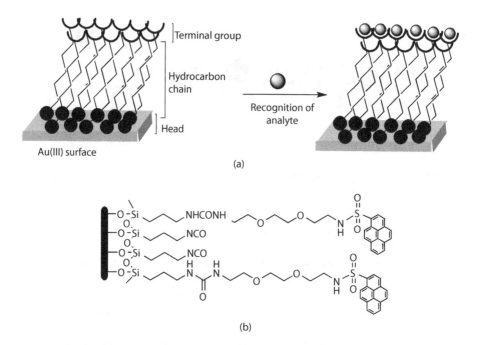

FIGURE 16.7 Molecular recognition through SAMs supported on gold surface (a) and Py-EOA-modified SAM film on glass surface (b).

A serpentine channel microfluidic chemosensor[33] patterned in a sol–gel film incorporating a cyclodextrin (CD) modified by a Tb^{3+}-containing macrocycle produced bright-green luminescence from the Tb^{3+} ions upon exposure to aqueous biphenyl solution. This resulted from recognition of the hydrocarbon biphenyl in the CD cavity that triggered Tb^{3+} emission through an absorption energy transfer–emission process.

Polydiacetylene-based conjugated vesicles (PDA vesicles) with alternating ene–yne backbone structures constitute self-assembled nanostructures that produce blue-to-red transitions (Figure 16.8a) under the influence of molecular recognition or surface perturbations.[34] Such self-assembled nanostructures efficiently integrate molecular recognition and signal transduction functions.

Sialic acid-β-glucoside (G1) and lactose-β-glucoside (G2) glycolipids embedded on the surfaces of PDA-phospholipid (dimyristoylphosphatidyl choline [DMPC]) vesicles model native glycolipids expressed on cell surfaces and are involved in the recognition of avian influenza H5N1 protein hemagglutinin (HA1). Upon recognizing HA1, clear color changes were observed (Figure 16.8b).[35]

A generic approach to the sensitive detection of noradrenalin and adrenalin catecholamines in a PDA–phospholipid matrix anchoring appropriate synthetic hosts was demonstrated using a fluorescence or visible color transition upon molecular recognition (Figure 16.9a)[36] of catecholamines with specific host–guest supramolecular interactions.

The combined effect of photo-controlled host–guest chemistry on photostimulus-responsive vesicles containing azobenzene allowed recognition of α-cyclodextrin

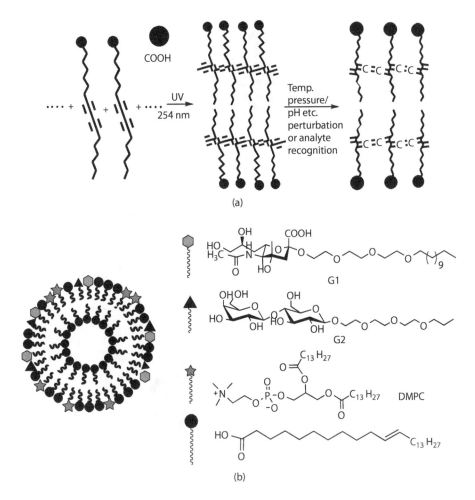

FIGURE 16.8 (a) PDA vesicles showing structural change upon surface perturbation.[34] (b) Recognition of HA1 by PDA-DMPC vesicles. (*Source:* Copyright (2009), American Chemical Society. With permission.)

(α-CD). Supramolecular interactions of azobenzene-loaded vesicles (Figure 16.9b) with α-CD cavities induced a chromatic transition that was reversible upon disassembly of the host–guest complex by an external photostimulus, thus allowing a light-triggered red-to-blue chromatic response.[37]

Gold nanoparticles (AuNPs) are interesting sensing materials because of their high sensitivity, unique distance-, size-, and shape-dependent optical properties, characteristic surface plasmon resonance (SPR), high extinction coefficients (e.g., 2.7×10^8 $M^{-1}cm^{-1}$ at ~520 nm for 13-nm spherical AuNPs), and super quenching capability. The detection limits for most AuNP-based colorimetric assays without signal amplification steps is in the range of nanometers to micrometers.

The key design strategy to AuNP-based colorimetric sensing is the switching of colloidal AuNP dispersion and aggregation stages using biological processes

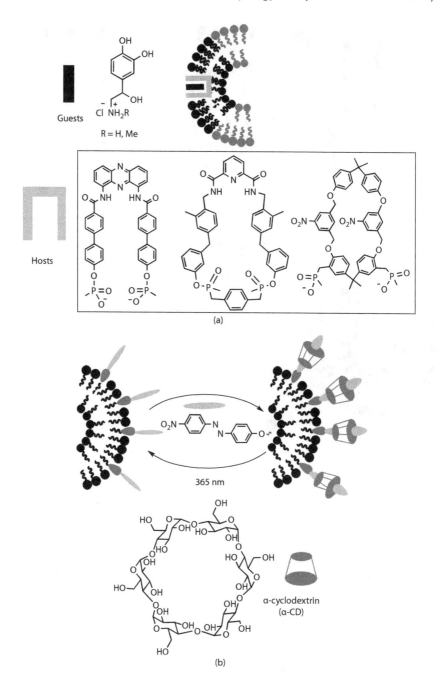

FIGURE 16.9 (a) Supramolecular recognition of catecholamines in PDA-phospholipid vesi-cles.[36] (b) Azobenzene-loaded photostimulus-responsive vesicle and host-guest recognition.[37]

(or analytes) of interest. The intense red-to-blue naked-eye color transition indicates analyte- and/or biotransformation-triggered aggregation of AuNPs that acts as the basis of most AuNP-based colorimetric sensing events.[38] The aggregation of AuNPs in these assays is due to an interparticle crosslinking mechanism or the controlled loss of colloidal stability in a non-crosslinking aggregation mechanism.

The design of nucleic-acid-functionalized nanomaterials, especially carbon, gold semiconductors, and magnetic nanoparticles, for bioimaging was reviewed recently.[39] Bioconjugation of AuNPs with readily available and stable DNA was achieved easily through covalent bonding or electrostatic attraction. However, the latter units are not as stable as the former in high ionic strength solutions. The DNA-AuNPs have high extinction coefficients and have been used for colorimetric, fluorescence, and scattering detection of different analytes.[40] Generally, when two complementary DNA-AuNPs are combined, they form DNA-linked aggregates that can dissociate reversibly with a concomitant purple-to-red color change.

The seminal work[41] that is the basis of most of the sensing events based on crosslinking aggregation of AuNPs involves self-assembling aggregation of two sets of AuNPs containing non-complementary DNA oligomers bound through thiol end groups. The event is accompanied by color transitions upon binding to an oligonucleotide duplex with "sticky ends" that are complementary to the two grafted sequences as shown in Figure 16.10a.

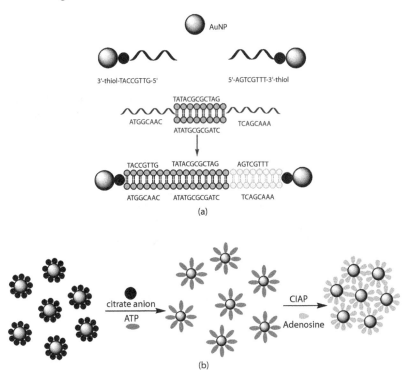

FIGURE 16.10 (a) Crosslinking AuNPs assembly approach[41] and (b) non-crosslinking approach using AuNP aggregation for sensing enzyme activity.[42]

In a rapid colorimetric enzyme sensing assay[42] using non-crosslinking AuNP aggregation, the loss of surface charge induced by aggregation led to reduced electrostatic repulsion and the onset of van der Waals and related forces of interaction. Upon treatment with calf intestine alkaline phosphatase (CIAP; Figure 16.10b), ATP adsorbed onto the AuNPs was dephosphorylated to ADP, leading to aggregation of AuNPs.

A similar naked-eye color change associated with aggregation of AuNPs containing folded aptamers was used as an efficient tool to test colloidal stability through weak interactions. The AuNPs containing folded aptamer target complexes were more stable and resisted salt-induced aggregation while remaining dispersed (red color) compared to those without the target that readily aggregated and induced a color transition to blue. The action of adenosine deaminase (ADA) on the dispersion of AuNPs converted entrapped adenosine to inosine and stabilized the AuNPs, resulting in a red-to-purple color change indicating aggregation.[43]

For a colorimetric approach to sensing kinase activity, peptide-capped AuNPs, in which 10% of the peptide ligands carried biotin–ATP substrates for kinase-aided biotinylation of the AuNPs were used.[44] When biotinylated AuNPs were treated with biotin-binding protein and avidin, immediate aggregation of AuNPs occurred as demonstrated by a color transition from red to blue (Figure 16.11a). However, when the process was repeated in the presence of an inhibitor (H89 and KN62), no color change to blue was observed due to lack of aggregation.

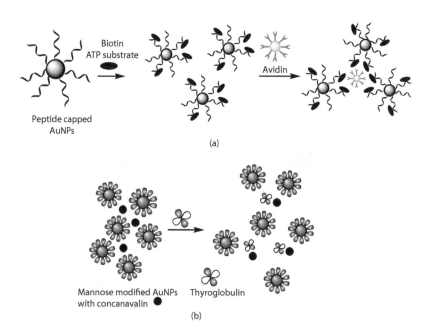

FIGURE 16.11 (a) Sensing kinase activity through aggregation of AuNPs. (*Source:* Copyright (2006) American Chemical Society. With permission.) (b) Detection of protein–protein interactions through AuNPs. (*Source:* Copyright (2005), Royal Society of Chemistry. With permission.)

An AuNP-based competitive colorimetric assay to detect protein–protein interactions was reported.[45] The mannose-modified AuNPs obtained by replacing citrate ions with thiol-appended mannose ligands recognized the carbohydrate-binding concanavalin A (Con A) protein that promotes agglomeration via multivalent ligand–protein interactions resulting in blue color (Figure 16.11b). The addition of thyroglobulin, a glycoprotein that binds Con A strongly, resulted in disaggregation of AuNPs and a burgundy color.

16.3 NAKED-EYE DETECTION AND SENSING VIA SUPRAMOLECULAR COORDINATION CHEMISTRY

Conceptually, a metal-based chemosensor for naked-eye detection also contains receptor and signaling units. The receptor unit selectively binds to analytes, thus relaying information to the signaling unit to change optical or other properties. The two strategies for designing a metal-based luminescent chemosensor are (1) a separated strategy and (2) an all-in-one strategy (Figure 16.12).

The separated strategy is the main approach for a phosphorescent transition metal-based chemosensor in which the receptor unit is separated from the signaling unit. When the substrates bind to the receptor unit, a phosphorescence signal from the organic signaling unit (Figure 16.12a) or the transition-metal signaling unit (Figure 16.12b) is turned on or off. For example, Duan and Bai designed a Ru(II)-based chromogenic and fluorogenic chemosensor 15 containing a quinone

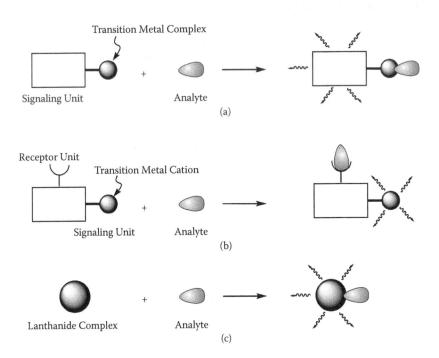

FIGURE 16.12   Separated (a and b) and all-in one (c) types of chromophores (c).

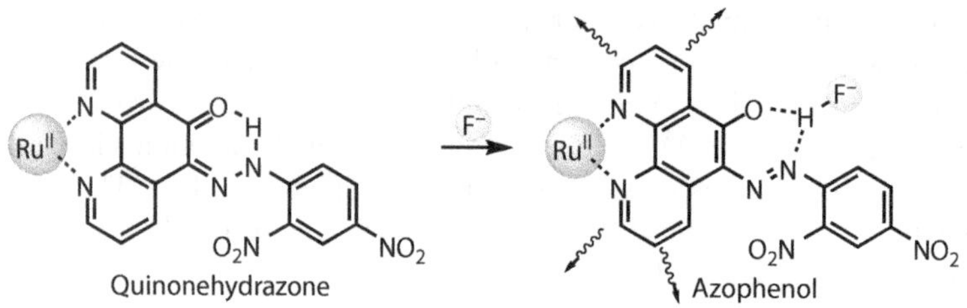

FIGURE 16.13   Naked-eye detection of F⁻ in water by Ru(II) complex.

hydrazone group for the naked-eye detection of fluoride ions. A hydrazone group was introduced into the ligand as an anion receptor by Schiff base formation with 1,10-phenanthroline-5,6-dione.[46] The presence of F⁻ anions significantly enhanced luminescence intensity at 630 nm of the Ru(II)-based luminophore by tautomerization of the quinonehydrazone to the azophenol (Figure 16.13).

The all-in-one strategy is based on lanthanide coordination chemistry and uses the ability of the lanthanide centers to function as both receptor and signaling units. Trivalent lanthanide centers have large ionic radii (0.89 to 1.16 Å in the octa-coordinated complexes)[47] and high coordination numbers (i.e., 8, 9, or 10) that offer coordination chemistries different from common transition metal complexes. When a ligand does not occupy all the coordination sites of the lanthanide center, counteranions and/or solvent molecules occupy the remaining coordination sites, allowing easy replacement with external guest molecules to form a ternary complex.[12] Since changes in the spectra and lifetimes of the lanthanide emissions are induced via ternary complexation, the proper combination of the lanthanide center and carefully designed ligand provides precise molecular recognition and luminescence sensing.

Lanthanide complexes with β-diketonate 16,[48–50] tripode 17,[51,52] and porphyrinate 18[53,54] (Figure 16.14) allowed detection of ternary complexation via changes in UV-Vis, luminescence,[50–52] and circular dichroism (CD) spectra,[53–56] as well as

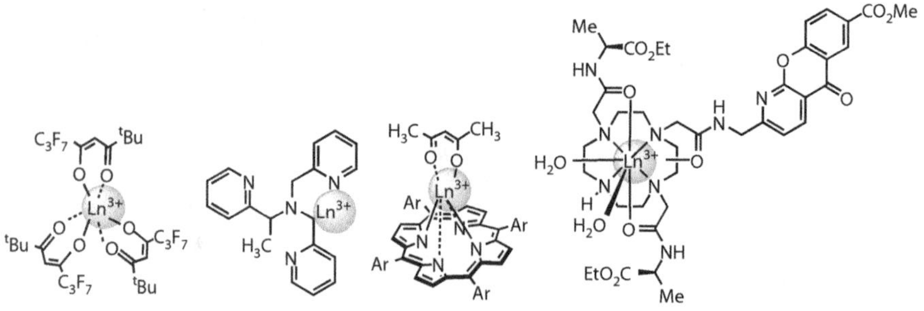

FIGURE 16.14   Several unsaturated lanthanide complexes for sensing external guest molecules.

electrochemical[48,49] and surface plasmon resonance (SPR) measurements.[57] The lanthanide tris-(β-diketonates) **16** selectively incorporate inorganic anions to form highly coordinated ternary complexes in which one anion directly coordinates with the lanthanide cation as a receptor.

When anion selectivity was examined using a poly(vinyl chloride) membrane electrode containing a lanthanide tris(β-diketonate) complex,[48,49] the Eu^{3+} complex exhibited high Cl^- anion-selective responses from various lanthanides in **16**. The lanthanide porphyrinate complexes **18** worked as CD probes for amino acids,[53,54] because the sign of the CD signals observed at the porphyrinate Soret bands was dependent on the stereochemistry of the amino acids. Naked-eye luminescence analysis is a promising tool in analytical chemistry, biochemistry, and cellular biology because its simplicity and sensitivity allow sensing and detection of biological guests.

16.3.1 NAKED-EYE DETECTION AND SENSING OF EXTERNAL ANIONS

Several types of unsaturated lanthanide complexes function as luminescent anion sensors. Parker et al. developed various lanthanide complexes with heptadentate cyclen ligands, such as **19.** The ligands selectively bind bidentate HCO_3^- guest anions that can be detected visually in mitochondria of cells.[58] This family of complexes with heptadentate cyclen exhibits greater dynamic displacement of coordinated water molecules by guest anions compared to octadentate cyclen complexes. Gunnlaugsson et al. reported interesting anion sensing using amide–cyclen lanthanide complexes and other environmental sensing and imaging techniques involving pH, metal cations, and damaged bone.[59]

Yamada et al. demonstrated CH_3 substitution of tetradentate tripod **17** that influenced the anion selective luminescent character of Eu^{3+} and Tb^{3+} complexes. Although the Eu^{3+} **17** complex showed NO_3^- anion selectivity, the Tb^{3+} **17** complex exhibited Cl^- anion selectivity.[51,52] Kataoka et al. further developed the CH_3 substitution effect into quinoline–amide mixed donor tripode **20a** for greater enhancement of anion selectivity using the Eu^{3+} complex (Figure 16.15).[60,66] Non-substituted tripode **20a** formed luminescent 1:1 (ligand:lanthanide) complexes with both $EuCl_3$ and $Eu(NO_3)_3$ salts that showed similar luminescent enhancement via UV irradiation of quinoline chromophores. In contrast, its disubstituted isomer **20b** formed a luminescent 1:1 complex with $EuCl_3$ and a 2:1 non-luminescent complex with $Eu(NO_3)_3$.

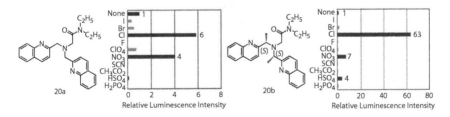

FIGURE 16.15   Anion-sensing profiles of Eu^{3+} complexes with tripod ligands **20a** and **20b**.

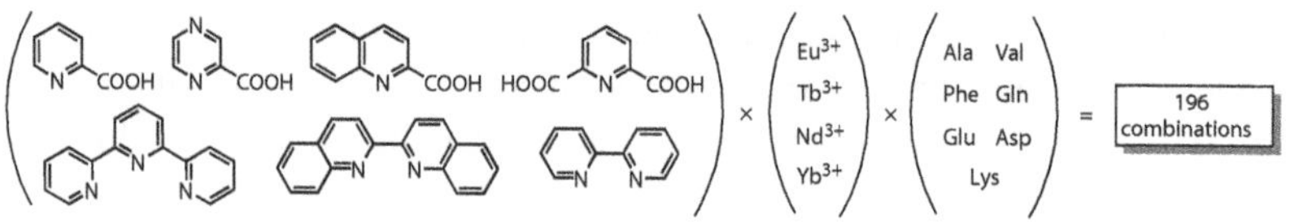

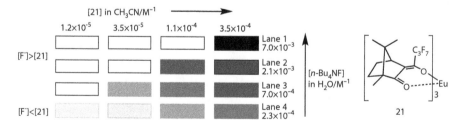

FIGURE 16.16   Combinatorial screening of amino acids.

The Eu^{3+} complex **20b** exhibited high Cl^- anion selectivity with a 63-fold luminescence enhancement. This anion-selective luminescence enhancement was observed easily with the naked eye using a hand-held UV lamp. In addition to optimization of donor combination and ligand geometry, simple CH_3 substitution on the ligand effectively changed the geometry of the lanthanide complex and adjusted luminescent anion selectivity.

Lanthanide coordination chemistry is still not completely understood, and many attempts are usually required to design specific luminescent lanthanide complexes. As an alternative to rational design, a combinatorial approach shows promise for the development of specific luminescent lanthanide materials. Shinoda et al. built a combinatorial library to optimize luminescent lanthanide complexes structurally for the selective detection of amino acids.[62] The lanthanide complex library included 196 combinations of 4 lanthanide centers, 7 pyridine ligands, and 7 amino acid substrates (Figure 16.16). The luminescence responses for amino acids depended on the nature of the ligand used. A series of Tb^{3+} complexes typically exhibited interesting luminescence responses. The Tb^{3+}–picolinic acid complex and Tb^{3+}–pyrazinecarboxylic acid complex preferred zwitterionic Ala, Val, Phe, and Gln, whereas the Tb^{3+} complex with dipicolinic acid favored anionic Glu and Asp.

Tsukube et al. developed a method for naked-eye sensing of F^- anions via ternary complexation with Eu^{3+}–tris-(β-diketonates) **21**. Luminescence images of mixed solutions containing aqueous samples of F^- anion and CH_3CN solutions of Eu^{3+} complex **21** are shown in Figure 16.17.[63] When an aqueous solution of F^- anions (7.0×10^{-4} mol L^{-1}, 0.5 mL) was added to a series of CH_3CN solutions containing complex **21** (9.5 mL; Figure 16.17, Lane 3), intense luminescence was observed

FIGURE 16.17   Naked-eye sensing of F^- anion concentration by red emission with Eu^{3+} complex **21**.

at [21]/[F⁻] ≥ 1. An excess of F⁻ anions did not enhance luminescence intensity at the same concentration of complex **21** due to decomposition of the Eu^{3+} complex (Figure 16.17, horizontal row). Since the use of complex **21** allowed F⁻ anion detection in an aqueous sample at parts-per-million (ppm) level (2.3×10^{-4} mol L^{-1} = ~4 ppm), the naked-eye observations produced an approximate estimate of the F⁻ anion concentrations in the aqueous sample. Since a hand-held UV lamp was used for the detection, the use of luminescent lanthanide complexes provided an effective method for in situ determination of a biologically important anion.

Properly designed unsaturated lanthanide complexes function as anion-specific luminescent receptors for ternary complexation with certain external anions that can be detected by naked-eye methods. Thus, such a sensing technique using lanthanide complexes may have practical applications.

16.3.2 Naked-Eye Detection and Sensing of Environmental Changes and Biological Substrates

Although several luminescent lanthanide complexes respond to environmental changes such as pH and temperature modifications, biological cage proteins containing lanthanide cations have been developed recently and provide environmental sensors in aqueous media.

Transferrin and lactoferrin belong to a family of Fe^{3+} cation buffer and carrier proteins. X-ray crystallography has shown that both proteins include two distorted octahedral Fe^{3+} sites, each of which is coordinated with one aspartate, one histidine, and two tyrosinate residues as well as one exogenous carbonate.[64,65] Luminescent lanthanides such as Tb^{3+} and Eu^{3+} cations can be incorporated along with Fe^{3+} cations into these binding sites[66] to provide water-soluble luminescent lanthanide complexes in sophisticated three-dimensional cages.

Complexation of Tb^{3+} occurred more rapidly and reversibly to induce Tb^{3+} luminescence changes at pH 6.0 to 8.0 (Figure 16.18a).[67] Apolactoferrin formed a more stable Tb^{3+} complex than the apotransferrin complex and its Tb^{3+} complex exhibited pH-dependent luminescence behavior at a lower pH. Such pH-dependent luminescence changes were detectable by the naked eye. Since the apotransferrin–Yb^{3+} complex permeated into the cell via a transferrin receptor mechanism,[68] incorporating luminescent lanthanide complexes into biological proteins may provide tools for sensors in intracellular environments.

Apoferritin **22a** is the protein component of iron-storage ferritin consisting of 24 subunits that form a hollow spherical shell with external and internal diameters of ~13 and ~7.5 nm, respectively (Figure 16.18b).[69] The inner cavity of apoferritin has been used recently as a nanoscale molecular vessel for generation of non-iron clusters,[70] non-covalent incorporation of organic guest molecules, and compartmentalization of drugs and MRI reagents.

Apoferritin also exhibits green luminescence in aqueous solution upon incorporation of Tb^{3+} cations (**22b**, Figure 16.18b).[71,72] Tsukube et al. demonstrated that the Tb^{3+} complex of apoferritin could be used in selective precipitation assays of

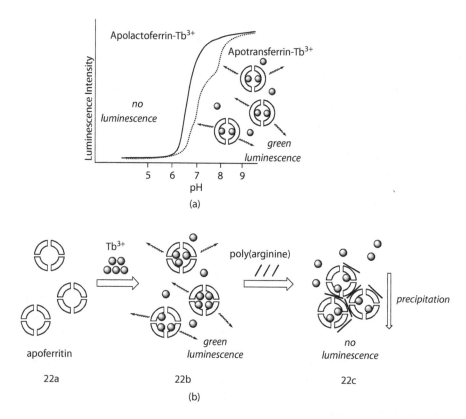

FIGURE 16.18 (a) pH-Dependent emission of apotransferrin-Tb^{3+} and apolactoferrin-Tb^{3+} complexes. (b) Self-assembly, Tb^{3+} complexation, and protein co-precipitation profiles **22a** through **c** of apoferritin at pH 7.0.

poly(arginines) and their tagged proteins (**22c**, Figure 16.18b).[73] The Tb^{3+}–apoferritin complex co-precipitated with a series of poly(arginines) in amounts dependent on the length of the added poly(arginines). The poly(arginine)-tagged albumin also co-precipitated with the Tb^{3+}–apoferritin complex, whereas cytochrome c, lysozyme, albumin, and other biological proteins did not precipitate or respond by producing luminescence. These co-precipitation phenomena were detected easily by naked-eye observation of the green luminescence, thus demonstrating promise as tools for specific detection of biopolycations.

16.4 CONCLUSIONS AND OUTLOOK

The promising approaches reviewed in this chapter demonstrate the opportunities offered by synergistic supramolecular approaches to applications of naked eye detectable chemosensing in chemical and biological systems. Future prospects for naked-eye sensing techniques using supramolecular concepts (assembling, disassembling, etc.) are promising for implementation in areas such as microelectronics and environmental science.

ACKNOWLEDGMENTS

K.S. thanks DST for the project SB/S1/OC-45/2013 and P.K. thanks CSIR and UGC, New Delhi. H.M. is grateful to the Japan Society for the Promotion of Science (Grants-in-Aid for Scientific Research 23350029 and 24655053).

REFERENCES

1. (a) Lehn, J.M., *Angew. Chem. Int. Ed. Engl.,* 1988, 27, 89. (b) Lehn, J.M., *Angew. Chem. Int. Ed. Engl.,* 1990, 29, 1304.
2. Cram, D.J., *Angew. Chem. Int. Ed. Engl.,* 1988, 27, 1009.
3. (a) Imahori, H., Umeyama, T., Kurotobi, K. et al., *Chem. Commun.,* 2012, 48, 4032. (b) Wasielewski, M.R., *Acc. Chem. Res.,* 2009, 42, 1910.
4. Bredas, J.L., Cornil, J., Beljonne, D. et al., *Acc. Chem. Res.,* 1999, 32, 267–276.
5. (a) Singh, K., Virk, T.S., Zhang, J. et al., *Chem. Commun.,* 2012, 48, 12174–12176. (b) Singh, K., Virk, T.S., Zhang, J. et al., *Chem. Commun.,* 2012, 48, 121–123. (c) Singh, K., Virk, T.S., Zhang, J. et al., *Chem. Commun.,* 2011, 47, 905–907.
6. (a) Zhang, J., Tan, J., Ma, Z. et al., *J. Am. Chem. Soc.,* 2013, 135, 558–561. (b) Zhang, J., Geng, H., Virk, T.S. et al., *Adv. Mater. Sci.,* 2012, 24, 2603–2607.
7. Zhu, L., Yi, Y., Li, Y. et al., *J. Am. Chem. Soc.,* 2012, 134, 2340–2347.
8. Hizume, Y., Tashiro, K., Charvet, R. et al., *J. Am. Chem. Soc.,* 2010, 132, 6628–6629.
9. Mativetsky, J.M., Kastler, M., Savage, R.C. et al., *Adv. Funct. Mater.,* 2009, 19, 2486–2494.
10. Ariga, K., Ito, H., Hill, J.P. et al., *Chem. Soc. Rev.,* 2012, 41, 5800–5835.
11. Mameri, S., Shinoda, S., and Tsukube, H., *Molecular Recognition with Designed Heterocycles and Their Lanthanide Complexes: Heterocyclic Supramolecules I.* Topics in Heterocyclic Chemistry Series, Vol. 17, 2008. Springer, Heidelberg, pp. 1–42.
12. (a) Singh, K., Sareen, D., Kaur, P. et al., *Chem. Eur. J.,* 2013, 19, 6914–6936. (b) Tsukube, H. and Shinoda, S., *Chem. Rev.,* 2002, 102, 2389–2404.
13. Altwood, A.L., Davies, J.E.D, MacNicol, D.D. et al., Eds., *Comprehensive Supramolecular Chemistry,* Vols. 1, 2, and 6, 1996. Pergamon, New York.
14. Xu, Z., Singh, N.J., Lim, J. et al. *J. Am. Chem. Soc.,* 2009, 131, 15528–15533.
15. Litvinchuk, S., Bollot, G., Mareda, J. et al., *J. Am. Chem. Soc.,* 2004, 126, 10067–10075.
16. Litvinchuk, S., Tanaka, H., Miyatake, T. et al., *Nat. Mater.* 2007, 6, 576–580.
17. Litvinchuk, S., Sorde, N., and Matile, S., *J. Am. Chem. Soc.,* 2005, 127, 9316–9317.
18. Yamana, K., Iwai, T., Ohtani, Y. et al., *Bioconjugate Chem.,* 2002, 13, 1266–1273.
19. Geiger, A., Burgstaller, P., von der Eltz, H. et al., *Nucleic Acids Res.* 1996, 24, 1029–1036.
20. Jenison, R.D., Gill, S.C., Pardi, A. et al., *Science,* 1994, 263, 1425–1429.
21. Stojanovic, M.N., de Prada, P., and Landry, D.W., *J. Am. Chem. Soc.,* 2001, 123, 4928–4931.
22. Seeman, N.C., *Angew. Chem., Int. Ed. Engl.,* 1998, 37, 3220–3238.
23. Hermann, T. and Patel, D.J., *Science,* 2000, 287, 820–825.
24. Li, J.J., Fang, X., and Tan, W., *Biochem. Biophys. Res. Commun.,* 2002, 292, 31–40.
25. Heyduk, E. and Heyduk, T., *Anal. Chem.,* 2005, 77, 1147–1156.
26. Nutiu, R. and Li, Y., *J. Am. Chem. Soc.,* 2003, 125, 4771–4778.
27. Huizenga, D.E. and Szostak, J.W. *Biochemistry* 1995, 34, 656–665.
28. Wilbur J.L. and Whitesides, G.M. In *Nanotechnology,* Timp, G., Ed. 1999. Springer, New York.
29. Prabhakaran, D., Yuehong, M., Nanjo, H. et al. *Anal Chem.,* 2007, 79, 4056–4065.
30. (a) Turyan, I. and Mandler, D., *Anal. Chem.,* 1997, 69, 894–897. (b) Moav, T., Hatzor, A., Cohen, H. et al., *Chem. Eur. J.,* 1998, 4, 502–506.

31. Ihara, T., Nakayama, M., Murata, M. et al., *Chem. Commun.* 1997, 1609–1613.
32. Ding, L., Liu, Y., Cao, Y. et al., *J. Mater. Chem.*, 2012, 22, 11574–11582.
33. Rudzinski, C.M., Young, A.M., and Nocera, D.G., *J. Am. Chem. Soc.*, 2002, 124 1723–1727.
34. Orynbayeva, Z., Kolusheva, S., Livneh, E. et al., *Angew. Chem., Int. Ed.*, 2005, 44, 1092–1096.
35. Deng, J., Sheng, Z., Zhou, K. et al., *Bioconjugate Chem.*, 2009, 20, 533–537.
36. Kolusheva, S., Molt, O., Herm, M. et al., *J. Am. Chem. Soc.*, 2005, 127, 10000–10001.
37. Chen, X., Hong, L., You, X. et al., *Chem. Commun.*, 2009, 1356–1358.
38. Lee, J.S., Han, M.S., and Mirkin, C. A., *Angew. Chem., Int. Ed.*, 2007, 46, 4093–4096.
39. Tyagi, A.K., Ramkumar, J., and Jayakumar, O.D., *Analyst*, 2012, 137, 760–764.
40. Wilson, R., *Chem. Soc. Rev.* 2008, 37, 2028–2045.
41. Mirkin, C.A., Letsinger, R.L., Mucic, R.C. et al., *Nature*, 1996, 382, 607–609.
42. Zhao, W., Chiuman, W., Lam, J.C.F. et al., *Chem. Commun.*, 2007, 3729–3731.
43. Zhao, W., Chiuman, W., Lam, J.C.F. et al., *J. Am. Chem. Soc.*, 2008, 130, 3610–3618.
44. Wang, Z., Levy, R., Fernig, D.G. et al., *J. Am. Chem. Soc.*, 2006, 128, 2214–2215.
45. Tsai, C.S., Yu, T.B., and Chen, C.T., *Chem. Comm.*, 2005, 4273–4275.
46. Lin, Z.H., Ou, S.J., Duan, C.Y. et al. *Chem. Commun.*, 2006, 624–626.
47. Shannon, R.D. *Acta Crystallogr. A*, 1976, A32, 751–767.
48. Mahajan, R.K., Kaur, I., Kaur, R. et al., *Chem. Commun.*, 2003, 2238–2239.
49. Mahajan, R. K., Kaur, I., Kaur, R. et al., *Anal. Chem.*, 2004, 76, 7354–7359.
50. Tsukube, H., Yano, K., and Shinoda, S., *Helv. Chim. Acta*, 2009, 92, 2488–2496.
51. Yamada, T., Shinoda, S., and Tsukube, H., *Chem. Commun.*, 2002, 1218–1219.
52. Yamada, T., Shinoda, S., Sugimoto, H. et al., *Inorg. Chem.*, 2003, 42, 7932–7937.
53. Tsukube, H., Wada, M., and Shinoda, S., *Chem. Commun.*, 1999, 1007–1008.
54. Tsukube, H., Tameshige, N., Shinoda, S. et al., *Chem. Commun.*, 2002, 2574–2575.
55. (a) Shinoda, S., *Chem. Soc. Rev.*, 2013, 42, 1825–1835. (b) Shinoda, S. and Tsukube, H., *Analyst*, 2011, 136, 431–435.
56. Shinoda, S., Terada, K., and Tsukube, H., *Chem. Asian J.*, 2012, 7, 400–405.
57. Tsukube, H., Yano, K., Ishida, A. et al., *Chem. Lett.*, 2007, 56, 554–555.
58. Butler, S.J. and Parker, D., *Chem. Soc. Rev.*, 2013, 42, 1652–1666.
59. (a) dos Santos, C.M.G., Harte, A.J., Quinn, S.J. et al., *Coord. Chem. Rev.*, 2008, 252, 2512–2527. (b) dos Santos, C. M. G. and Gunnlaugsson, T., *Dalton Trans.*, 2009, 4712–4721. (c) McMahon, B., Mauer, P., McCoy, C.P. et al., *J. Am. Chem. Soc.*, 2009, 131, 17542–17543.
60. Kataoka, Y., Paul, D., Miyake, H., Shinoda, S. and Tsukube, H., *Dalton Trans.*, 2007, 2784–2791.
61. Kataoka, Y., Paul, D., Miyake, H. et al., *Chem. Eur. J.*, 2008, 14, 5258–5266.
62. Shinoda, S., Yano, K., and Tsukube, H., *Chem. Commun.*, 2010, 46, 3110–3112.
63. Tsukube, H., Onimaru, A., and Shinoda, S., *Bull. Chem. Soc. Jpn.*, 2006, 79, 725–730.
64. Sun, H., Li, H., and Sadler, P., *Chem. Rev.*, 1999, 99, 2817–2842.
65. Hall, D.R., Hadden, J.M., Leonard, G.A. et al., *Acta Crystallogr. D*, 2002, 58, 70–80.
66. Pecoraro, V.L., Harris, W.R., Carrano, C.J. et al., *Biochemistry*, 1981, 20, 7033–7039.
67. Kataoka, Y., Shinoda, S., and Tsukube, H., *J. Nanosci. Nanotechnol.*, 2009, 9, 655–657.
68. Du, X.L., Zhang, T.L., Yuan, L. et al., *Eur. J. Biochem.*, 2002, 269, 6082–6090.
69. Hempstead, P.D., Yewdall, S.J., Fernie, A.R. et al., *J. Mol. Biol.*, 1997, 268, 424–448.
70. Ueno, T., Abe, M., Hirata, K. et al., *J. Am. Chem. Soc.*, 2009, 131, 5094–5100.
71. Wardeska, J.G., Viglione, B., and Chasteen, N. D., *J. Biol. Chem.*, 1986, 261, 6677–6683.
72. Barnés, C.M., Petoud, S., Cohen, S. M. et al., *J. Biol. Inorg. Chem.*, 2003, 8, 195–205.
73. Tsukube, H., Noda, Y., and Shinoda, S., *Chem. Eur. J.*, 2010, 16, 4273–4278.

17 Self-Assembled Catenanes, Knots, and Links and Synergistic Effects of Weak Interactions

Loïc J. Charbonnière and Ali Trabolsi

CONTENTS

17.1 INTRODUCTION

17.1.1 MOLECULAR TOPOLOGY AND NEED FOR MULTIPLE WEAK INTERACTIONS IN SYNERGY

Molecular topology was introduced in 1961 by Edel Wasserman and Harry Frisch[1] following the work of Wasserman. Nonmechanically interlocked rings were obtained from macrocyclic hydrocarbons called catenanes[2] (from the Latin *catena* which

$$2_1^2 \qquad 3_1 \qquad 4_1^2 \qquad 5_1 \qquad 6_2^3$$

FIGURE 17.1   Most common knots and links and their mathematical descriptions. Left to right: [2]catenane, trefoil knot, Solomon link, pentafoil (or Solomon's seal) knot, and Borromean ring.

means *chain*). Their concept of topological isomers was summarized in a recent comprehensive review on the subject[3] as "two objects containing the same atoms and chemical bond connectivities, but that cannot be interconverted by any deformation action in three-dimensional space." Figure 17.1 represents some of the most common examples of knots and links such as the [2]catenate, trefoil knot, Solomon link, pentafoil knot, and Borromean ring.

Knots and links are commonly found in our daily life, in fashion, in architecture, and even in natural[4] and unnatural[5] DNA strands. Knot theory has long been a matter of interest and a well-studied area for mathematicians who developed a theory around them and devised a specific nomenclature to describe knots and links. A knot or a link is denoted according to the theory developed by James Alexander and Garlan Briggs[3] as X_z^y, in which X is the minimum number of nodes, y is the number of components or strands, and z is the order within the links with the same number of components and crossings. Selected examples are shown in Figure 17.1.

Because of their importance in naturally occurring biological systems and their aesthetic appeal, the synthesis of these topological structures also became the focus of interest of chemists. After the first synthesis of a [2]catenane[2] (Hopf or 2_1^2 link) and subsequent relevant works based on the Möbius ring concept,[7,8] it was realized early that the step-by-step chemical approaches yielded only poor quantities of the expected knots and an alternative approach was necessary.

One of the strategies employed in the preparation of such structures consisted of introducing molecular information as chemical functions into the strands to increase the number of non-covalent weak interactions among these molecular codons. The objective was to favor some of the topological isomers among others. It was anticipated that the newly added weak interactions acting synergistically would improve the pre-organization of the desired structure and consequently increase the yield of the final knotting step while avoiding the formation of oligomers.

For example, Sauvage and coworkers pioneered the use of metal coordination to template the preparation of a metallo-catenane assembled around a Cu(I) ion with a chemical yield of 42% for the cyclization step (see Section 17.2.1).[9] By utilizing lessons from the synthesis of [2]catenanes, the preparation of higher-order knots and links soon met with success. This pattern in which [2]catenanes precede the formation of more complex structures has been repeated throughout the short history of molecular topology involving other non-covalent interactions such as H bonding, π–π stacking, and hydrophobic interactions.

Rather than providing a comprehensive review, we will first focus on some prototypical examples of simple knotted structures obtained by these various types of non-covalent interactions, followed by a section describing selected examples of knots and links of increasing complexity. We will also describe the rules of the combined interactions in these complex molecular assemblies.

17.1.2 Self-Assembly of Topological Structures

In chemistry, self-assembly is defined as a phenomenon by which molecules order themselves into a particular arrangement without external intervention. The process occurs under the guidance of numerous weak non-covalent bonding interactions that represent the core of supramolecular chemistry.[10] It is not surprising that chemists engaged in the synthesis of topologically non-trivial structures are interested in utilizing self-assembly to direct the templation of chemical components into intermediates prone to cyclization, followed by the formation of covalent bonds.

Nevertheless, self-assembling processes by nature rely on weak interactions and the isolation of chemically stable knots and links requires a final knotting step in which covalent bonds are formed. While irreversible reactions such as Williamson ether reactions[11] and ring closing methathesis[12] are employed successfully to form knots and links, their template-directed syntheses have recently been enriched by the concept of dynamic covalent chemistry (DCC).[3] DCC employs reversible covalent bond formation that allows equilibration of a system toward the most thermodynamically stable structures dictated by the sum of the non-covalent bonding interactions.

Different syntheses have applied the DCC approach to knots and links, among which the most successful are the formation of imine bonds allowing for the isolation of Solomon links[14] and pentafoil knots,[15] or disulfide bridges produced by aqueous oxidation of two dithiols.[16] Template-directed strategy coupled to DCC (which takes advantage of multiple weak interactions working in synergy to achieve the thermodynamically controlled synthesis of topologically non-trivial structures) has became one the best ways to synthesize knots and links efficiently.

17.2 INTERACTIONS GOVERNING ASSEMBLIES OF KNOTS AND LINKS

This section aims to describe the repertoire of non-covalent bonding interactions ranging from coordination to weak hydrogen bonding that drive the self-assemblies of topological molecules by presenting illustrative examples.

17.2.1 Metal Coordination

The use of metal coordination in the template-directed synthesis of knots and links is one of the earliest successful strategies employed by chemists. The first example of metal-templated synthesis of a [2]catenane was reported in Strasbourg by Sauvage et al.[9] who first used the method to assemble two phenanthroline-based strands around a tetrahedral copper(I) cation (Scheme 17.1).

After assembly around Cu(I), the terminal hydroxyl groups of the phenyl substituents of each 2,9-diphenyl-1,10-phenanthroline were joined by a Williamson

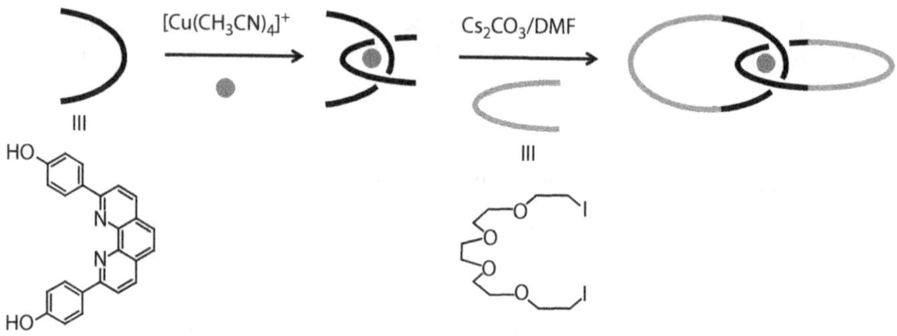

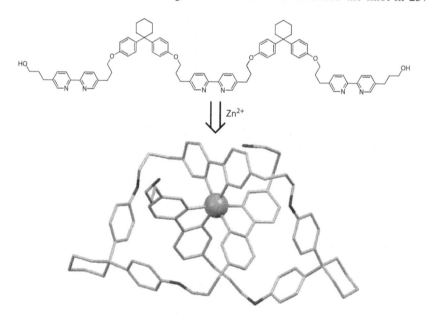

SCHEME 17.1 Metal-assisted template synthesis of a [2]catenane developed by Sauvage et al.[9]

ether macrocyclization reaction using a diiodopentathylene glycol spacer. This novel strategy allowed a favourable orientation of the termini and the final catenate was isolated in 42% yield.

As a template for the formation of a trefoil knot, Hunter et al. employed the self-assembly of three bipyridine units of a single linear strand around an octahedral zinc(II) metal center. The templation of the knotted structure was evidenced by an x-ray crystal structure of the intermediate Zn complex (Scheme 17.2, CCDC163073).[18]

The two terminal hydroxyl groups on the strand of the open-knot zinc complex were connected with 3,6,9-trioxaundecanedioic acid using EDCI coupling in an esterification reaction under high dilution conditions to afford the knot in 25%

SCHEME 17.2 Tris-bidentate ligand developed by Hunter et al.[18] and x-ray crystal structure. H atoms are omitted for clarity. Redrawn from CCDC163073) of its Zn complex obtained by metal templation and forming a trefoil knot.

yield.[19] The authors also demonstrated that the closing of the trefoil knot may be accomplished in 68% yield by ring closing metathesis when terminal alkynes are present in the structure of the strand.

The intent of metal coordination is to provide both a robust template platform and a large diversity of stereochemical geometries defined by the coordination properties of the metal. In addition to tetrahedral[9] and octahedral[19] coordination geometries obtained, for example, with Cu(I) or Zn(II), respectively, linear arrangements can be obtained using Au(I) centers coordinated to alkyne- and phosphine-terminated strands to prepare different types of catenates.[20,21] Square planar coordination around Pd(II) was also used to generate a doubly interlocked [2]catenane, known as a Solomon link.[22]

17.2.2 Anion Coordination

Although the chemistry of anion directed assemblies is a flourishing field[23] that can profit from the geometrical properties of anions similar to those of cations,[24] examples of templated synthesis assisted by anion coordination are far less common than cationic examples. A pioneer example came from Beer's group.[25,26] The authors took advantage of the tetrahedral organization of a sulfate anion coordinated to two hydrogen atoms of secondary amides (Scheme 17.3).

The ligands were designed around 3-substituted pyridinium[26] or 1,3-disubstituted phenyl rings[25] that pre-organize them for efficient bidentate coordination of each strand to the sulfate anion. In the example in Scheme 17.3, the catenation step was realized by ring closing metathesis using Grubb's catalyst in 80% chemical yield.

17.2.3 Hydrogen Bonding Interactions

Although hydrogen bonding interactions are rather weak (2 to 10 KJ/mol), the combination of many of these interactions can be used as an efficient template-directed strategy to prepare knotted compounds. This strategy was successfully employed by Vögtle et al. on the directed condensation of a variety of rigid diamines with diacyl or disulfonyl chlorides for the construction of a trefoil knot.[27]

For the synthesis of the trefoil knot depicted in Scheme 17.4,[28] the condensation in high dilution conditions led to a mixture of conventional macrocycles resulting from the [2 + 2] and [3 + 3] condensations of the diamine and dipicolinic acid in15 and

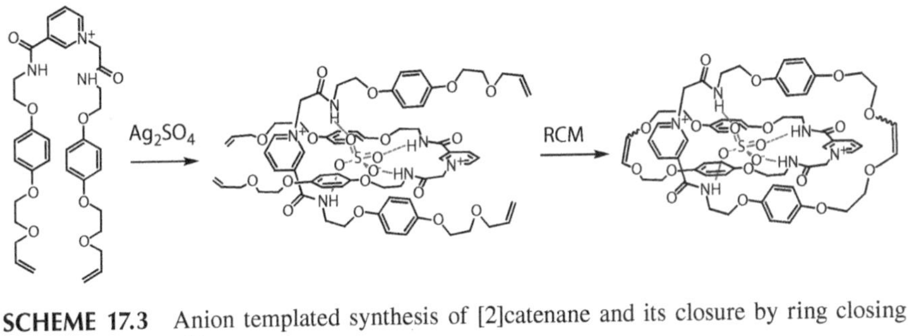

SCHEME 17.3 Anion templated synthesis of [2]catenane and its closure by ring closing metathesis (RCM).[26]

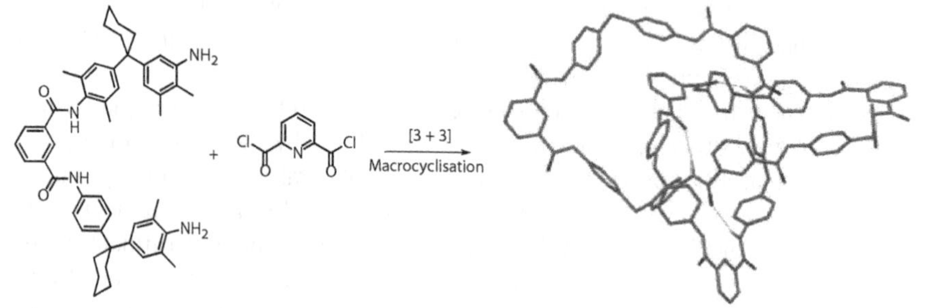

SCHEME 17.4 Hydrogen templated synthesis of trefoil knot obtained by [3 + 3] macrocyclization.[28] Right: Crystal structure (adapted from CCDC 139484) of trefoil knot with intrastrand H-bonding interactions. H atoms, cyclohexyl, and methyl groups are omitted for clarity.

23% yields, respectively. A third product was obtained in 20% yield and identified on the basis of mass spectrometry and x-ray crystal structure as a trefoil knot.

The spontaneous formation of the trefoil knot is a result of strong intramolecular hydrogen bonding interactions serving as a template that can be enhanced by the use of non-polar aprotic solvents such as dichloromethane. The efficacy of hydrogen bonding interactions in the mechanism of the trefoil knot formation was also confirmed through a step-by-step approach in which some of the precursors of the final structure were prepared independently and were shown to self-assemble into an open trefoil knot structure.[29]

17.2.4 π-Stacking Interactions

The introduction of π-stacking interactions in supramolecular assemblies was driven mainly by the tremendous work of Stoddart and co-workers who took advantage of the presence of electron-poor bipyridinium dications and electron-rich aromatic polyethers to template assemblies by planar aromatic interactions leading to a library of supermolecules such as rotaxanes[30] and catenanes.[31] Another pioneering effort was carried out by Sanders et al.[32,33] who used neutral aromatic electron-poor and -rich entities instead of charged species (Scheme 17.5).

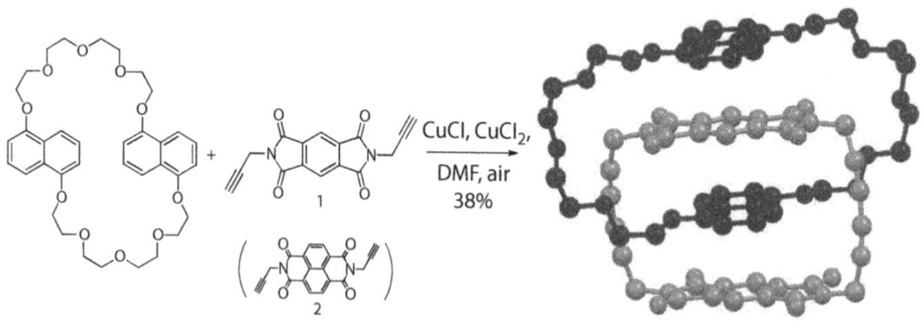

SCHEME 17.5 Methodology developed by Sanders et al. for high yield chemical knotting of electron poor dialkynes (light grey) in cavity of electron-rich bisnaphthalene (dark gray) forming a [2]catenane. H atoms are omitted for clarity. (Adapted from CCDC 182443.)[32,33]

Starting from a 1:2 mixture of an electron-rich bis-1,5-(dinaphtho)-38-crown-10 and the electron-poor bis-acetylenic compound **1**, Sanders used the template effect obtained by the inclusion of the pyromellitic diimide **1** in the crown ether to direct the oxidative coupling of the terminal alkynes toward the formation of the cyclized structure, the [2]catenane, with a 38% chemical yield. When **1** was replaced by a more electron-poor moiety obtained from 1,4,5,8-naphthalene tetracarboxylic diimide (**2**, Scheme 17.5) the yield of the catenane improved to 52%.

The influence of the stacking interaction was also demonstrated by an experiment in which the mechanical bond formation in the [2]catenane was carried out in the presence of equimolar amounts of the crown ether and diacetylene **1** or **2**. While a statistical knotting would have led to equivalent yields for the homocatenanes (obtained from the condensation of similar diacetylenes) and a double yield for the heterocatenane, it was found that the latter was obtained with a chemical yield four times higher. This was the result of a higher content of inclusion precursor resulting from the more electron-poor compound **2**, thereby inducing a larger amount of hetero-coupling.

This example clearly demonstrates that multiple weak π-stacking interactions can be combined synergistically to obtain strongly pre-organized precursors allowing for the knotting steps to be performed with high chemical yields.

17.3 TEMPLATE-DIRECTED SELF-ASSEMBLY TOWARD COMPLEX MOLECULAR KNOTS AND LINKS

The template-directed synthesis of knots and links can be governed by different types of interactions in which a large change in free energy caused by the addition of multiple weak interactions such as hydrogen bonds may compensate for the lack of strong interactions such as coordination bonds. Nevertheless, the formation of these knots and links is most often a combination of such interactions resulting in the final assembly and it is only due to a subtle balance of such interactions that one can achieve the synthesis of knots and links of high complexity.

Since many examples of catenanes were described in the previous section, this section will focus on representative examples of more complex molecular knots and links.

17.3.1 TREFOIL KNOT (TK)

Recently, we reported the synthesis of a metal-templated trefoil knot using dynamic covalent chemistry of imine bonds.[34] Inspired by the work of Stoddart on the synthesis of Borromean rings[35] and with the aid of DFT calculations, we anticipated that the combination of a 5,5′-dimethyl-2,2′-bipyridine functionalized with *para*-aminomethyl-phenyl ether (Scheme 17.6) would provide adequate geometry for the bidentate coordination to a Zinc(II) cation.

The structure of the trefoil knot was demonstrated by a combination of high resolution mass spectrometry, NMR experiments, and computational modeling. Interestingly, the synthesis of the trefoil knot was accomplished with a consequent 56% yield, but it was shown to be accompanied by the formation of a [2]catenane species isolated in 22% yield as well as by the larger Solomon link structure detected only in minor amounts by mass spectrometry during the course of the reaction.

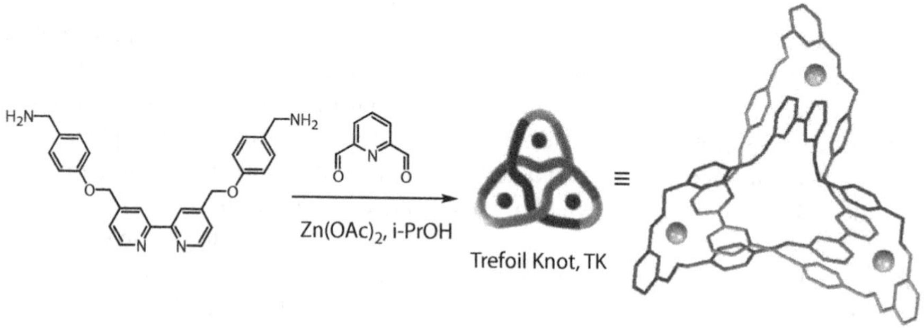

SCHEME 17.6 Synthesis of molecular trefoil knot by dynamic covalent chemistry of imine bond formation.[34]

Based on the x-ray crystal structure of the [2]catenane and the DFT-calculated structure of the trefoil knot, the formation of the knotted structures resulted from a strong zinc templated chelation combined with the formation of dynamic covalent imine bonds catalyzed by Zn(II) acting as a Lewis acid and templated by strong π–π stacking interactions between the bipyridine of one strand and the phenoxy rings of another—an organization that predisposes the amine functions close to the formyl groups.

In summary, metal templation and π-π stacking interactions were employed cooperatively along with the formation of reversible covalent imine bonds to template the three different links obtained when the two ligands were mixed with zinc.

17.3.2 SOLOMON LINK

A step forward in the complexity of topological molecular objects is reached with the synthesis of molecular Solomon links that are composed of two distinct strands intertwined with four alternate crossings (Figure 17.1). As a probable consequence of their increased complexity, examples of Solomon links are less common than catenates or trefoil knots. Peinador et al. reported a series of three Solomon links obtained in a single step by the self-assembly of five components: two diazapyrene based strands; two Pd(en)(OTf)$_2$ coordination complexes (en = ethylene diamine, OTf = triflate); and an electron rich cyclophane containing four phenylene groups (Scheme 17.7).[22]

The self-assembly was governed by a combination of metal templation using Pd(II) metal cations implying a square planar coordination that affords the closing of the strand made by the diazapyrene ligands, templated by π-stacking interactions between the electron poor diazapyrenium aromatic cycles and the electron-rich 1,4-disubstituted alkoxyphenyl rings. The assembly process was generalized to other Solomon links obtained with square planar Pt(II) cations or by replacing the diazapyrene by 4,4'-bipyridine strands.

In conclusion, donor–acceptor π–π stacking interactions were utilized in synergy with metal coordination in the preparation of the described Solomon link.

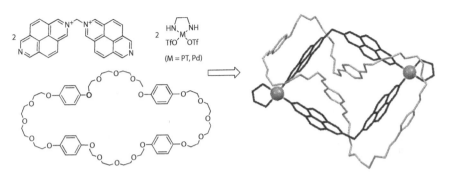

SCHEME 17.7 Formation of Solomon link by self-assembly of two diazapyrene-based ligands coordinated to two Pd(II) cations and intertwined with tetraphenylenecyclophane (redrawn from CCDC 728058).[22]

17.3.3 Pentafoil Knot

Inspired by the work of Lehn et al. on pentanuclear circular helicates,[36] and taking advantage of their work on the formation of metal-assisted catenanes using dynamic covalent chemistry,[37] Leigh et al. developed a strategy to generate a pentafoil knot, also called Solomon's seal knot.[15] Lehn noted that the introduction of ethylene bridges between three coordinating 2,2'-bipyridine units linked in 5 and 5' positions led to the formation of circular helicates. Interestingly, the sizes of such assemblies correlated to the counter ions used; chloride anions afforded isolation of the penta-nuclear species. Leigh et al. replaced the terminal pyridines by formyl groups that allowed the closing of structures by imine bond formation in the presence of appro-priately sized diamine linkers (Scheme 17.8).

Their strategy was successful and they were able to isolate the pentafoil knot in 44% yield. In addition to the metal-assisted templation of the strands to form a cir-cular helicate precursor, the chloride anion was shown to assist the formation of the pentanuclear adduct. The use of 3,6-dioxo-1,8-diaminooctane allowed the knots to be closed while simple aliphatic diamines failed to produce a clean closing proce-dure. The synergy of metal and anion coordinations was employed simultaneously to template the formation of a pentafoil knot structure.

17.3.4 Borromean Ring

The total synthesis of molecular Borromean rings, three interlocked but non-catenated rings, employing metal templation and dynamic covalent chemistry was reported by Stoddart et al. in 2004.[35] Fascinated by the topology of the Borromeo family crest and with the aid of molecular modeling, the group combined the chemistry of imine bonds with the templation power of Zinc(II) ions to achieve in an all-in-one synthetic strat-egy: the formation of molecular Borromean rings in a nearly quantitative conversion.

The strategy relies on the use of a pair of Zinc(II) chelating ligands described as endo-ligands and consisting of 2,6-diformylpyridine (DFP) and exo-diaminebipyri-dyl derivative (DAB). When combined in a 1:1:1 ratio of DAB:DFP:Zn(II), molecular Borromean rings are formed from 18 precursors (Scheme 17.9). The synthesis of

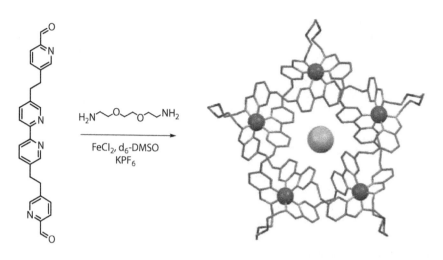

SCHEME 17.8 Metal-assisted synthesis of pentafoil knot of Leigh et al.[15] and its crystal structure. Fe atoms are shown as dark grey balls, chloride anion as light gray. H atoms are omitted for clarity (adapted from CCDC 845599).

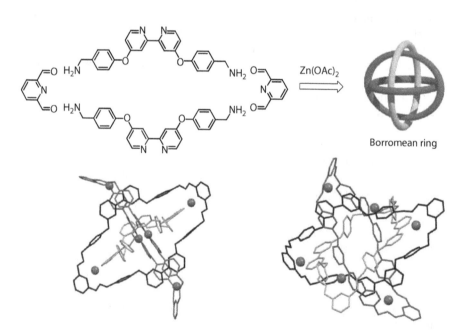

SCHEME 17.9 Synthesis of Borromean ring by Stoddart et al.[35] using dynamic covalent chemistry with a bipyridine-containing ligand, diformyl pyridine and Zn(II) (top) and two representations of corresponding x-ray crystal structure. Zn atoms are shown as balls. H atoms are omitted for clarity (adapted from CCDC 231701). (*Source:* Meyer, C. D. et al., *Chem. Eur. J.* 16 (2010): 12570–12581. With permission.)

the rings in an all-in-one strategy combining metal templation and DCC in a nearly quantitative conversion clearly demonstrates the efficacy of this strategy and sets a precedent for the continued development of their complex structures.[38] Metal templation and $\pi-\pi$ stacking effectively template the formation of the three-interlocked rings of the Borromean structure.

17.4 CONCLUSION AND PERSPECTIVES

The examples illustrated in this chapter clearly demonstrate the wide variety of covalent, coordinative, and weak interactions utilized by chemists in the preparation of molecular knots and links. Strategies employing a diversity of weak interactions acting in synergy and coupled to strong covalent bonds proved effective in the preparation of knots and links.

Despite the numerous molecular topologies reported to date, it is clear that very few of the intrinsic properties of these assemblies such as the topologically unconditional chirality of trefoil knots and Solomon links or the incredible kinetic inertness of Cu(I)-based [2]catenates have been exploited. A step forward for a better understanding and use of these supramolecular assemblies resides in the availability of researchers to exploit such properties.

Among the possible perspectives, the geometrical properties of these objects may be particularly appealing for their nanometer scale ordering on surfaces. Additionally, as numerous knots and links are based on metal complexes, the tuning of their redox properties may be adapted to perform the write–read–erase processes mandatory for molecular data storage in microelectronics.

Finally, the molecular architectures of knots and links often present well-defined cavities with hydrophilic or lipophilic characters, favoring the inclusion of adequately shaped molecules and may be exploited as specific binding sites for selective catalytic processes.[39]

REFERENCES

1. Frisch, H. L. and Wasserman, E. *J. Am. Chem. Soc.* 83 (1961): 3789–3795.
2. Wasserman, E. *J. Am. Chem. Soc.* 82 (1960): 4433–4434.
3. Forgan-Ross, S., Sauvage, J. P., and Stoddart, J. F. *Chem. Rev.* 111 (2011): 5434–5464.
4. Spengler, S. J., Stasiak, A., and Cozzarelli, N. R., *Cell* 42 (1985): 325–334.
5. Lu, C. H., Cecconelo, A., Elbaz, J. et al. *Nano Lett.* 13 (2013): 2303–2308.
6. Stoddart, J. F. *Chem. Soc. Rev.* 38 (2009): 1802–1820.
7. Ben Efraim, D. A., Christopher, B., and Wasserman, E. *J. Am. Chem. Soc.* 92 (1970): 2133–2135.
8. Walba, D. M., Richards, R. M., and Haliwanger, R. C. *J. Am. Chem. Soc.* 104 (1982): 3219–3221.
9. Dietrich-Buchecker, C. O., Sauvage, J. P., and Kintzinger, J. P. *Tetrahedron Lett.* 24 (1983): 5095–5098.
10. Lehn, J. M. *Supramolecular Chemistry: Concepts and Perspectives.* VCH, New York (1995).
11. Dietrich-Buchecker, C. O. and Sauvage, J. P. *Angew. Chem. Intl. Ed.* 28 (1989): 189–192.
12. Guo, J., Mayers, P. C., Breault, G. A. et al., *Nat. Chem.* 2 (2010): 218–222.

13. Rowan, S. J., Cantrill, S. J., Cousins, G. et al. *Angew. Chem. Intl. Ed.* 41 (2002): 898–952.
14. Pentecost, C. D., Chichak, K. S., Peters, A. J. et al. *Angew. Chem. Intl. Ed.* 46 (2007): 218–222.
15. Ayme, J. F., Beves, J. E., Leigh, D. A. et al. *Nat. Chem.* 4 (2012): 15–20.
16. Ponnuswamy, N., Cougnon, F. B. L., Clough, J. M. et al. *Science* 338 (2012): 783–785.
17. Ayme, J. F., Beves, J. E., Campbell, C. J. et al. *Chem. Soc. Rev.* 42 (2013): 1700–1712.
18. Adams, H., Ashworth, E., Breault, G. A. et al. *Nature* 411 (2001): 763.
19. Guo, J. Mayers, P. C., Breault, G. A. et al. *Nat. Chem.* 2 (2010): 218–222.
20. McArdle, C. P., Jennings, M. C., Vittal, J. J. et al., *Chem. Eur. J.* 7 (2001): 3572–3583.
21. McArdle, C. P., Vittal, J. J., and Puddephatt, R. J. *Angew. Chem. Intl. Ed.* 39 (2000): 3819–3822.
22. Peinador, C., Blanco, V., and Quintela, J. M. *J. Am. Chem. Soc.* 131 (2009): 920–921.
23. Gale, P. A. and Quesada, R. *Coord. Chem. Rev.* 250 (2006): 3219–3244.
24. Bowman-James, K. *Acc. Chem. Res.* 38(2005): 671–678.
25. Sambrook, M. R., Beer, P. D., Wisner, J. A. et al. *J. Am. Chem. Soc.* 126 (2004): 15364–15365.
26. Huang, B. Q., Santos, S. M., Felix, V. et al. *Chem. Commun.* (2008): 4610–4612.
27. Vögtle, F., Dünwald, T., and Schmidt, T. *Acc. Chem. Res.* 29 (1996): 451–460.
28. Safarowsky, O., Nieger, M., Frölich, R. et al., *Angew. Chem. Intl. Ed.* 39 (2000): 1616-1618.
29. Brüggemann, J., Bitter, S., Müller, S. et al. *Angew. Chem. Intl. Ed.* 46 (2007): 254–259.
30. Cantrill, S. J., Rowan, S. J., and Stoddart, J. F. *Org. Lett.* 1 (1999): 1363–1366.
31. Griffiths, K. E. and Stoddart, J. F. *Pure Appl. Chem.* 80 (2008): 485–506.
32. Hamilton, D. G., Davies, J. E., Prodi, L. et al. *Chem. Eur. J.* 4 (1998): 608–620.
33. Hamilton, D. G., Sanders, J. K., Davies, J. E. et al. *Chem. Commun.* (1997): 807–898.
34. Prakasam, T., Lusi, M., Elhabiri, M. et al. *Angew. Chem. Intl. Ed.* (2013) 52(2013): 9956–9960.
35. Chichak, K. S., Cantrill, S. J., Pease, A. R. et al. *Science* 304 (2004): 1308–1312.
36. Hasenknopf, B., Lehn, J. M., Boumediene, N. et al. *J. Am. Chem. Soc.* 119 (1997): 10956–10962.
37. Leigh, D. A., Lusby, P. J., Teat, S. J. et al. *Angew. Chem. Intl. Ed.* 40 (2001): 1538–1543.
38. Meyer, C. D., Forgan, R. S., Chichak, K. S. et al. *Chem. Eur. J.* 16 (2010): 12570–12581.
39. Kang, J. M. and Rebeck, J., Jr. *Nature* 385 (1997): 50–52.

Index